PROPE...

OF

ATLANTIC TECHNOLOGY

Chip Scale Package (CSP)

Electronic Packaging and Interconnection Series

ALVINO • *Plastics for Electronics*

CLASSON • *Surface Mount Technology for Concurrent Engineering and Manufacturing*

DI GIACOMO • *Reliability of Electronic Packages*

GARROW AND TURLIK • *Multichip Module Technology Handbook*

GINSBERG AND SCHNORR • *Multichip Modules and Related Technologies*

HARMON • *Wire Bonding, 2/E*

HARPER • *Electronic Packaging and Interconnection Handbook, Second Edition*

HARPER AND MILLER • *Electronic Packaging, Microelectronics, and Interconnection Dictionary*

HARPER AND SAMPSON • *Electronic Materials and Processing Handbook, 2/e*

HWANG • *Modern Solder Technology for Competitive Electronics Manufacturing*

LAU • *Ball Grid Array Technology*

LAU • *Electronic Packaging*

LAU • *Flip Chip Technologies*

LAU • *Solder Joint Reliability of BGA, CSP, Flip Chip*

LICARI • *Multichip Module Design, Fabrication, and Testing*

SERGENT • *Thermal Management Hardbolt*

SMITH AND WHITEHALL • *Optimizing Quality in Electronics Assembly*

Related Books of Interest

BOSWELL • *Subcontracting Electronics*

BOSWELL AND WICKAM • *Surface Mount Guidelines for Processing Control, Quality, and Reliability*

CHEN • *Computer Engineering Handbook*

COOMBS • *Printed Circuits Handbook, 4/E*

CHRISTIANSEN • *Electronics Engineers' Handbook, 4/E*

GINSBERG • *Printed Circuits Design*

JURGEN • *Automotive Electronics Handbook*

MANKO • *Solders and Soldering, 3/E*

VAN ZANT • *Microchip Fabrication, 2/e*

Chip Scale Package (CSP)

Design, Materials, Processes, Reliability, and Applications

John H. Lau
Express Packaging Systems, Inc.

Shi-Wei Ricky Lee
The Hong Kong University of Science & Technology

McGraw-Hill

New York   San Francisco   Washington, D.C.   Auckland   Bogotá
Caracas   Lisbon   London   Madrid   Mexico City   Milan
Montreal   New Delhi   San Juan   Singapore
Sydney   Tokyo   Toronto

Library of Congress Cataloging-in-Publication Data

Lau, John H.
    Chip scale package, CSP : design, materias, processes, and
applications / John H. Lau, Shi-Wei Ricky Lee.
        p.   cm.
    Includes index.
    ISBN 0-07-038304-9
    1. Integrated circuits—Design and construction.
    2. Microelectronic packaging.   I. Lee, Shi-Wei Ricky.   II. Title.
    TK7874.L3167   1999
    621.3815—dc21   98-53224
                                                                CIP

McGraw-Hill

A Division of The McGraw-Hill Companies

1 2 3 4 5 6 7 8 9 0   DOC/DOC   9 0 4 3 2 1 0 9

ISBN 0-07-038304-9

*The sponsoring editor for this book was Stephen S. Chapman, the editing
supervisor was Penny Linskey, and the production supervisor was Sherri
Souffrance. It was set in Century Schoolbook by Victoria Khavkina of
McGraw-Hill's Professional Book Group composition unit.*

Printed and bound by R. R. Donnelley & Sons Company.

McGraw-Hill books are available at special quantity discounts to use
as premiums and sales promotions, or for use in corporate training programs. For more information, please write to the Director of Special
Sales, McGraw-Hill, 11 West 19th Street, New York, NY 10011. Or contact your local bookstore.

This book is printed on recycled, acid-free paper containing
a minimum of 50% recycled, de-inked fiber.

To our parents
for their encouragement

To our wives, Teresa Lau and Jackie Lee,
for their support

To our daughters, Judy Lau, Ingrid Lee, and
Denise Lee,
for their understanding

To those who published peer-reviewed papers
for sharing their knowledge

Contents

Preface xvii
Acknowledgmetns xxi

Part 1   Flip Chip and Wire Bond for CSP

**Chapter 1 Solder-Bumped Flip Chip and Wire-Bonding
Chip on CSP Substrate**   1

  1.1   Introduction   1
  1.2   Cost Analysis: Solder-Bumped Flip Chip Versus Wire Bonding   2
     1.2.1   Assembly Process   4
     1.2.2   Major Equipment   5
     1.2.3   Materials/Labor/Operation   5
     1.2.4   Cost Comparison   15
     1.2.5   Summary   18
  1.3   How To Select Underfill Materials   19
     1.3.1   Underfill Materials and Applications   19
     1.3.2   Curing Conditions   21
     1.3.3   Underfill Material Properties   22
     1.3.4   Underfill Flow Rate   30
     1.3.5   Mechanical Performance   32
     1.3.6   Electrical Performance   34
     1.3.7   Ranking of Underfill Encapsulants   34
     1.3.8   Summary   37
  1.4   Summary   37
  1.5   Acknowledgments   39
  1.6   References   39

Part 2   Customized Leadframe Based CSPs

**Chapter 2 Fujitsu's Small Outline No-Lead/C-Lead Package
(SON/SOC)**   43

  2.1   Introduction and Overview   43
  2.2   Design Concepts and Package Structure   43
  2.3   Material Issues   48

2.4 Manufacturing Process   48
2.5 Electrical and Thermal Performance   49
2.6 Qualifications and Reliability   52
2.7 Applications and Advantages   54
2.8 Summary and Concluding Remarks   58
2.9 References   59

Chapter 3 Fujitsu's Bump Chip Carrier (BCC)   61

3.1 Introduction and Overview   61
3.2 Design Concepts and Package Structure   61
3.3 Material Issues   63
3.4 Manufacturing Process   64
3.5 Electrical and Thermal Performance   66
3.6 Qualifications and Reliability   67
3.7 Applications and Advantages   69
3.8 Summary and Concluding Remarks   70
3.9 References   70

Chapter 4 Fujitsu's MicroBGA and Quad Flat Nonleaded
Package (QFN)   71

4.1 Introduction and Overview   71
4.2 Design Concepts and Package Structure   71
4.3 Material Issues   74
4.4 Manufacturing Process   75
4.5 Qualifications and Reliability   80
4.6 Applications and Advantages   83
4.7 Summary and Concluding Remarks   85
4.8 References   86

Chapter 5 Hitachi Cable's Lead-on-Chip Chip Scale
Package (LOC-CSP)   87

5.1 Introduction and Overview   87
5.2 Design Concepts and Package Structure   87
5.3 Material Issues   89
5.4 Manufacturing Process   91
5.5 Electrical Performance and Package Reliability   93
5.6 Applications and Advantages   94
5.7 Summary and Concluding Remarks   95
5.8 References   96

Chapter 6 Hitachi Cable's Micro Stud Array Package (MSA) 97

6.1 Introduction and Overview   97
6.2 Design Concepts and Package Structure   97

6.3   Material Issues and Manufacturing Process   100
6.4   Performance and Reliability   103
6.5   Applications and Advantages   104
6.6   Summary and Concluding Remarks   105
6.7   References   105

Chapter 7 LG Semicon's Bottom-Leaded Plastic Package (BLP)   107

7.1   Introduction and Overview   107
7.2   Design Concepts and Package Structure   107
7.3   Material Issues   110
7.4   Manufacturing Process   112
7.5   Electrical and Thermal Performance   115
7.6   Qualifications and Reliability   118
7.7   Applications and Advantages   124
7.8   Summary and Concluding Remarks   125
7.9   References   127

Chapter 8 TI Japan's Memory Chip Scale Package with LOC (MCSP) 129

8.1   Introduction and Overview   129
8.2   Design Concepts and Package Structure   129
8.3   Material Issues   131
8.4   Manufacturing Process   134
8.5   Electrical and Thermal Performance   136
8.6   Qualifications and Reliability   137
8.7   Applications and Advantages   140
8.8   Summary and Concluding Remarks   141
8.9   References   141

Part 3   CSPs with Flexible Substrate

Chapter 9 3M's Enhanced Flex CSP   143

9.1   Introduction and Overview   143
9.2   Design Concepts and Package Structure   143
9.3   Material Issues   145
9.4   Manufacturing Process   148
9.5   Performance and Reliability   150
9.6   Applications and Advantages   155
9.7   Summary and Concluding Remarks   156
9.8   References   156

Chapter 10 General Elecrtric's Chip-on-Flex Chip Scale
Package (COF-CSP)   157

10.1   Introduction and Overview   157
10.2   Design Concepts and Package Structure   157

10.3   Material Issues   163
10.4   Manufacturing Process   164
10.5   Performance and Reliability   168
10.6   Applications and Advantages   169
10.7   Summary and Concluding Remarks   172
10.8   References   172

Chapter 11 Hitachi's Chip Scale Package for Memory Devices   173

11.1   Introduction and Overview   173
11.2   Design Concepts and Package Structure   173
11.3   Material Issues   176
11.4   Manufacturing Process   177
11.5   Qualifications and Reliability   179
11.6   Applications and Advantages   180
11.7   Summary and Concluding Remarks   181
11.8   References   181

Chapter 12 IZM's *flex*PAC   183

12.1   Introduction and Overview   183
12.2   Design Concepts and Package Structure   183
12.3   Material Issues   186
12.4   Manufacturing Process   188
12.5   Qualifications and Reliability   194
12.6   Applications and Advantages   196
12.7   Summary and Concluding Remarks   198
12.8   References   198

Chapter 13 NEC's Fine-Pitch Ball Grid Array (FPBGA)   201

13.1   Introduction and Overview   201
13.2   Design Concepts and Package Structure   201
13.3   Material Issues   206
13.4   Manufacturing Process   208
13.5   Performance and Reliability   213
13.6   Applications and Advantages   216
13.7   Summary and Concluding Remarks   218
13.8   References   218

Chapter 14 Nitto Denko's Molded Chip Size Package (MCSP)   219

14.1   Introduction and Overview   219
14.2   Design Concepts and Package Structure   219
14.3   Material Issues   220
14.4   Manufacturing Process   225
14.5   Qualifications and Reliability   226
14.6   Applications and Advantages   231

14.7 Summary and Concluding Remarks   232
14.8 References   232

Chapter 15 Sharp's Chip Scale Package   233

15.1 Introduction and Overview   233
15.2 Design Concepts and Package Structure   233
15.3 Material Issues   237
15.4 Manufacturing Process   238
15.5 Performance and Package Qualifications   240
15.6 Solder Joint Reliability   244
15.7 Applications and Advantages   255
15.8 Summary and Concluding Remarks   257
15.9 References   257

Chapter 16 Tessera's Micro-Ball Grid Array (μBGA)   259

16.1 Introduction and Overview   259
16.2 Design Concepts and Package Structure   259
16.3 Material Issues   263
16.4 Manufacturing Process   265
16.5 Electrical and Thermal Performance   266
16.6 Qualifications and Reliability   271
16.7 Applications and Advantages   277
16.8 Summary and Concluding Remarks   281
16.9 References   281

Chapter 17 TI Japan's Micro-Star BGA (μStar BGA)   283

17.1 Introduction and Overview   283
17.2 Design Concepts and Package Structure   283
17.3 Material Issues and Manufacturing Process   284
17.4 Qualifications and Reliability   285
17.5 Applications and Advantages   291
17.6 Summary and Concluding Remarks   292
17.7 References   292

Chapter 18 TI Japan's Memory Chip Scale Package with
Flexible Substrate (MCSP)   293

18.1 Introduction and Overview   293
18.2 Design Concepts and Package Structure   293
18.3 Material Issues   294
18.4 Manufacturing Process   297
18.5 Qualifications and Reliability   299
18.6 Applications and Advantages   302
18.7 Summary and Concluding Remarks   303
18.8 References   303

Part 4   CSPs with Rigid Substrate

Chapter 19 Amkor/Anam's ChipArray Package   305

19.1   Introduction and Overview   305
19.2   Design Concepts and Package Structure   305
19.3   Material Issues   307
19.4   Manufacturing Process   309
19.5   Performance and Reliability   310
19.6   Applications and Advantages   311
19.7   Summary and Concluding Remarks   311
19.8   References   312

Chapter 20 EPS's Low-Cost Solder-Bumped Flip Chip NuCSP   313

20.1   Introduction and Overview   313
20.2   Design Concepts and Package Structure   313
20.3   Material Issues   315
       20.3.1 Wafer Bumping and Solder-Bump Characterizations 317
       20.3.2 Solder-Bump Height Measurements   318
       20.3.3 Solder-Bump Strength Measurements   319
20.4   NuCSP Substrate Design and Fabrication   320
20.5   NuCSP Assembly Process   325
       20.5.1 Fluxing and Pick and Place   325
       20.5.2 Solder Reflow   325
       20.5.3 Inspection   325
       20.5.4 Underfill Application   327
20.6   NuCSP Mechanical and Electrical Performance   328
       20.6.1 Mechanical Performance of NuCSP   328
       20.6.2 Electrical Performance of NuCSP   329
20.7   NuCSP Solder Joint Reliability on PCB   329
20.8   Applications and Advantages   334
20.9   Summary and Concluding Remarks   334
20.10 Acknowledgments   335
20.11 References   335

Chapter 21 IBM's Ceramic Mini-Ball Grid Array Package (Mini-BGA)   337

21.1   Introduction and Overview   337
21.2   Design Concepts and Package Structure   337
21.3   Material Issues   339
21.4   Manufacturing Process   341
21.5   Performance and Reliability   346
21.6   Applications and Advantages   347
21.7   Summary and Concluding Remarks   347
21.8   References   348

Chapter 22 IBM's Flip Chip–Plastic Ball Grid Array Package (FC/PBGA) 349

22.1   Introduction and Overview   349
22.2   Design Concepts and Package Structure   349

22.3 Material Issues   351
22.4 Manufacturing Process   351
22.5 Qualifications and Reliability   353
22.6 Summary and Concluding Remarks   356
22.7 References   357

Chapter 23 Matsushita's MN-PAC   359

23.1 Introduction and Overview   359
23.2 Design Concepts and Package Structure   359
23.3 Material Issues   361
23.4 Manufacturing Process   363
23.5 Electrical and Thermal Performance   365
23.6 Qualifications and Reliability   368
23.7 Applications and Advantages   371
23.8 Summary and Concluding Remarks   373
23.9 References   375

Chapter 24 Motorola's SLICC and JACS-Pak   377

24.1 Introduction and Overview   377
24.2 Design Concepts and Package Structure   378
24.3 Material Issues   381
24.4 Manufacturing Process   383
24.5 Electrical and Thermal Performance   384
24.6 Qualifications and Reliability   389
24.7 Summary and Concluding Remarks   395
24.8 References   397

Chapter 25 National Semiconductor's Plastic Chip Carrier (PCC)   399

25.1 Introduction and Overview   399
25.2 Design Concepts and Package Structure   399
25.3 Material Issues   401
25.4 Manufacturing Process   402
25.5 Performance and Reliability   403
25.6 Applications and Advantages   404
25.7 Summary and Concluding Remarks   405
25.8 References   406

Chapter 26 NEC's Three-Dimensional Memory Module (3DM) and CSP   407

26.1 Introduction and Overview   407
26.2 Design Concepts and Package Structure   407
26.3 Material Issues   414
26.4 Manufacturing Process   418
26.5 Performance and Reliability   422
26.6 Applications and Advantages   424
26.7 Summary and Concluding Remarks   426
26.8 References   427

Chapter 27 Sony's Transformed Grid Array Package (TGA)   429

  27.1 Introduction and Overview   429
  27.2 Design Concepts and Package Structure   429
  27.3 Material Issues   431
  27.4 Manufacturing Process   432
  27.5 Qualifications and Reliability   433
  27.6 Applications and Advantages   435
  27.7 Summary and Concluding Remarks   437
  27.8 References   438

Chapter 28 Toshiba's Ceramic/Plastic Fine-Pitch BGA Package (C/P-FBGA)   439

  28.1 Introduction and Overview   439
  28.2 Design Concepts and Package Structure   439
  28.3 Material Issues   441
  28.4 Manufacturing Process   443
  28.5 Qualifications and Reliability   448
  28.6 Applications and Advantages   452
  28.7 Summary and Concluding Remarks   452
  28.8 References   453

Part 5   Wafer-Level Redistribution CSPs

Chapter 29 ChipScale's Micro SMT Package (MSMT)   455

  29.1 Introduction and Overview   455
  29.2 Design Concepts and Package Structure   455
  29.3 Material Issues   459
  29.4 Manufacturing Process   460
  29.5 Performance and Reliability   461
  29.6 Applications and Advantages   462
  29.7 Summary and Concluding Remarks   464
  29.8 References   465

Chapter 30 EPIC's Chip Scale Package   467

  30.1 Introduction and Overview   467
  30.2 Design Concepts and Package Structure   467
  30.3 Material Issues   471
  30.4 Manufacturing Process   471
  30.5 Performance and Reliability   473
  30.6 Applications and Advantages   474
  30.7 Summary and Concluding Remarks   475
  30.8 References   475

Chapter 31 Flip Chip Technologies' *Ultra*CSP   477

31.1 Introduction and Overview   477
31.2 Design Concepts and Package Structure   477
31.3 Material Issues   479
31.4 Manufacturing Process   482
31.5 Performance and Reliability   482
31.6 Applications and Advantages   485
31.7 Summary and Concluding Remarks   486
31.8 References   486

Chapter 32 Fujitsu's *Super* CSP (SCSP)   487

32.1 Introduction and Overview   487
32.2 Design Concepts and Package Structure   487
32.3 Material Issues   489
32.4 Manufacturing Process   489
32.5 Qualifications and Reliability   490
32.6 Summary and Concluding Remarks   493
32.7 References   494

Chapter 33 Mitsubishi's Chip Scale Package (CSP)   495

33.1 Introduction and Overview   495
33.2 Design Concepts and Package Structure   496
33.3 Material Issues   499
33.4 Manufacturing Process   505
33.5 Performance and Reliability   512
33.6 Applications and Advantages   516
33.7 Summary and Concluding Remarks   517
33.8 References   518

Chapter 34 National Semiconductor's μSMD   519

34.1 Introduction and Overview   519
34.2 Design Concepts and Package Structure   519
34.3 Material Issues and Manufacturing Process   520
34.4 Solder Joint Reliability   521
34.5 Summary and Concluding Remarks   526
34.6 References   527

Chapter 35 Sandia National Laboratories' Mini Ball Grid Array Package (mBGA)   529

35.1 Introduction and Overview   529
35.2 Design Concepts and Package Structure   529
35.3 Material Issues   533

35.4   Manufacturing Process   534
35.5   Electrical and Thermal Performance   536
35.6   Bump Shear Strength and Assembly Joint Reliability   540
35.7   Applications and Advantages   542
35.8   Summary and Concluding Remarks   543
35.9   References   544

Chapter 36 ShellCase's Shell-PACK/Shell-BGA   545

36.1   Introduction and Overview   545
36.2   Design Concepts and Package Structure   545
36.3   Material Issues   547
36.4   Manufacturing Process   548
36.5   Performance and Reliability   548
36.6   Applications and Advantages   553
36.7   Summary and Concluding Remarks   555
36.8   References   555

Index   557

Preface

Theoretically speaking, one of the most cost-effective packaging techniques is direct chip attach (DCA) which is directly attaching the chip on the printed circuit board (PCB), or on the flexible printed circuit (FPC), or on the glass (COG) "without" a package. However, because of the cost and infrastructure in supplying the known good die (KGD) and the corresponding fine line and spacing of PCB, or FPC, or glass, most in the industry are still working on these issues.

In the meantime, a class of new technology called chip scale package (CSP) has emerged. As a matter of fact, CSP is not just an emerging technology—it quickly is becoming the package of choice for packaging the memory IC devices. In the last few years, the electronics packaging industry has witnessed an explosive growth in the research and development efforts devoted to CSP. One of the unique features of most CSPs is using a substrate (interposer) or metal layer to redistribute the very find-pitch (as small as 0.075 mm) peripheral (or staggered) pads on the chip to a much larger-pitch (1mm, 0.8 mm, 0.75 mm, and 0.5 mm) area-array pads on the PCB, or FPC, or glass.

The definition of CSP given by IPC is that the package area is less than 1.5 times that of the chip area. However, don't be too hung up on this definition. If someone finds a cost-effective and reliable CSP that does not satisfy this definition, use it!

The advantages of CSP versus DCA are that with the substrate the CSP is easier to test-at-speed and burn-in for KGD, to handle, to assemble, to rework, to standardize, to protect the die, to deal with die shrink and expand, and it is subject to less infrastructure constraints. On the other hand, the advantages of DCA are that it has better electrical performance, more direct heat path, less weight, smaller size, and lower cost.

There are more than 40 different CSPs reported at this writing and most of them are designed to be used for static random access memories (SRAM), dynamic random access memories (DRAM), flash memories, and not-so-high-pin-count nor high-power application-specific

ICs (ASIC) and microprocessors. The important CSP parameters such as the ICs, substrates, packaging, routing capabilities, thermal and electrical management, assembly processes, reliability, qualification, applications, and infrastructure have been studied by many experts. Their conclusions have already been disclosed in diverse journals or, more incidentally, in the proceedings of many conferences, symposia, and workshops whose primary emphases are materials science or electronic packaging and interconnection. Consequently, there is no single source of information devoted to the state of the art of CSP technology. This book aims to remedy this deficiency and to present, in one volume, a timely summary of progress in all aspects of this fascinating field. It is written for everyone who can learn quickly about the basics and problem-solving methods, understand the trade-offs, and make system-level decisions with the CSPs.

This book is organized into five basic parts. Chapter 1 (Part 1) briefly discusses the two most popular interconnect technologies, namely, solder-bumped flip chip and wire-bonding chip on CSP substrate. The major equipment and materials for these technologies are compared and analyzed. Also, the curing conditions, material properties, mechanical and thermal performance, and selection of underfill encapsulant for the applications of solder-bumped flip chip on CSP substrates are presented.

The design concepts and package structure, material issues, manufacturing process, electrical and thermal performance, qualification and reliability test data, and applications of Customized Leadframe Based CSPs are examined in Part 2 (Chapters 2–8); CSPs with Flexible Substrates are discussed in Part 3 (Chapters 9–18); CSPs with Rigid Substrates are presented in Part 4 (Chapters 19–28); and Wafer-Level Redistribution CSPs are provided in Part 5 (Chapters 29–36) of this book.

For whom is this book intended? Undoubtedly, it will be of interest to three groups of specialists: 1) those who are active or intend to become active in research and development of CSPs; 2) those who have encountered practical CSP problems and wish to understand and learn more methods of solving such problems; and 3) those who have to choose a reliable, creative, high performance, robust, and cost effective packaging technique for their interconnect system. This book also can be used as a text for college and graduate students who could become our future leaders, scientists, and engineers in the electronics industry.

We hope this book will serve as a valuable reference source to all those faced with the challenging problems created by the ever-increasing IC speed and density and reducing product size and weight. We also hope that it will aid in stimulating further research and development on electrical and thermal designs, materials, process, manufacturing, electrical and thermal management, testing, and reliability, and more sound applications of CSPs in electronics products.

The organizations that learn how to design CSPs in their interconnect systems have the potential to make major advances in the electronics industry and to gain great benefits in cost, performance, quality, size, and weight. It is our hope that the information presented in this book may assist in removing roadblocks, avoiding unnecessary false starts, and accelerating design, materials, and process development of CSPs. We are strongly against the notion that electronics packaging and interconnection are the bottle neck of high-speed computing. Rather, we want to consider this as the golden opportunity to make a major contribution to the electronics industry by developing innovative, cost-effective, and reliable CSPs. It is an exciting time for CSP!

John H. Lau, Ph.D., P.E., IEEE Fellow
Express Packaging Systems, Inc.

Shi-Wei Ricky Lee, Ph.D.
The Hong Kong University of Science & Technology

Acknowledgments

Development and preparation of *Chip Scale Package: Design, Materials, Process, Reliability, and Applications* was facilitated by the efforts of a number of dedicated people at McGraw-Hill. We would like to thank them all, with special mention to Penny Linskey, Regina Frappolli, Sherri Souffrance, and Victoria Khavkina for their unswerving support and advocacy. Our special thanks to Steve Chapman (Executive Editor of Electronics and Optical Engineering) who made our dream of this book come true by effectively sponsoring the project and solving many problems that arose during the book's preparation. It has been a great pleasure to work with them in transforming our messy manuscript into a very attractive printed book.

The material in this book clearly has been derived from many sources including individuals, companies, and organizations, and we have attempted to acknowledge, in the appropriate parts of the book, the assistance that we have been given. It would be quite impossible for us to list the names of everyone we thank for their cooperation in producing this book, but we extend our gratitude to them all.

Also, we want to thank several professional societies and publishers for permitting us to reproduce some of their illustrations and information in this book. For example, the American Society of Mechanical Engineers (ASME) Conference Proceedings (e.g., *International Intersociety Electronic Packaging Conference*) and Transactions (e.g., *Journal of Electronic Packaging*), the Institute of Electrical and Electronic Engineers (IEEE) Conference Proceedings (e.g., *Electronic Components & Technology Conference and International Electronic Manufacturing and Technology Symposium*) and Transactions (e.g., *Components, Packaging, and Manufacturing Technology*), the International Microelectronics and Packaging Society (IMAPS) Conference Proceedings (e.g., *International Symposium on Microelectronics*), and Transactions (e.g.,*International Journal of Microcircuits & Electronic Packaging*), American Society of Metals (ASM), Conference Proceedings and books (e.g., *Electronic Materials Handbook*, Vol. 1, *Packaging*), the Surface Mount Technology

Association (SMTA) Conference Proceedings (e.g., *Surface Mount International Conference & Exposition*), and Journals (e.g., *Journal of Surface Mount Technology*), the National Electronic Packaging Conferences (NEPCON) and Proceedings, the *IBM Journal of Research and Development, Electronic Packaging & Production, Advanced Packaging, Circuits Assembly, Surface Mount Technology, Connection Technology, Solid State Technology, Circuit World, Microelectronics International,* and *Soldering and Surface Mount Technology.*

John Lau wants to thank his former employer, Hewlett-Packard Company, for providing him an excellent working environment that has nurtured him as a human being, fulfilled his desire for job satisfaction, and enhanced his professional reputation.

He also wants to thank Mr. Terry T. M. Gou (Chairman and C.E.O. of Hon Hai Precision Industry Co., Ltd.) for his trust, respect, and support of his work at EPS. Furthermore, he wants to thank his eminent colleagues at Hewlett-Packard Company, Express Packaging Systems, and throughout the electronics industry for their useful help, strong support, and stimulating discussions. Working and socializing with them have been a privilege and an adventure. He learned a lot about life and electronics packaging from them.

Shi-Wei Ricky Lee wishes to express his gratitude to his colleagues at Hong Kong University of Science and Technology. Without their efforts to establish a pro-electronic packaging environment, he probably would not have begun his endeavor in this discipline. Special thanks are also due to the Research Grany Council (RGC) of Hong Kong for their financial support to part of his research activities in electronic packaging.

Lastly, John Lau wants to thank his daughter, Judy, and his wife, Teresa, for their love, consideration, and patience by allowing him to work many weekends on the book. Their simple belief that he is making a small contribution to the electronics industry was a strong motivation for him. During this holiday season, thinking that Judy had just survived her first semester at UC Berkeley and that Teresa and he are in good health, he wants to thank God for his generosity and blessings.

Ricky Lee shares the same feeling about the support of his family. In a certain period of time, he worked for more than sixteen hours a day on this book. The only time he could see his two young daughters, Ingrid and Denise, was at 7 a.m. when he took the girls to their school bus. He still recalls clearly that, after getting on the bus his daughters waved to him and said, "See you tomorrow, Dad!"

John H. Lau, EPS, Inc., Palo Alto, California
Shi-Wei Ricky Lee, Visiting Scholar at EPS, Inc.

Solder-Bumped Flip Chip and Wire-Bonding Chip on CSP Substrate

1.1 Introduction

Figure 1.1 schematically shows a printed circuit board (PCB) with three different packaging technologies, namely the ball grid array (BGA), chip scale package (CSP), and direct chip attach (DCA). BGA is an "old" technology because many ceramic BGAs, plastic BGAs, tape BGAs, and metal BGAs are already in mass production in various products.

Theoretically speaking, the most cost-effective packaging technology is DCA, which attaches the chip directly to the PCB without a package such as the solder-bumped flip chip or wire-bonding chip on board. However, because of the infrastructure and cost of supplying the known good die (KGD) and the corresponding fine line and spacing PCB, most in the industry are still working on these issues. It should be noted that with the sequential buildup PCB technologies with micro vias, such as the DYCOstrate, plasma-etched redistribu-

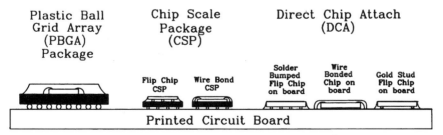

Plastic Ball Grid Array (PBGA) Package — Chip Scale Package (CSP) — Direct Chip Attach (DCA)

Flip Chip CSP — Wire Bond CSP — Solder Bumped Flip Chip on board — Wire Bonded Chip on board — Gold Stud Flip Chip on board

Printed Circuit Board

Figure 1.1 BGA, CSP, and DCA on printed circuit board.

tion layers (PERLs), surface laminar circuits (SLCs), film redistribution layer (FRL), interpenetrating polymer buildup structure system (IBSS), high-density interconnect (HDI), conductive adhesive bonded flex, sequential bonded films, sequential bonded sheets, and micro-filled via technology, PCBs with 0.05-mm (2-mil) line width and space should be available at reasonable cost by the turn of this century.

In the meantime, a class of new technology called CSP has surfaced. The unique feature of most CSPs is the use of a substrate (interposer, or carrier or substrate, or metal layer) to redistribute the very fine pitch (as small as 0.075 mm) peripheral array pads on the chip to much larger pitch (1 mm, 0.8 mm, 0.75 mm, and 0.5 mm) area array pads on the PCB. The definition of CSP given by IPC is that the package area is less than 1.5 times the chip area. However, don't be too hung up on this definition. If someone finds a cost-effective and reliable CSP which does not satisfy this definition, use it!

The advantages of CSP versus DCA are that with the substrate (interposer), CSP is easier to test at speed and burn in for KGD, to handle, to assemble, to rework, and to standardize; protecting the die and dealing with die shrink are easier; and CSP is subject to fewer infrastructure constraints. On the other hand, the advantages of DCA are that it has better electrical and thermal performance and less weight, size, and cost.

There are more than 40 different CSPs reported today, and most of them are designed to be used for static random-access memories (SRAM), dynamic random-access memories (DRAM), flash memories, and not-so-high-pin-count application-specific ICs (ASICs) and microprocessors. Higher-pin-count and performance CSPs are being explored in laboratories. Most of these CSPs use either wire-bonding or solder-bumped flip-chip technologies to connect the IC chip to the organic or tape substrate. In this chapter, the cost analysis of these two technologies and the selection criteria for underfill materials for flip-chip technology are presented.

1.2   Cost Analysis: Solder-Bumped Flip Chip versus Wire Bonding

Wire-bonding chips on organic substrate have been used for many years. Today, more than 90 percent of the IC chips used in the world use wire bonds because high-speed automatic wire bonders meet most of the needs for interconnecting the semiconductor device to the next-level packaging.

The past decade witnessed an explosive growth in the research and development efforts devoted to solder-bumped flip-chip technology as

TABLE 1.1   Comparison between solder-bumped Flip-Chip and Wire-Bonding
Technologies

	Flip-chip technology	Wire-bonding technology
Advantages	▪ High density ▪ High I/Os ▪ High performance ▪ Noise control ▪ Thin profile ▪ SMT compatible ▪ Area array technology ▪ Small device footprints ▪ Self-alignment	▪ Mature technology ▪ Infrastructure exists ▪ Flexible for new devices ▪ Flexible for new bonding patterns
Disadvantages	▪ Availability of wafers ▪ Availability of dies ▪ Availability of KGD ▪ Wafer bumping ▪ Test and burn-in ▪ Underfill encapsulation ▪ Additional equipment ▪ Additional processes ▪ Rework after encapsulation difficult ▪ Die shrink	▪ Availability of wafers ▪ Availability of dies ▪ Availability of KGD ▪ Tests and burn-in ▪ I/O limitation ▪ Peripheral technology ▪ Sequential process ▪ Additional equipment ▪ Additional processes ▪ Rework difficult ▪ Glob-top encapsulation

a direct result of the smaller form factor higher package density, performance, and interconnection requirements and the limitations of wire-bonding technology. In comparison to the popular wire-bonding technology, flip-chip technology provides higher packaging density (more I/Os) and performance (shorter possible leads, lower inductance, and better noise control), smaller device footprints, and lower packaging profile. The advantages and disadvantages of flip-chip and wire-bonding technology are shown in Table 1.1.

What is the cost comparison between solder-bumped flip-chip and wire-bonding technology? This is a frequently asked question, and answering it is almost impossible. First of all, either flip chip or wire bonding is just an interconnect technology for connecting the IC chip to the next level of interconnection. Second, the cost impact of these interconnect technologies on a final product depends on the upstream (e.g., semiconductor design, materials, and manufacturing) and downstream (e.g., substrate and printed circuit board) elements of this level of interconnect, and the functions of the final product. However, with some assumptions, the cost of solder-bumped flip chip and wire bonding can be analyzed.

1.2.1 Assembly process

Figure 1.2 shows a very simplified assembly process for wire-bonding chips and solder-bumped flip chips on organic substrates. More detailed design, materials, process, reliability, and applications for wire-bonding and flip-chip technologies can be found in Ref. [1–34]. It should be noted that there are two ways to do the screening test for solder-bumped flip-chip technology. One is to test before wafer bumping. In this case, the probe marks (damages) will be on the pad, which could affect the integrity of the under-bump metallurgy and have po-

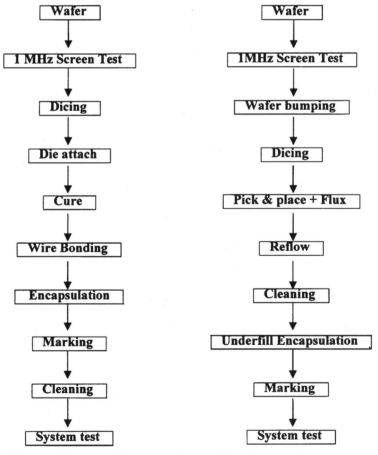

Wire Bonding Chip On Organic Substrate **Solder Bumped Flip Chip On Organic Substrate**

Figure 1.2 Assembly process for wire-bonding and flip-chip technologies.

tential effects on long-term reliability. The other is to test after wafer bumping. In this case, the test will contaminate the probe needles and result in shorts. Also, solder bumps may be damaged. However, better yield can be obtained because there will be better electrical contact between the probe needles and the solder bumps. For a mature wafer-bumping process, the bump yield is usually very high (>99 percent).

From a cost point of view, the most important difference between wire bonding and solder-bumped flip chip is that the wire-bonding technology needs gold wire and the solder-bumped flip-chip technology needs wafer bumping. The second most important difference between these two technologies is the effect on IC chip yields of the 1-MHz screening test and at-speed/burn-in system tests. Finally, the major equipment needed for these interconnect methods differs.

1.2.2 Major equipment

The major equipment needed by wire-bonding chips and solder-bumped flip chips on a CSP organic substrate is shown in Table 1.2. It should be noted that equipment required by both technologies is not shown. Also, all the equipment is assumed to be utilized 300 days (24 hours a day) a year, and the capacity is assumed to be 12 million chips per year.

It can be seen from Table 1.2 that expensive pick and place and fluxing machines are necessary for flip-chip technology. Even though the wire bonder is cheaper, however, because its throughput is much lower (it depends on the number of pads on a chip) than that of the pick and place machine (which performs gang bonding on a chip), more wire bonders are needed. Consequently, the cost of major equipment for flip chip ($2,700,000) is lower than that for wire bonding ($4,790,000). Also, the manufacturing floor space for flip chip should be smaller.

1.2.3 Materials/labor/operation

As mentioned earlier, the largest cost differences between wire bonding and flip chip are those for materials and IC chip yields. The labor/operation costs for these two assembly processes are assumed to be the same.

1.2.3.1 The wafer. The physically possible number of undamaged chips N_c stepped from a wafer (Fig. 1.2) may be given by

$$N_c = \pi \frac{[\phi - (1 + \theta)\sqrt{A/\theta}]^2}{4A} \tag{1.1}$$

TABLE 1.2   Major Equipment for Flip-Chip and Wire-Bonding Technologies

Major equipment for flip chip	
Equipment for	Cost
Pick & place + flux	3 @ $500K = $1500K
Reflow oven	$70K
Clean	$130K
Underfill dispenser	5 @ 120K = $600K
Vertical cure oven	$100K
X-ray	$300K
Total	**$2,700K**
Major equipment for wire bonding	
Equipment for	Cost
Die attach	$220K
Cure oven	$20K
Wire bonders	35 @ $110K = $3850K
Encapsulation	$700K (Transfer Mold)
	or
	5 @ $120K = $600K
	(glob-top dispenser) plus
	Verticle cure oven = $100K
Total	**$4,790K**

- Machine utilization is 7200 h/year.
- Machine capacity is 12 million chips per year.
- Pick and place + fluxing is assumed to require 6.5 s per chip.
- Underfill dispensing is assumed to require 10 s per chip (to completely fill the chip).
- Wire bonders' speed is assumed to be 8 bonds (4 wires) per s. Also, it is assumed that there are 300 pads per chip.
- Glob-top dispensing dam and encapsulant are assumed to require 10 s per chip.
- Equipment required by both techniques is not shown.

where

$$A = xy \qquad (1.2)$$

and

$$\theta = \frac{x}{y} \geq 1 \qquad (1.3)$$

where x and y are the dimensions of a rectangular chip (in millimeters), with x no less than y; θ is the ratio between x and y; ϕ is the wafer diameter (in millimeters); and A is the area of the chip (in square millimeters). For example, for a 200-mm wafer with a square chip $A = 100$ mm^2, $N_c = 255$ chips.

For a given chip size, the physically possible number of pads N_p depends on the pad pitch p and pad configuration. The N_p on a chip surface is given as follows:

For area-array pads:

$$N_p = \left(\frac{x}{p} - 1\right)\left(\frac{y}{p} - 1\right) \tag{1.4}$$

For peripheral-array pads:

$$N_p = 2\left(\frac{x}{p} - 1\right) + 2\left(\frac{y}{p} - 1\right) \tag{1.5}$$

For peripheral-staggered-array pads:

$$N_p = 4\left(\frac{x}{p} - 1\right) + 4\left(\frac{y}{p} - 1\right) - 4 \tag{1.6}$$

For example, for a chip with $x = 10$ mm, $y = 8$ mm, and $p = 0.25$ mm, $N_p = 1209$ area-array pads, $N_p = 140$ peripheral-array pads, and $N_p = 276$ peripheral-staggered-array pads. Obviously, area array is the choice for high-density ICs and packaging. It is interesting to note that the solder joints on an area-array flip-chip assembly are more reliable than those on a peripheral-array flip-chip assembly, since the interior solder bumps provide resistance to the relative displacement of the chip and the substrate.

1.2.3.2 Wafer-bumping cost per die. Wafer bumping is the heart of solder-bumped flip-chip technology. The cost of wafer bumping is affected by the true yield Y_T of the IC chips. The wafer-bumping cost per die ($C_{B/D}$) can be determined by

$$C_{B/D} = \frac{C_B}{Y_T N_C} \tag{1.7}$$

where C_B is the wafer-bumping cost (ranging from \$25 to \$250 per wafer), N_c is given in Eq. (1.1), and Y_T is the true IC chip yield after at-speed/burn-in system tests.

Figures 1.3 and 1.4, respectively, show the wafer-bumping cost per die $C_{B/D}$ for various Y_T for the 200-mm and 150-mm wafers. It can be seen that for both wafers, the $C_{B/D}$ increases as Y_T decreases. Also, it

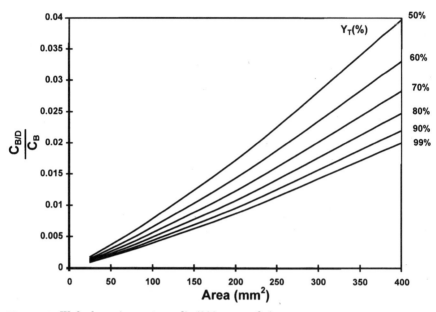

Figure 1.3 Wafer-bumping cost per die (200-mm wafer).

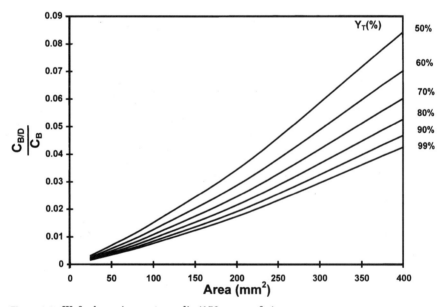

Figure 1.4 Wafer-bumping cost per die (150-mm wafer).

can be shown from Eqs. (1.1) and (1.7) that if the wafer-bumping cost C_B of the 200-mm wafer is no more than two times that of the 150-mm wafer, then the wafer-bumping cost per die is lower for the larger wafer (assuming Y_T is the same for both wafers). For example, for a 200-mm wafer with $A = 100$ mm^2, $N_c \sim 255$. Assuming $C_B = \$100$, then $C_{B/D} = \$0.44$ if $Y_T = 0.9$ and $C_{B/D} = \$0.65$ if $Y_T = 0.6$. Thus Y_T and C_B play important roles in flip-chip technology.

1.2.3.3 Wafer-bumping cost per pad. The wafer-bumping cost per pad $C_{B/P}$ can be determined by

$$C_{B/P} = \frac{C_B}{Y_T N_C N_P} \tag{1.8}$$

where N_c is given in Eq. (1.1); N_p is given in Eq. (1.4), (1.5), or (1.6); C_B is the wafer-bumping cost; and Y_T is the true IC chip yield.

Figures 1.5 and 1.6, respectively, show the $C_{B/P}$ for the 200-mm and 150-mm wafers with peripheral-array pads. It can be seen that $C_{B/P}$ depends on the chip size, pad pitch, C_B, and Y_T. For both wafers, the larger the chip size, the higher the $C_{B/P}$, and the larger the pad pitch, the higher the $C_{B/P}$. Also, $C_{B/P}$ is lower for the 200-mm wafer if the C_B of the larger wafer is no more than twice that of the smaller wafer.

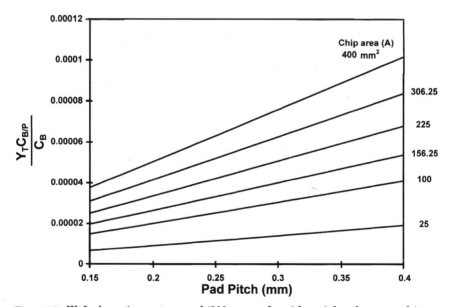

Figure 1.5 Wafer-bumping cost per pad (200-mm wafer with peripheral-array pads).

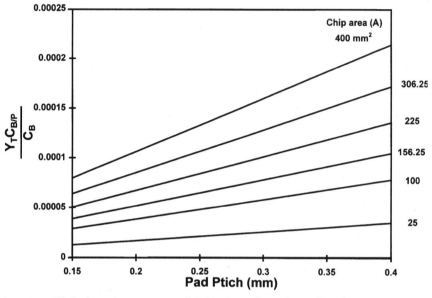

Figure 1.6   Wafer-bumping cost per pad (150-mm wafer with peripheral-array pads).

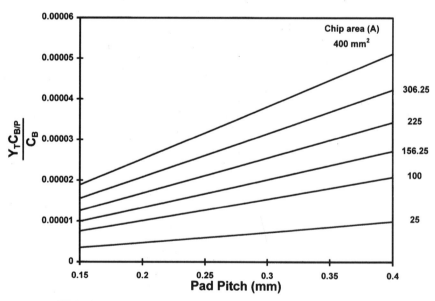

Figure 1.7   Wafer-bumping cost per pad (200-mm wafer with peripheral-staggered-array pads).

Figures 1.7 and 1.8, respectively, show the $C_{B/P}$ for the 200-mm and 150-mm wafers with peripheral-staggered-array pads. It can be seen from Figs. 1.5 and 1.7 and Figs. 1.6 and 1.8 that $C_{B/P}$ is cut almost by half, since there are almost twice as many pads on the peripheral-staggered-array chip. Figures 1.9 and 1.10 show the $C_{B/P}$ for the 200-mm and 150-mm wafers with area-array pads. It can be seen that the saving in $C_{B/P}$ is very compelling, since there are many more pads on the chip than in the peripheral-array cases.

Consider a 200-mm wafer with $A = 100$ mm^2 and $p = 0.25$ mm. Assuming $Y_T = 0.9$, then $C_{B/P} = \$0.00036$ for area-array pads, $C_{B/P} = \$0.00311$ for peripheral-array pads, and $C_{B/P} = \$0.00158$ for peripheral-staggered-array pads. It can be seen that the cost saving for chips with area-array pads is very compelling.

1.2.3.4 Wire-bonding cost per die. The wire-bonding cost per die $C_{W/D}$ can be determined by

$$C_{W/D} = \frac{Y_I L_A C_W N_P}{Y_T} \qquad (1.9)$$

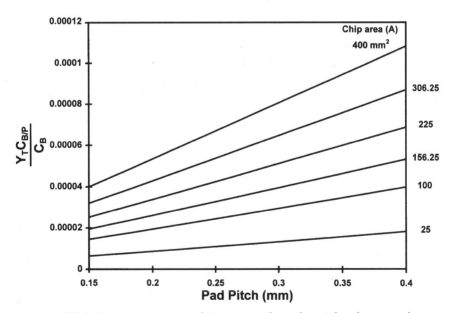

Figure 1.8 Wafer-bumping cost per pad (150-mm wafer with peripheral-staggered-array pads).

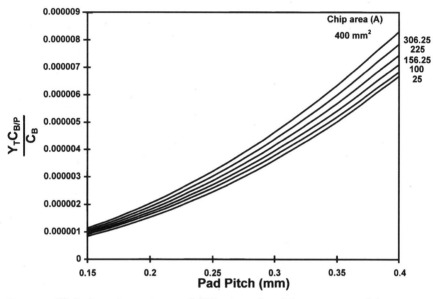

Figure 1.9 Wafer-bumping cost per pad (200-mm wafer with area-array pads).

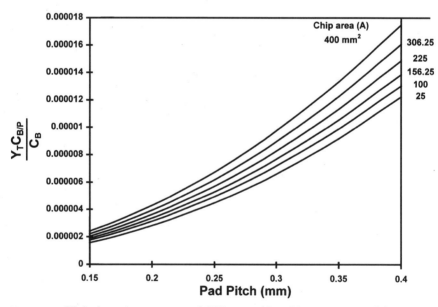

Figure 1.10 Wafer-bumping cost per pad (150-mm wafer with area-array pads).

where C_W is the wire cost (Table 1.3), L_A is the average wire length per pad, Y_I is the initial screening IC chip yield (after 1-MHz screening test), Y_T is the true IC chip yield, and N_p is given by either Eq. (1.5) or (1.6).

Figures 1.11 and 1.12 show, respectively, the $C_{W/D}$ for the chips with peripheral-array pads and for those with peripheral-staggered-array pads ($x = y$ has been assumed). It can be seen that the $C_{W/D}$ decreases with pad pitch increase and decreases with chip-size decrease. Also, the $C_{W/D}$ with peripheral-staggered-array pads is almost half of that with peripheral-array pads.

Consider a chip with $x = 10$ mm, $y = 8$ mm, and $p = 0.25$ mm. Assuming $L_A = 4.572$ mm, $C_W = \$70.30/\text{K ft} = \$0.000231/\text{mm}$, $Y_I = 0.9$, and $Y_T = 0.6$, then $C_{W/D} = \$0.22$ for peripheral-array pads and

TABLE 1.3 Cost of 99.99% Gold Wire (as of September 1998)

(Spool: 2 in)

Diameter (in)	1000 ft	5000 ft	10,000 ft	25,000 ft
0.0007	\$485.00/lot	\$143.70/1000 ft	\$93.70/1000 ft	\$60.00/1000 ft
0.001	\$375.00/lot	\$135.30/1000 ft	\$96.50/1000 ft	\$70.30/1000 ft
0.0015	\$410.00/lot	\$192.10/1000 ft	\$149.70/1000 ft	\$121.10/1000 ft

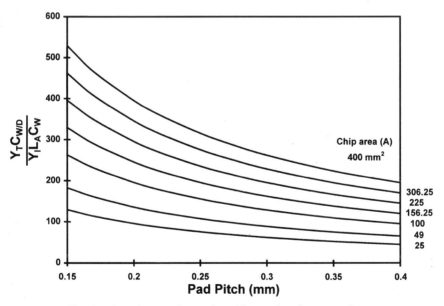

Figure 1.11 Wire-bonding cost per die (wafer with peripheral-array pads).

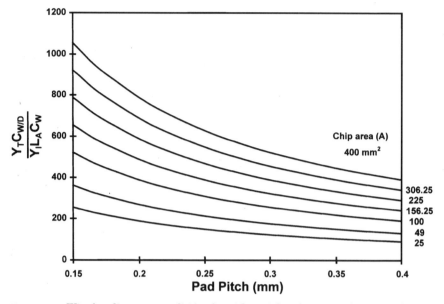

Figure 1.12 Wire-bonding cost per die (wafer with peripheral-staggered-array pads).

$C_{W/D}$ = \$0.44 for peripheral-staggered-array pads. If Y_I = 0.7 and Y_T = 0.6, then $C_{W/D}$ = \$0.17 for peripheral-array pads and $C_{W/D}$ = \$0.34 for peripheral-staggered-array pads. It can be seen that for wire-bonding technology, it is desirable to have $Y_I \sim Y_T$; otherwise, the expensive gold wire will be bonded on the bad dies.

1.2.3.5 Wire-bonding cost per pad. The wire-bonding cost per pad $C_{W/P}$ can be determined by

$$C_{W/P} = \frac{Y_I L_A C_W}{Y_I} \tag{1.10}$$

Equation (1.10) is shown in Fig. 1.13 with various values of Y_I and Y_T. It can be seen that the higher the Y_T, the lower the $C_{W/P}$. Also, for a given Y_T, the $C_{W/P}$ increases with a higher Y_I. For example, for a 200-mm wafer with A = 100 mm² and p = 0.25 mm, assuming L_A = 4.572 mm and C_W = \$0.000231/mm, then $C_{W/P}$ = \$0.001115 if Y_T = 0.9 and Y_I = 0.95, and $C_{W/P}$ = \$0.001056 if $Y_I = Y_T$ = 0.9. By comparing this example with that in Sec. 1.2.3.3, it can be seen that $C_{W/P}$ is smaller than $C_{B/P}$. However, flip-chip technology provides the possibility for area-array pads and in this case $C_{B/P}$ is a few times smaller than $C_{W/P}$.

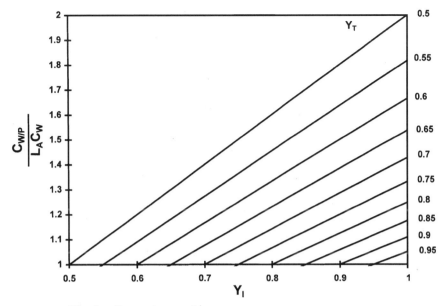

Figure 1.13 Wire-bonding cost per pad.

1.2.4 Cost comparison

The cost ratio ß between the wafer-bumping cost per die, Eq. (1.7), or per pad, Eq. (1.8), and the wire-bonding cost per die, Eq. (1.9), or per pad, Eq. (1.10), is given by

$$ß = \frac{C_B}{Y_I L_A C_W N_P N_C} \qquad (1.11)$$

Equation (1.11) is plotted in Figs. 1.14 and 1.15, respectively, for 200-mm and 150-mm wafers with peripheral-array pads. It can be seen that the ß is larger for larger chip size and larger pad pitch. Also, the ß is larger for smaller Y_I. Figures 1.16 and 1.17 show the plots of Eq. (1.11), respectively, for the 200-mm and 150-mm wafers with peripheral-staggered-array pads. It can be seen that the ß drops to almost half of that with peripheral array pads.

For a 200-mm wafer with $x = y = 10$ mm and $p = 0.25$ mm on a peripheral-staggered-array pad configuration, assuming that $L_A = 4.572$ mm, $C_W = \$0.000231/mm$, $C_B = \$100$, and $Y_I = 0.9$, then ß = 1.35. This means that from the materials, labor, and operation points of view, flip-chip technology is more expensive than wire-bonding technology. It should be remembered that the major equipment cost for wire bonding is higher than that for flip chip (1.77 times). Thus, flip chip is still

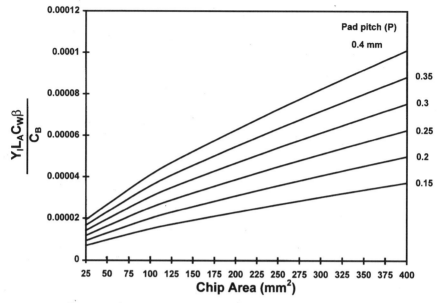

Figure 1.14 Cost ratio between wafer-bumping cost per pad and wire-bonding cost per pad (200-mm wafer with peripheral-array pads).

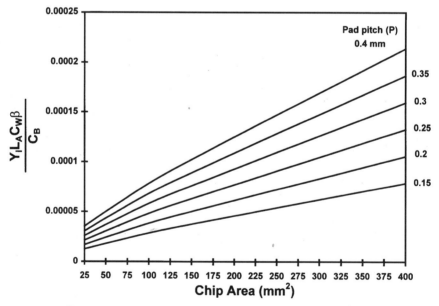

Figure 1.15 Cost ratio between wafer-bumping cost per pad and wire-bonding cost per pad (150-mm wafer with peripheral-array pads).

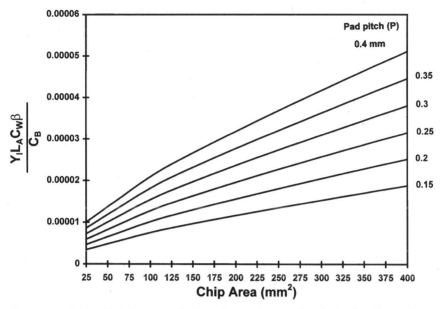

Figure 1.16 Cost ratio between wafer-bumping cost per pad and wire-bonding cost per pad (200-mm wafer with peripheral-staggered-array pads).

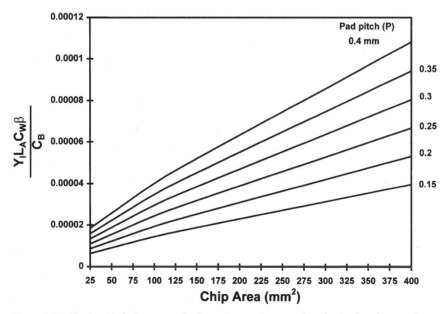

Figure 1.17 Cost ratio between wafer-bumping cost per pad and wire-bonding cost per pad (150-mm wafer with peripheral-staggered-array pads).

cheaper. If we reduce the pitch to $p = 0.16$ mm and keep everything the same, then ß $= 0.85$. That means that flip chip is cheaper even without the major equipment cost advantage. However, in this case, for a peripheral-array pad configuration, wire bonding is cheaper.

If we reduce the chip size to $x = y = 5$ mm, then ß $= 0.63$ for $p = 0.25$ mm on peripheral-staggered-array pads and ß $= 0.39$ for $p = 0.16$ mm on peripheral-staggered-array pads. Thus, for this case, flip chip is cheaper even with peripheral-array pads. It should be pointed out, as shown in Sec. 1.2.3.3, that flip chip could be even cheaper if an area-array pad configuration were used. Unfortunately, it cannot be compared with wire bonding, since the latter usually does not use area-array pads.

1.2.5 Summary

The cost analysis of wire-bonding chip and solder-bumped flip chip on a CSP organic substrate has been presented. Useful equations for determining the wafer-bumping and wire-bonding costs per die and per pad have also been provided. Furthermore, useful charts relating the important variables such as the wafer size, IC chip yields, chip dimensions, and costs are provided for engineering practice convenience. Some important results are summarized as follows:

- The major equipment cost for wire bonding is higher than that for flip chip.
- Wire bonding occupies more manufacturing floor space than flip chip does.
- The wafer-bumping cost per die increases as Y_T decreases.
- The wafer-bumping cost per die is lower for larger wafers.
- The wafer-bumping cost per pad is higher for larger chip sizes.
- The wafer-bumping cost per pad is higher for larger pad pitches.
- The wafer-bumping cost per pad is lower for larger wafers.
- The wafer-bumping cost per pad is very low for area-array chips.
- The wire-bonding cost per die decreases as pad pitch increases.
- The wire-bonding cost per die decreases as chip size decreases.
- The wire-bonding cost per pad increases with a higher ratio of Y_I and Y_T.
- Area-array solder-bumped flip chip is cheaper than wire bonding.
- For peripheral-array chips, wire bonding is cheaper than flip chip if the chip size and pad pitch are very large. Otherwise, flip chip is cheaper.

1.3 How to Select Underfill Materials

One of the major reasons why solder-bumped flip chip on low-cost organic CSP substrates works is because of the underfill epoxy encapsulant [1–6, 10–29]. It reduces the effect of the global thermal expansion mismatch between the silicon chip and the organic substrate, i.e., it reduces the stresses and strains in the flip-chip solder bumps (since the chip and the substrate are tightly held by the underfill) and redistributes over the entire chip area the stresses and strains that would otherwise be increasingly concentrated near the corner solder bumps of the chip. Other advantages of underfill encapsulant are that it protects the chip from moisture, ionic contaminants, radiation, and hostile operating environments such as thermal [3, 5, 18], mechanical pull [3], shear [3, 22, 26–28], and twist [3, 22], and shock/vibration [24].

In this chapter, eleven different underfill encapsulants with different filler size and content and different epoxies are studied. Their curing conditions, such as time and temperature, are measured by a differential scanning calorimeter (DSC) unit. Their material properties, such as the TCE (thermal coefficient of expansion), T_g (glass transition temperature), dynamic storage modulus, tangent delta, and moisture content, are measured using thermal mechanical analysis (TMA), dynamic mechanical analysis (DMA), and thermal gravimetric analysis (TGA). Their flow rate and mechanical (shear) strength in a solder-bumped flip chip on board are measured. Their effects on the electrical performance (voltage) of a functional flip-chip device are determined experimentally. For each test configuration, the sample size is three and the values reported herein are the average.

1.3.1 Underfill materials and applications

There are eleven different encapsulant materials under consideration, namely, Underfills A, B, C, D1, D2, D3 (three different lots), E, F, G, H, and I (Table 1.4). All of the underfill encapsulants are premixed at the supplier sites, packed in plastic syringes (5 to 10 mL), and then frozen packed at $-40°C$ to prevent curing. In shipping, these underfill materials require special handling to maintain the low temperature continuously. Upon receiving the package, one needs to unpack the package, take out the syringes quickly, and store them in a freezer at an uninterrupted temperature of $-40°C$. Under these conditions, most of the underfill encapsulants would have approximately one year storage life.

The filler content and size for Underfills A, B, C, D1, D2, D3, E, F, G, H, and I are shown in Table 1.4. For all these underfills, the filler is silica. The resin of Underfills A, B, C, D1, D2, and D3 is bisphenol-type epoxy; that of Underfill H is a mixture of bisphenol epoxy and a

TABLE 1.4   Average Curing Conditions for Underfills A, B, C, D1, D2, D3, E, F, G, H, and I

Underfill	Filler content and size	Type of epoxy	Curing temperature (°C)	Curing condition temperature (°C)/time (min)			
A	40%, <20 μm	Bis	138.3	150/15	160/8	165/6	170/4
B	60%, <20 μm	Bis	139.7	150/15	160/7	165/5	170/4
C	60%, <20 μm	Bis	135.1	150/30	160/13	165/9	170/6
D1	60%, <20 μm	Bis	140.5	150/33	160/13	165/9	170/6
D2	60%, <20 μm	Bis	148.5	150/35	160/15	165/10	170/7
D3	60%, <10 μm	Bis	140.9	150/19	160/9	165/7	170/5
E	67%, <5 μm	Cyc*	158.3	150/190	160/90	165/64	170/45
F	67%, <20 μm	Cyc*	163.4	150/272	160/128	165/90	170/63
G	67%, <5 μm	Cyc*	157.2	150/528	160/236	165/160	170/109
H	68%, <12 μm	Bis*	141.2	150/295	160/68	165/34	170/17
I	67%, <11 μm	Cyc*	145.0	150/140	160/39	165/21	170/12

Bis means bisphenol; Bis* is a mixture of bisphenol and trade secret resin; Cyc* is a mixture of cycloaliphatic and trade secret resin.

trade secret resin; and that of Underfills E, F, G, and I is a mixture of cycloaliphatic epoxy and a trade secret resin. Before the underfill encapsulants are applied, they are removed from the freezer and thawed at room temperature. It takes about an hour to thaw a 5- to 10-mL syringe. Once thawed, most of the underfill materials have a pot life of 14 to 16 h.

Dispensing the underfill encapsulants through a syringe is accomplished with a version system (to locate the edges of the chip) and a pumping system (to control the amount of underfill material). The temperature of the heating plate is around 90°C. After the underfill materials are dispensed on either one side or two adjacent sides of the chip, the chip and substrate should remain on the heating plate for about 30 to 90 s (depending on the chip, and the gap between the chip and the substrate and the temperature of the hating plate) to enhance the flow rate of the underfill. After the underfill has flowed out from the other sides of the chip, the chip on board will be put into a curing oven for 10 to 60 min at 130 to 170°C depending on the curing conditions of the underfill materials.

The most desired features of underfill materials are (1) low viscosity (fast flow), which can increase throughput, (2) low curing temperature/fast curing time, which can reduce cost and be less harmful to other components, (3) low TCE, which can reduce thermal expansion mismatch between the chip, solder bumps, and substrate, (4) high modulus, which leads to good mechanical properties, (5) high glass transition temperature T_g, which enables the material to endure higher-temperature environments, (6) low moisture absorption, which can extend shelf life, and (7) good adhesion between the underfill and the solder mark on the PCB and between the underfill and the passivation on the chip, which can improve product lifetime.

1.3.2 Curing conditions

To determine the curing conditions of underfill materials, they are put into an aluminum pan (which will form a disc sample with dimensions 6.4 ± 0.2 mm in diameter and 1.6 ± 0.1 mm in height), weighed, and then put into a DSC unit. The objective of DSC is to measure the amount of energy (heat) absorbed or released by a sample as it is heated, cooled, or held at a constant (isothermal) temperature. The instrument design consists of two independent furnaces, one for the sample and one for the reference (Fig. 1.18). When an exothermic or endothermic change occurs in the sample material, energy is applied to or removed from one or both furnaces to compensate for the energy change occurring in the sample. Since the system is always directly measuring energy flow to or from the sample, DSC can directly measure melting temperature, T_g, temperature onset of crystallization, and temperature onset of curing. The kinetic software enables analysis of a DSC peak to obtain specific kinetic parameters that characterize a reaction process.

Any material reaction can be represented by the following equation:

$$A \xrightarrow{k} B + \Delta H \tag{1.12}$$

Figure 1.18 Specimen setup for DSC (differential scanning calorimeter).

where A is the material before reaction, B is the material after reaction, ΔH is the heat absorbed or released, and k is the Arrhenius rate constant. The Arrhenius equation is given by

$$k = Z \exp(-E_a/RT) \qquad (1.13)$$

where Z is the pre-exponential constant, E_a is the activation energy of the reaction, R is the universal gas constant [8.314 J/(°C · mol)], and T is the absolute temperature in kelvins.

The rate of reaction dx/dt can be directly measured by DSC and is expressed as

$$dx/dt = k(1 - x)^n \qquad (1.14)$$

where dx/dt is the rate of reaction, x is the fraction reacted, t is the time, k is the Arrhenius rate constant, and n is the order of reaction. Combining the above equations and assuming an nth-order reaction kinetics and constant program rate, activation energy, and pre-exponential constant, we have

$$dx/dt = Z \exp(-E_a/RT)(1 - x)^n \qquad (1.15)$$

The fraction reacted x is directly related to the fractional area of the DSC reaction peak. The kinetic parameters Z, E_a, and n are determined with an advanced multilinear regression method (MLR). Thermal scan is carried out at a 5°C/min heating rate ranging from 40 to 250°C.

Figure 1.19 shows a typical DSC thermal scan curve, and Fig. 1.20 shows typical degree of conversion (reaction) versus time curves for Underfill D2. It can be seen that the average curing temperature of Underfill D2 is 148.5°C and that Underfill D2 cannot be 100 percent cured if the applied temperature is less than 148.5°C. Table 1.4 summarizes the average curing conditions for Underfills A, B, C, D1, D2, D3, E, F, G, H, and I. It can be seen that the curing conditions vary with different underfill materials and that the materials cure faster at higher temperatures. Also, Underfills A, B, C, D1, D2, and D3 cure faster than the others.

1.3.3 Underfill material properties

1.3.3.1 TCE. The TCE of Underfills A, B, C, D1, D2, D3, E, F, G, H, and I (with sample dimensions 6.4 ± 0.2 mm diameter and 1.6 ± 0.1 mm height) is determined by TMA in an expansion quartz system (50 to 200°C) at a 5°C/min heating rate (Fig. 1.21). The objective of TMA is to measure the change in the dimensions of a sample (such as ex-

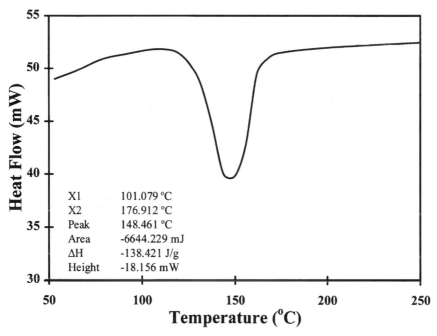

Figure 1.19 DSC thermal scan curve for Underfill D2.

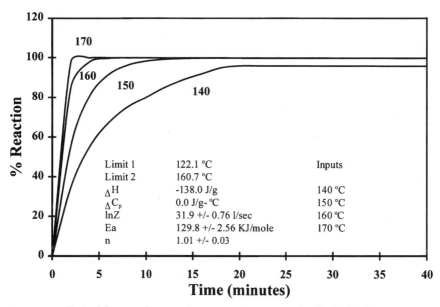

Figure 1.20 Typical degree of conversion versus time curves for Underfill D2.

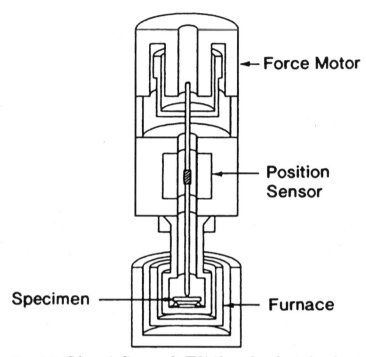

Figure 1.21 Schematic diagram of a TMA (thermal mechanical analyzer) and DMA (dynamic mechanical analyzer). The TMA force motor applies static load, whereas the DMA force motor applies dynamic load on the specimen.

pansion or contraction) as the sample is heated, cooled, or held at a constant (isothermal) temperature. The instrument design consists of a platinum-wound furnace system which can be operated from -170 to $1000°C$. The scan rate is from 0.1 to $100°C/min$. The specimen is mounted between a quartz platform and a probe, and then a static load F_S (Fig. 1.22) is applied to the specimen; the dimensions of the specimen are monitored by the linear variable differential transducer (LVDT) throughout the analysis. The TCE of the material is obtained from the slope of the dimension change versus temperature curve, and the T_g of the material is obtained from the onset of the two different slopes of the curve.

Figure 1.23 shows a typical expansion curve for Underfill D2, and Table 1.5 summarizes the average results for Underfills A, B, C, D1, D2, D3, E, F, G, H, and I. It can be seen that the TCE of Underfill A (40 percent filler content) is the largest ($46.5 \times 10^{-6}/°C$) and the TCE of Underfill E (67 percent filler content) is the smallest ($21 \times$

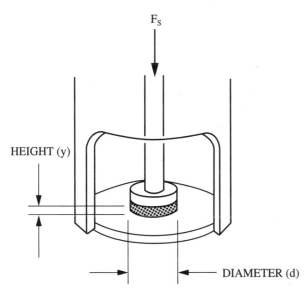

Figure 1.22 Outlook of quartz tube, probe, and specimen
with height = y and diameter = d of TMA for TCE measurement. (F_s = static force)

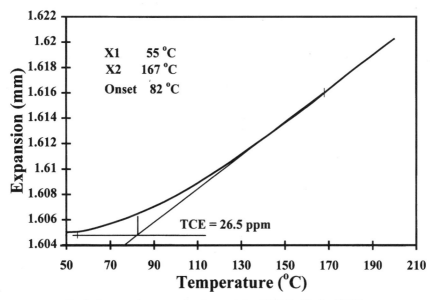

Figure 1.23 Typical expansion curve for determining TCE for Underfill D2.

TABLE 1.5   Average Material Properties of Underfills A, B, C, D1, D2, D3, E, F, G, H, and I and Their Effects on Moisture Content

Underfill	Filler content and size	TCE (ppm/°C)	T_g (°C)	Storage modulus (GPa)			Moisture content (%)	
				25°C	55°C	110°C	Dry	20 h of steam aging
A	40%, <20 μm	46.5	142	3.88	3.8	3.0	0.09	0.33
B	60%, <20 μm	36.2	142	3.3	3.3	2.73	0.05	0.51
C	60%, <20 μm	26.1	158	4.4	3.6	3.42	0.04	0.28
D1	60%, <20 μm	23.4	150	6.45	6.38	5.62	0.07	0.38
D2	60%, <20 μm	26.6	140	6.41	6.24	5.06	0.05	0.28
D3	60%, <10 μm	24.2	150	6.73	6.6	5.84	0.09	0.22
E	67%, <5 μm	21.0	137	4.8	3.3	2.86	0.04	0.52
F	67%, <20 μm	38.6	120	5.4	5.3	2.0	0.08	0.58
G	67%, <5 μm	39.3	140	7.2	6.9	4.2	0.05	0.42
H	68%, <12 μm	28.6	151	7.8	7.67	6.67	0.07	0.28
I	67%, <11 μm	30.0	186	5.09	5.09	4.79	0.09	0.28

$10^{-6}/°C$). In general, the higher the filler content, the lower the TCE. This is because the filler (made from silica) expands less than the epoxy resin. For solder joint thermal-fatigue reliability, it is preferable to have lower-TCE ($<27 \times 10^{-6}/°C$) underfill materials to reduce the thermal expansion mismatch among the chip, solder bumps, and substrate.

1.3.3.2 Storage modulus. The modulus for Underfills A, B, C, D1, D2, D3, E, F, G, H, and I is measured with a three-point bending specimen (3.0 ± 0.3 mm $\times$ 2.9 ± 0.3 mm $\times$ 19 ± 3 mm) in a DMA unit (55 to 200°C) at a heating rate of 5°C/min (Fig. 1.21). The objective of DMA is to measure mechanical properties such as modulus as a function of time, temperature, frequency, stress, or combinations of these parameters. The force motor of the instrument can be programmed to apply constant stress, dynamic stress, or combinations of both. The core rod applies stress to the sample and is held in place using an electromagnetic suspension. The ceramic furnace with platinum furnace element is capable of heating and cooling at a very high rate and can be heated up to 1000°C. For electronics packaging materials, flexural properties such as the storage modulus, loss modulus, and tangent delta (tan δ) can be obtained with DMA.

The storage modulus E_S is a measure of the energy stored per cycle of deformation and can be expressed as

$$E_S = \frac{F_d x^3 \cos \delta}{4y^3 z \Delta} \tag{1.16}$$

where F_d is the dynamic load, δ is the phase angle, and Δ is the maximum dynamic deflection of the specimen (Fig. 1.24). Figure 1.25 shows a typical flexural storage modulus curve for Underfill D2 as a function of temperature. The average results for the storage modulus of Underfills A, B, C, D1, D2, D3, E, F, G, H, and I are reported in Table 1.5. It can be seen that the storage modulus of all the underfill materials is temperature-dependent: The higher the temperature, the

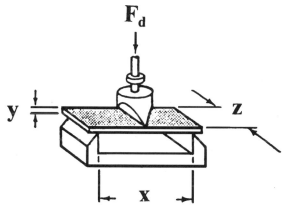

Figure 1.24 Three-point bending for dynamic modulus and tangent delta. F_d = dynamic force.

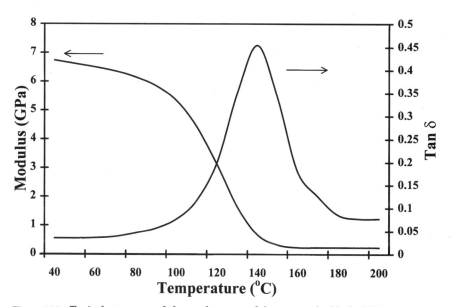

Figure 1.25 Typical storage modulus and tangent delta curves for Underfill D2.

lower the modulus. Also, in general, the higher the filler content (which acts like reinforcement in filler-reinforced composites), the higher the storage modulus.

The flexural loss modulus E_l is a measure of the energy lost per cycle of deformation and can be expressed as

$$E_l = \frac{F_d x^3 \sin \delta}{4y^3 z \Delta} \tag{1.17}$$

1.3.3.3 Tan δ and T_g. Tangent delta (tan δ), which is a measure of material-related damping property, for Underfills A, B, C, D1, D2, D3, E, F, G, H, and I can be obtained by dividing E_l by E_s, or

$$\tan \delta = E_l/E_s \tag{1.18}$$

The temperature at the peak of a tan δ curve is often reported in literature as the glass transition temperature T_g. Figure 1.25 shows a typical tangent delta curve for Underfill D2, and the average values of T_g for all the underfills are shown in Table 1.5. In general, the T_g is higher for underfills with bisphenol epoxy than for those with cycloaliphatic epoxy; the exception is Underfill I, which uses a high-T_g resin.

1.3.3.4 Moisture content. Two sets of tests are carried out to determine the moisture content; one is for dry specimens and the other is for steam aging specimens. The steam aging specimens are prepared under steam evaporation for 20 h in a closed hot water bath. All the specimens are 6.4 ± 0.2 mm in diameter and 1.6 ± 0.1 mm in height. Weight changes for Underfill A, B, C, D1, D2, D3, E, F, G, H, and I are measured with TGA equipment (Fig. 1.26) at 110°C for 4 h. The objective of TGA is to measure the change in the mass of a sample as the sample is heated, cooled, or held at a constant (isothermal) temperature. The instrument consists of a microbalance which allows the sensitive measurement of weight changes as small as a few micrograms, furnace, and sample holder area. The vertical analytical design also serves to isolate the balance mechanism from the furnace, eliminating temperature fluctuations which cause drifts and nonlinearity. This system can measure curie point, decomposition temperature, moisture uptake, and component separation.

The change in mass during thermal scan can be expressed as

$$\frac{W_f - W_i}{W_i} \times 100\% \tag{1.19}$$

where W_f is the final weight after thermal scan and W_i is the initial weight before thermal scan. Figure 1.27 shows a typical percent

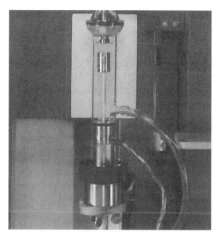

Figure 1.26   High-sensitivity weight measurements with TGA (thermal gravimetric analyzer).

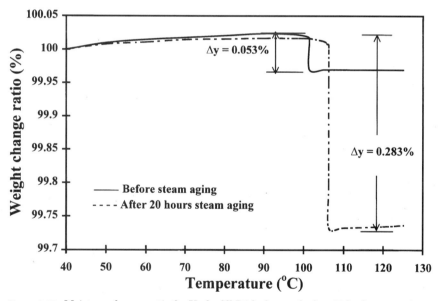

Figure 1.27   Moisture change ratio for Underfill D2 before and after 20 h of steam aging.

weight change ratio (moisture content) for Underfill D2 before and after 20 h of steam aging. The average moisture content of all the underfills is shown in Table 1.5. It can be seen that the average moisture content of all the underfills after 20 h of steam aging is at least three times that before (the dry condition). In general, this is especially true for higher-filler-content underfills, since porosity increases when the amount of filler increases.

1.3.4 Underfill flow rate

The flow rate of Underfills A, B, C, D1, D2, D3, E, F, G, H, and I is measured underneath a functional solder-bumped flip chip on a BT (bismaleimide triazine) substrate. The dimensions of the chip are 6.3×3.6 mm, and it has 32 bumps (Fig. 1.28). The BT substrate has two metal layers (Figs. 1.29 and 1.30 show, respectively, the top and bottom layers) and 32 vias (0.254 mm in diameter). The standoff height of the solder bumps is 0.07 ± 0.02 mm. The assembled flip chip is placed on a hot plate at 80°C. Approximately 0.025 mL of room-temperature underfill is placed around two sides of the chip in an L

Figure 1.28 UMC's chip with 32 solder bumps on two opposite sides.

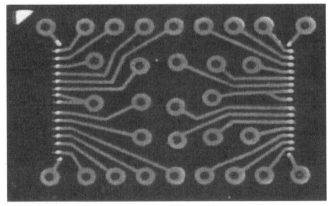

Figure 1.29 Top side of the low-cost substrate with 32 vias and 32 Cu pads on two opposite sides.

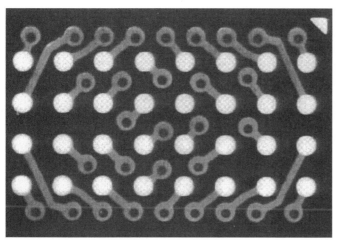

Figure 1.30 Bottom side of the low-cost substrate with 32 vias and 32 1-mm pitch land grid array (LGA) Cu pads.

TABLE 1.6   Average Flow Rate of Solder-Bumped Flip Chip on Board with Underfills A, B, C, D1, D2, D3, E, F, G, H, and I

Underfill	Filler content and size	Flow rate (mm/s)
A	40%, <20 μm	0.167
B	60%, <20 μm	0.133
C	60%, <20 μm	0.003
D1	60%, <20 μm	0.3
D2	60%, <20 μm	0.3
D3	60%, <10 μm	0.3
E	67%, <5 μm	0.118
F	67%, <20 μm	0.105
G	67%, <5 μm	0.105
H	68%, <12 μm	0.63
I	67%, <11 μm	0.42

pattern. The time for the underfill to completely fill the gap is recorded. It should be pointed out that there is no underfill material flow to the bottom side of the substrate.

The average flow rate for Underfills A, B, C, D1, D2, D3, E, F, G, H, and I is shown in Table 1.6. It can be seen that, in general, the flow rate is strongly affected by the filler content and size: The greater the filler content and the larger the filler size (i.e., the higher the viscosity), the lower the flow rate.

1.3.5 Mechanical performance

The Royce Instruments Model 550 is used to perform the mechanical shear tests. The shear wedge is placed against one edge of the solder-bumped flip chip with underfill on the BT substrate, which is clamped on the stage. A push of the wedge is applied to shear the chip/bumps/underfill away. This is a destructive test.

The average test results for Underfills A, B, C, D1, D2, D3, E, F, G, H, and I are shown in Table 1.7, and a typical load-deflection curve for a solder-bumped flip chip on board with Underfill D2 is shown in Fig. 1.31. It can be seen from Table 1.7 that the average shear force varies for the different underfills under consideration. It can also be seen that, in most of the cases, the shear force for the solder-bumped flip chip with underfill drops about 25 percent after 20 h of steam aging. Thus, the moisture content has significant influence on the mechanical performance of underfilled flip-chip assemblies.

There are two failure modes. In one, the chip is broken into pieces and some of the broken pieces are left on the substrate (Fig. 1.32). Most of the dry underfills exhibit this type of failure. In the other failure mode, the whole chip is peeled off from the substrate and some underfill is left on the chip (Fig. 1.33). Most of the underfills exhibit this type of failure after 20 h of steam aging. The measured shear forces for the first failure mode are larger than those for the second failure mode. It

TABLE 1.7 **Effects of Moisture on the Mechanical Shear Test Performance of Solder-Bumped Flip Chip on Board with Underfills A, B, C, D1, D2, D3, E, F, G, H, and I**

| | Shear force (kgf) | | |
Underfill	Dry condition	20 h of steam aging	Change (%)
A	43.2	33.8	27.8
B	40.5	31.3	29.4
C	43.5	34.2	27.2
D1	55.2	42.3	30.5
D2	65.5	62.2	5.3
D3	62.4	31.2	100
E	41.6	33.2	25.3
F	47.9	32.3	48.3
G	21.7	18.3	18.6
H	43.1	39.9	8.0
I	31.2	25.8	20.9

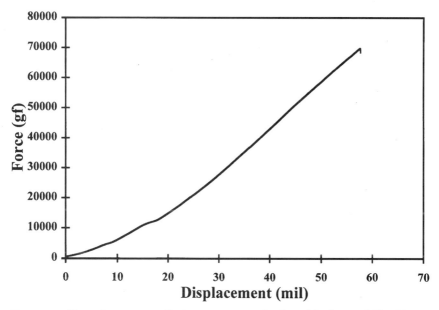

Figure 1.31 Shear force versus displacement curve for the solder-bumped flip-chip assembly with Underfill D2.

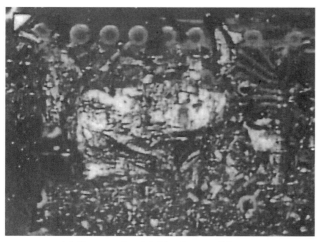

Figure 1.32 Failure mode of solder-bumped flip chip on board with underfill at dry condition. The chip is broken into many small pieces, some of which remain on the substrate.

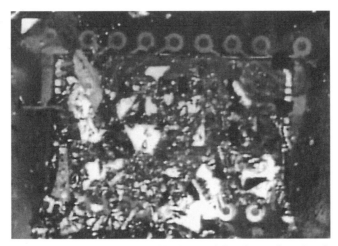

Figure 1.33 Failure mode of solder-bumped flip chip on board with underfill after 20 h of steam aging. The whole chip has peeled off from the substrate.

seems that moisture degrades interface adhesion and leads to debonding of the interface between the chip and the underfill encapsulant.

1.3.6   Electrical performance

The electrical performance of Underfills A, B, C, D1, D2, D3, E, F, G, H, and I is measured under 5 V dc by using a Kepco power supply. The average voltage readout for the solder-bumped flip chip on board with and without underfill is summarized in Table 1.8. It can be seen that, for all the underfills under consideration, there is almost no difference in voltage readout between chips on board with underfill at dry condition and those without underfill. Also, for solder-bumped flip chips on board with underfills, the difference in voltage readout between those before and those after 20 h of steam aging is insignificant.

1.3.7   Ranking of underfill encapsulants

How to choose an underfill encapsulant? The answer is not easy, since the performance of underfill depends on many factors, such as, among others, the curing condition, TCE, storage modulus, T_g, moisture uptake, flow rate, and shear strength (adhesion). However, based on the measured results for the eleven different underfills under consideration, a simple method is proposed to rank (choose among) the underfills.

Table 1.9 shows the grading criteria for nine important properties of underfills: (1) curing time at 165°C, (2) TCE, (3) T_g, (4) storage modulus at 25°C, (5) storage modulus at 110°C, (6) moisture uptake,

TABLE 1.8   Average Electrical Performance of Solder-Bumped Flip Chip on Board with Underfills A, B, C, D1, D2, D3, E, F, G, H, and I

| | Voltage readout (V) | | | | | |
| | Without underfill | | With Underfill at dry condition | | With underfill after 20 h of steam aging | |
Underfill	Low	High	Low	High	Low	High
A	3.88	4.05	3.88	4.05	3.87	4.03
B	3.92	4.06	3.94	4.07	3.93	4.05
C	3.91	4.05	3.88	4.05	3.85	4.04
D1	3.90	4.05	3.89	4.05	3.87	4.04
D2	3.91	4.05	3.91	4.06	3.89	4.05
D3	3.91	4.06	3.90	4.06	3.88	4.05
E	,3.93	4.08	3.94	4.10	3.92	4.10
F	3.92	4.07	3.93	4.07	3.90	4.08
G	3.94	4.09	3.95	4.10	3.93	4.09
H	3.89	4.05	3.90	4.05	3.88	4.04
I	3.92	4.07	3.92	4.08	3.90	4.06

TABLE 1.9   Underfill Material Properties Grading Criteria

| | Grading (1 = poor; 3 = average; 6 = excellent)* | | | | | |
Properties	1	2	3	4	5	6
1. Curing time (min) at 165°C	>60	40–59	30–39	20–29	10–19	<10
2. TCE (ppm/°C)	45–49	40–44	35–39	30–34	25–29	20–24
3. T_g (°C)	120–129	130–139	140–149	150–159	160–169	>170
4. Storage modulus (GPa) at 25°C, 1 Hz	<3	3.1–3.9	4–4.9	5–5.9	6–6.9	7–7.9
5. Storage modulus (GPa) at 110°C, 1 Hz	<2	2.1–2.9	3–3.9	4–4.9	5–5.9	6–6.9
6. Moisture uptake (%)	0.55–0.6	0.5–0.54	0.4–0.49	0.35–0.39	0.3–0.34	<0.3
7. Flow rate (mm/s)	<0.1	0.1–0.19	0.2–0.29	0.3–0.39	0.4–0.49	>0.5
8. Shear force (kgf) before steam aging	20–29	30–39	40–44	45–49	50–59	>60
9. Shear force (kgf) after steam aging	<20	20–29	30–34	35–39	40–59	>60

*The grading is based on the present test results of Underfills A, B, C, D1, D2, D3, E, F, G, H, and I.

(7) flow rate, (8) shear force at dry condition, and (9) shear force after 20 h of steam aging. For example, if the curing time at 165°C is 15 min, then the grading is 5. Also, for example, if the flow rate is 0.1 mm/s, then the grading is 2.

The rationales for establishing this table are: (1) The faster the curing time, the lower the manufacturing cost and the less harm may be done to other components; (2) the lower the TCE, the less the mismatch stresses; (3) the higher the T_g, the higher the temperature the underfill can withstand; (4) and (5) the higher the storage modulus, the better the mechanical performance; (6) the lower the moisture uptake, the better the shelf life; (7) the faster the flow rate, the higher the throughput; and (8) and (9) the higher the shear force, the better the adhesion strength.

Table 1.10 shows the ranking of Underfills A, B, C, D1, D2, D3, E, F, G, H, and I based on their measurement results. The way this table was constructed is as follows: For Underfill A, the measured curing time at 165°C is 6 min (Table 1.4), and thus in column 1 it is given a grade of 6 (<10 min in Table 1.9); the measured TCE is 46.5 ppm/°C (Table 1.5), and thus in column 2 it is given a grade of 1 (45–49 ppm/°C in Table 1.9); the measured T_g is 142°C (Table 1.5), and thus in column 3 it is given a grade of 3 (140–149°C in Table 1.9); etc.

It can be seen from Table 1.10 that, based on the total grading, Underfills D2, D3, D1, and H are ranked above the others. Thus, they are chosen for the solder-bumped flip chip on board applications. However, it should be noted that these underfills are not perfect. For example, Underfill D2 has the best shear force results; however, its flow rate is less than half that of Underfill H.

TABLE 1.10   Ranking of Underfills A, B, C, D1, D2, D3, E, F, G, H, and I

Underfill	Properties grading (from measured results and Table 1.9)									Sum	Rank
	1	2	3	4	5	6	7	8	9		
A	6	1	3	2	3	5	2	3	3	28	5
B	6	3	3	2	2	2	2	3	3	26	6
C	6	5	4	3	3	6	1	3	3	34	4
D1	6	6	4	5	5	4	4	5	5	44	2
D2	5	5	3	5	5	6	4	6	6	45	1
D3	6	6	4	5	5	6	4	6	3	45	1
E	1	6	2	3	2	2	2	3	3	24	7
F	1	3	1	4	1	1	2	4	3	20	8
G	1	3	3	6	4	3	2	1	1	24	7
H	3	5	5	6	6	6	6	3	4	44	2
I	4	4	6	4	4	6	5	2	2	37	3

1.3.8 Summary

Eleven different underfills with different resin epoxies and different filler contents and sizes have been studied. Their curing conditions and material properties have been measured using DSC, TMA, DMA, and TGA equipment. Furthermore, their flow rates and effects on mechanical (shear) and electrical (voltage) performance in a solder-bumped functional flip chip on an organic substrate have been experimentally determined. Some important results are summarized as follows:

- A simple method has been presented for choosing (ranking) underfill materials.

- For all the underfills considered, curing time is temperature-dependent; they cured faster at higher temperatures.

- For all the underfills considered, the higher the filler content, the lower the TCE.

- In general, the T_g is higher for underfill materials with bisphenol epoxy than for those with cycloaliphatic epoxy; the exception is Underfill I, which uses a high-T_g resin.

- For all the underfills considered, the storage modulus is temperature-dependent: The higher the temperature, the lower the modulus.

- For all the underfills considered, the moisture content after 20 h of steam aging is at least three times that in the dry condition.

- In general, the flow rate is strongly affected by the filler content and size: the greater the filler content and the larger the size, the lower the flow rate.

- In general, the moisture content in the underfill significantly affects the mechanical shear strength of solder-bumped flip-chip assemblies.

- For all the underfills considered, the effects of underfill on the electrical (voltage readout) performance of the functional chip are insignificant.

1.4 Summary

As mentioned earlier, there are more than 40 different CSPs reported in the literature. Even though they are different in design, materials, and applications, however, they can be classified into four groups, namely (1) customized-lead-frame-based CSPs, (2) flexible interposer

TABLE 1.11   Summary of CSP Groups, Companies and Their CSP Names, and the
Corresponding Chapter Number in this Book

CSP groups	Companies and their CSP names	Chapter no.
Customized-lead-frame-based CSP	• Fujitsu's Small Outline No-Lead/C-Lead Package (SON/SOC)	2
	• Fujitsu's Bump Chip Carrier (BCC)	3
	• Fujitsu's MicroBGA and Quad Flat Nonleaded (QFN) Package	4
	• Hitachi Cable's Lead-on-Chip Chip Scale Package (LOC-CSP)	5
	• Hitachi Cable's Micro Stud Array (MSA)	6
	• LG Semicon's Bottom-Leaded Plastic Package (BLP)	7
	• TI Japan's Memory Chip Scale Package with LOC (MCSP)	8
CSP with flexible substrate	• 3M's Enhanced Flex CSP	9
	• General Electric's Chip-On-Flex Chip Scale Package (COF-CSP)	10
	• Hitachi's Chip Scale Package for Memory Devices	11
	• IZM's *flex*PAC	12
	• NEC's Fine-Pitch Ball Grid Array (FPBGA)	13
	• Nitto Denko's Molded Chip Size Package (MCSP)	14
	• Sharp's Chip Scale Package	15
	• Tessera's Micro-Ball Grid Array (μBGA)	16
	• TI Japan's Micro-Star BGA (μStar BGA)	17
	• TI Japan's Memory Chip Scale Package with Flexible Substrate (MCSP)	18
CSP with rigid substrate	• Amkor/Anam's ChipArray Package	*19*
	• EPS's Low-Cost Solder-Bumped NuCSP	20
	• IBM's Ceramic Mini-Ball Grid Array Package (Mini-BGA)	21
	• IBM's Flip Chip–Plastic Ball Grid Array Package (FC-PBGA)	22
	• Mitsushita's Stud Bump Bonding Package (MN-PAC)	23
	• Motorola's SLICC and JACS-Pak	24
	• National Semiconductor's Plastic Chip Carrier (PCC)	25
	• NEC's Three-Dimensional Memory Module (3DM) and CSP	26
	• Sony's Transformed Grid Array Package (TGA)	27
	• Toshiba's Ceramic/Plastic Fine-Pitch Ball Grid Array Package (C/P-FBGA)	28
Wafer-level redistribution CSP	• ChipScale's Micro SMT Package (MSMT)	29
	• EPIC's Chip Scale Package	30
	• Flip Chip Technologies' *Ultra*CSP	31
	• Fujitsu's *Super*CSP (SCSP)	32
	• Mitsubishi's Chip Scale Package (CSP)	33
	• National Semiconductor's MSMD	34
	• Sandia National Laboratories' Mini Ball Grid Array (μBGA)	35
	• ShellCase's Shell-PACK/Shell-BGA	36

CSPs, (3) rigid substrate CSPs, and (4) wafer-level redistribution CSPs. The first group of CSPs [also called lead on chip (LOC)] is developed mainly for die expand and the system houses wanting to keep the same footprints on the PCB purposes. The other three groups use a substrate (either flexible or rigid) or a metal layer (on the wafer) to redistribute the very fine pitch peripheral-array pads on the chip to the larger-pitch (0.5, 0.75, 0.8, and 1 mm) area-array pads on the PCB.

In this book, 35 different CSPs are studied (Table 1.11). The focus is to examine and comment on the design concepts and package structure, material issues, manufacturing process, electrical and thermal performance, qualifications and reliability, and applications of each CSP. Some noted CSPs are shown in [35–49].

1.5   Acknowledgments

The first author (JL) would like to thank Chris Cheng of Express Packaging Systems, Inc., for his useful help and constructive comments on this chapter.

1.6   References

1. R. Tummala, E. Rymaszewski, and A. Klopfenstein, *Microelectronics Packaging Handbook*, 2d ed., Chapman-Hall, New York, 1997.
2. J. H. Lau, C. Wong, J. Prince, and W. Nakayama, *Electronic Packaging: Design, Materials, Process, and Reliability*, McGraw-Hill, New York, 1998.
3. J. H. Lau and Y. H. Pao, *Solder Joint Reliability of BGA, CSP, Flip Chip, and Fine Pitch SMT Assemblies*, McGraw-Hill, New York, 1997.
4. J. H. Lau, *Ball Grid Array Technology*, McGraw-Hill, New York, 1995.
5. J. H. Lau, *Flip Chip Technologies*, McGraw-Hill, New York, 1996.
6. J. H. Lau, *Chip on Board Technologies for Multichip Modules*, Van Nostrand Reinhold, New York, 1994.
7. G. Harman, *Wire Bonding in Microelectronics, Materials, Processes, Reliability, and Yield*, McGraw-Hill, New York, 1997.
8. P. E. Garrou and I. Turlik, *Multichip Module Technology Handbook*, McGraw-Hill, New York, 1998.
9. A. Elshabini-Riad and F. D. Barlow III, *Thin Film Technology Handbook*, McGraw-Hill, New York, 1998.
10. J. Sweet, D. Peterson, J. Emerson, and R. Mitchell, "Liquid Encapsulant Stress Variations as Measured with the ATC04 Assembly Test Chip," *Proceedings of the IEEE Electronic Components & Technology Conference*, May 1995, pp. 300–304.
11. C. P. Wong, S. H. Shi, and G. Jefferson, "High Performance No Flow Underfills for Low-Cost Flip-Chip Applications," *Proceedings of the IEEE Electronic Components & Technology Conference*, May 1997, pp. 850–858.
12. C. P. Wong, J. M. Segelken, and C. N. Robinson, "Chip On Board Encapsulation," in J. H. Lau (ed.), *Chip on Board Technologies for Multichip Modules*, Van Nostrand Reinhold, New York, 1994, pp. 470–503.
13. T. Y. Wu, Y. Tsukada, and W. T. Chen, "Materials and Mechanics Issues in Flip-Chip Organic Packaging," *Proceedings of the IEEE Electronic Components & Technology Conference*, May 1996, pp. 524–534.

14. S. Han and K. K. Wang, "Analysis of the Flow of Encapsulant during Underfill Encapsulation of Flip-Chips," *IEEE Transactions on CPMT, Part B, Advanced Packaging,* vol. 20, no. 4, 1997, pp. 424–433.
15. S. Han and K. K. Wang, "Study on the Pressurized Underfill Encapsulation of Flip-Chips," *IEEE Transactions on CPMT, Part B, Advanced Packaging,* vol. 20, no. 4, 1997, pp. 434–442.
16. D. M. Shi and J. W. Carbin, "Advances in Flip-Chip Underfill Flow and Cure Rates and Their Enhancement of Manufacturing Processes and Component Reliability," *Proceedings of the IEEE Electronic Components & Technology Conference,* May 1996, pp. 1025–1031.
17. D. Suryanarayana, J. Varcoe, and J. Ellerson, "Repairability of Underfill Encapsulated Flip Chip Packages," *Proceedings of the IEEE Electronic Components & Technology Conference,* May 1995, pp. 524–528.
18. J. H. Lau, "Thermal Fatigue Life Prediction of Encapsulated Flip Chip Solder Joints for Surface Laminar Circuit Packaging," ASME Paper No. 92W/EEP-34, ASME Winter Annual Meeting, 1992.
19. J. H. Lau, M. Heydinger, J. Glazer, and D. Uno, "Design and Procurement of Eutectic Sn/Pb Solder-Bumped Flip Chip Test Die and Organic Substrates," *Circuit World,* vol. 21, no. 4, March 1995, pp. 20–24.
20. W. Wun and J. H. Lau, "Characterization and Evaluation of the Underfill Encapsulants for Flip Chip Assembly," *Circuit World,* vol. 21, no. 4, March 1995, pp. 25–32.
21. M. Kelly and J. H. Lau, "Low Cost Solder Bumped Flip Chip MCM-L Demonstration," *Circuit World,* vol. 21, no. 4, July 1995, pp. 14–17.
22. J. H. Lau, T. Krulevitch, W. Schar, M. Heydinger, S. Erasmus, and J. Gleason, "Experimental and Analytical Studies of Encapsulated Flip Chip Solder Bumps on Surface Laminar Circuit Boards," *Circuit World,* vol. 19, no. 3, 1993, pp. 18–24.
23. J. H. Lau, "Solder Joint Reliability of Flip Chip and Plastic Ball Grid Array Assemblies Under Thermal, Mechanical, and Vibration Conditions," *IEEE Transactions on CPMT, Part B, Advanced Packaging,* vol. 19, no. 4, 1996, pp. 728–735.
24. J. H. Lau, E. Schneider, and T. Baker, "Shock and Vibration of Solder Bumped Flip Chip on Organic Coated Copper Boards," *ASME Transactions, Journal of Electronic Packaging,* vol. 118, no. 2, June 1996, pp. 101–104.
25. E. Zakel and H. Reichl, "Flip Chip Assembly Using the Gold, Gold-Tin and Nickel-Gold Metallurgy," in J. H. Lau (ed.), *Flip Chip Technologies,* McGraw-Hill, New York, 1996, pp. 415–490.
26. J. H. Lau, C. Chang, and R. Chen, "Effects of Underfill Encapsulant on the Mechanical and Electrical Performance of a Functional Flip Chip Device," to be published by *Journal of Electronics Manufacturing,* 1998.
27. J. H. Lau and C. Chang, "Characterization of Underfill Materials for Functional Solder Bumped Flip Chip on Board Applications," *IEEE 48th ECTC Proceedings,* May 1998, pp. 1361–1371.
28. J. H. Lau and C. Chang, "How to Select Underfill Materials for Solder Bumped Flip Chips on Low Cost Substrates?" *Proceedings of the IMAPS 31st International Symposium on Microelectronics,* November 1998, pp. 693-700.
29. J. H. Lau, C. Chang, T. Chen, D. Cheng, and E. Lao, "A Low-Cost Solder-Bumped Chip Scale Package—NuCSP," *Circuit World,* vol. 24, no. 3, 1998, pp. 11–25.
30. J. H. Lau, "Flip Chip on Printed Circuit Board (FCOB) with Anisotropic Conductive Film (ACF)," *Advanced Packaging,* July–August 1998, pp. 44–48.
31. J. H. Lau and T. Chou, "A Low-Cost Chip Size Package—NuCSP," *Circuit World,* vol. 24, no. 1, 1997, pp. 34–38.
32. J. H. Lau, "The Roles of DNP (Distance to Neutral Point) on Solder Joint Reliability of Area Array Assemblies," *Soldering & Surface Mount Technology,* no. 20, July 1997, pp. 58–60.
33. J. H. Lau, "Electronics Packaging Technology Update: BGA, CSP, DCA, and Flip Chip," *Circuit World,* vol. 23, no. 4, 1997, pp. 22–25.
34. J. H. Lau and J. Wei, "Temperature-Dependent Elasto-Plastic-Creep Analysis of NuCSP's Solder Joints," ASME Paper No. 97WA/EEP, November 1997.

35. J. Kasai, M. Sato, T. Fujisawa, T. Uno, M. Waki, K. Hayashida, and T. Kawahara, "Low Cost Chip Scale Package for Memory Products," *Proceedings of the SMI Conference*, August 1995, pp. 6–17.
36. G. Murakami, "Rationale for Chip Scale Packaging (CSP) Rather Than Multichip Modules (MCM)," *Proceedings of the SMI Conference*, August 1995, pp. 1–5.
37. S. Lee, J. Lee, S. Oh, and H. Chung, "Passivation Cracking Mechanism in High Density Memory Devices Assembled in SOJ Packages Adopting LOC Die Attach Technique," *Proceedings of the IEEE Electronic Components & Technology Conference*, May 1995, pp. 455–462.
38. M. Amagai, "The Effect of Stress Intensity of Package Cracking in Lead-On-Chip (LOC) Packages," *Proceedings of the IEEE Japan International Electronics Manufacturing Technology Symposium*, December 1995, pp. 415–420.
39. Y. Okugawa, T. Yoshida, T. Suzuki, and H. Nakayoshi, "New Tape LOC Adhesive Tapes," *Proceedings of the IEEE Electronic Components & Technology Conference*, May 1994, pp. 570–574.
40. H. Nakayoshi, N. Izawa, T. Ishikawa, and T. Suzuki, "Memory Package with LOC Structure Using New Adhesive Material," *Proceedings of the IEEE Electronic Components & Technology Conference*, May 1994, pp. 575–579.
41. Y. Kunitomo, "Practical Chip Size Package Realized by Ceramic LGA Substrate and SBB Technology," *Proceedings of the SMI Conference*, August 1995, pp. 18–25.
42. R. Master, R. Jackson, S. Ray, and A. Ingraham, "Ceramic Mini-Ball Grid Array Package for High Speed Device," *Proceedings of the IEEE Electronic Components & Technology Conference*, May 1995, pp. 46–50.
43. P. Lall, G. Gold, B. Miles, K. Banerji, P. Thompson, C. Koehler, and I. Adhihetty, "Reliability Characterization of the SLICC Package," *Proceedings of the IEEE Electronic Components & Technology Conference*, May 1996, pp. 1202–1210.
44. H. Iwasaki, "CSTP: Chip Scale Thin Package," *Proceedings of SEMICON Japan*, November 1994, pp. 488–495.
45. G. Forman, R. Fillion, R. Kole, R. Wojnarowski, and J. Rose, "Development of GE's Plastic Thin-Zero Outline Package (TZOP) Technology," *Proceedings of the IEEE Electronic Components & Technology Conference*, May 1995, pp. 664–668.
46. S. Matsuda, K. Kata, and E. Hagimoto, "Simple-Structure, Generally Applicable Chip-Scale Package," *Proceedings of the IEEE Electronic Components & Technology Conference*, May 1995, pp. 218–223.
47. S. Tanigawa, K. Igarashi, M. Nagasawa, and N. Yeti, "The Resin Molded Chip Size Package (MCSP)," *Proceedings of the IEEE Japan International Electronics Manufacturing Technology Symposium*, December 1995, pp. 410–415.
48. T. Kayo, K. Be, N. Sakaguchi, and S. Wakabayashi, "Reliability of mBGA Mounted on a Printed Circuit Board," *Proceedings of the SMI Conference*, August 1995, pp. 43–56.
49. M. Yasunaga, S. Baba, M. Matsuo, H. Matsushima, S. Nako, and T. Tachikawa, "Chip Scale Package (CSP) a Lightly Dressed LSI Chip," *Proceedings of the IEEE Japan International Electronics Manufacturing Technology Symposium*, December 1994, pp. 169–176.

Fujitsu's Small Outline No-Lead/C-Lead Package (SON/SOC)

2.1 Introduction and Overview

The small outline no-lead (SON) package was developed by Fujitsu Ltd. in 1995 based on the company's multiframe lead-over-chip (MF-LOC) technology [1]. Later on this package evolved to become the small outline C-lead (SOC) package, which can be stacked to form a three-dimensional packaging module (3DPM) [2]. These packages belong to the category of lead-frame-based CSP and are aimed at applications for low-pin-count (<50) ICs such as memory devices. Compared to conventional TSOP, Fujitsu's SON and SOC have smaller form factors and possess electrical and thermal performance superior to that of TSOP. Good package and solder joint reliability through qualification tests were reported as well. Currently Fujitsu produces more than 2 million SON packages per month for flash memories and DRAMs. The SOC package and associated 3DPM find wide application in PC memory cards.

2.2 Design Concepts and Package Structure

The lead-on-chip (LOC) configuration is a packaging structure that increases the die-to-package area ratio for lead-frame-based packages. This technology has been widely applied to SOIC packages (Fig. 2.1) and can be implemented in two ways. Since the inner leads are above the IC, the conventional die paddle cannot be used. Instead, the die may be mounted on the lead fingers by double-sided adhesive tape (Tape-LOC), or an additional lead frame may be used to hold the die (MF-LOC), as shown in Fig. 2.2. LG Semicon adopted the former configuration to develop a CSP named bottom-leaded plastic (BLP) pack-

Figure 2.1 SOJ packages using LOC technology.

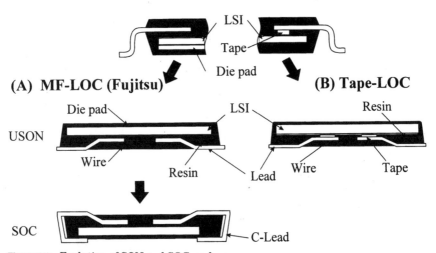

Figure 2.2 Evolution of SON and SOC packages.

age [3], see Chapter 7 of this book. On the other hand, Fujitsu used its MF-LOC technology to implement a chip scale package called SON. Both BLP and SON have a face-down chip and conform to the JEDEC MO-196 standard for ultra-thin small outline nonleaded (USON) packages. However, while the BLP is a paddleless package, the SON has a die mounting pad because of the existence of an additional lead frame. In 1997, Fujitsu extended its SON to a C-lead version (SOC). Figure 2.2 presents the evolution of SON and SOC and the cross section of corresponding package structures. For both packages, the first-level and second-level interconnects are wire bonds and plated flat lands, respectively.

Figure 2.3 Top and bottom views of SON-26.

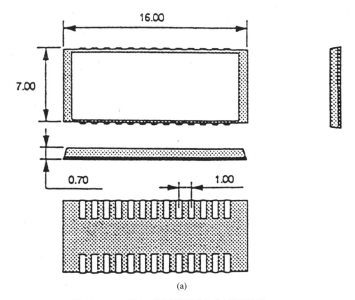

(a)

Figure 2.4a Package outline: (a) SON-26, (b) SON-40.

The first version of SON was developed to package 16M DRAM, as shown in Fig. 2.3. It has 26 bottom lands with a pitch of 1.0 mm. The outline of SON-26 is specified in Fig. 2.4a. The current lineup of SON is for 40 and 46 pins with a reduced pitch of 0.5 mm. These modules are mainly to package 16M flash ROM, as shown in Fig. 2.5. The dimension outline and internal LOC structure of SON-40 are presented in Figs. 2.4b and 2.6, respectively [4].

The SOC is basically a "flipped" SON with the bottom leads extended and wrapped around to the back of the package. Therefore, this package has a face-up chip instead of one that is face down. Since the SOC has terminals at both the top and the bottom surfaces, it be-

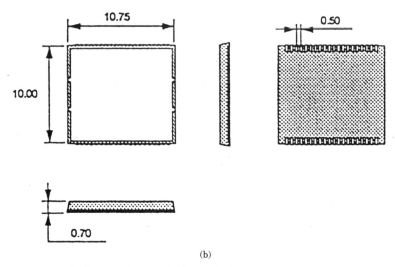

(b)

Figure 2.4b   Package outline: (*a*) SON-26, (*b*) SON-40.

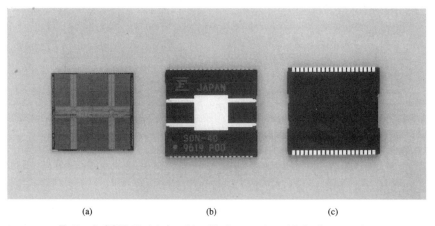

(a)   (b)   (c)

Figure 2.5   Fujitsu's SON-40: (*a*) the chip, (*b*) the top view, (*c*) the bottom view.

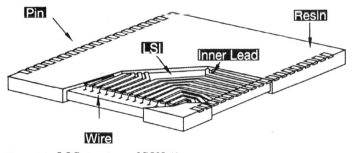

Figure 2.6   LOC structure of SON-40.

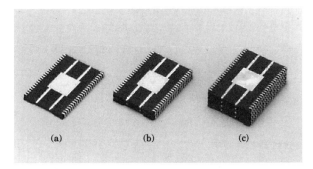

Figure 2.7 SOC-46 packages: (*a*) single module, (*b*) two-stack, (*c*) four-stack.

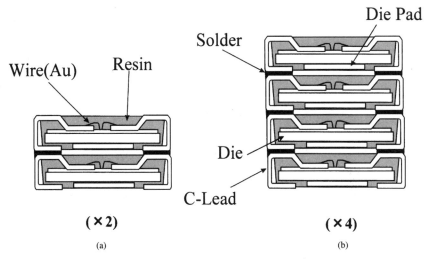

Figure 2.8 Cross-sectional view of SOC-3DPM structure: (*a*) two-stack, (*b*) four-stack.

comes an ideal candidate for building stacked memory devices, as shown in Fig. 2.7. The current lineup of SOC is 46 pins with a pitch of 0.5 mm. The body size and package height of a single module are 8×12 mm and 0.75 mm. The main application is for packaging 16M SDRAM. The stack-up version of SOC is called 3DPM. Currently there are two configurations, namely, two-stack and four-stack SOC-3DPM. The total package heights are 1.5 and 3.0 mm, respectively. The cross-sectional view of SOC-3DPM structure is presented in Fig. 2.8 [5].

TABLE 2.1   Properties of New Molding Compound for MF-LOC Packages

	Unit	Developed mold resin	Conventional mold resin
Epoxy type		Biphenyl	Biphenyl
Filler content	wt%	90	80
Stress-relieving agent		—	Added
Epoxy bromide		—	Added
Antimony oxide		—	Added
Viscosity (175°C)	poise	200–300	200–300
C.T.E.	ppm/°C	6–8	10–12
Oxidant index		>40	35
Flammability	UL94	V-0(1/16″)	V-0(1/16″)

2.3   Material Issues

Both SON and SOC are leadless plastic-encapsulated packages based on the MF-LOC technology. The major package constituents in addition to the silicon chip include the gold bonding wires, the die-attach adhesive, the custom-designed lead frames, and the encapsulation molding compound (EMC). According to Fujitsu [2], both Cu and alloy 42 were evaluated for the lead-frame material. From their testing results, the Cu lead frames have much better electrical and thermal performance.

It was reported that, in the MF-LOC technology, the EMC is the element that most significantly affects the package reliability. To ensure the performance of SON and SOC packages, a new EMC was developed, as shown in Table 2.1 [1]. This EMC has a highly filled resin to reduce the thermal expansion and moisture absorption. Because of the amount of filler (90 percent by weight), the fire retardant agent and the stress-relieving agent are not needed. With this newly developed EMC, the package reliability at high temperature has been substantially improved.

2.4   Manufacturing Process

The fabrication of SON and SOC is similar to that of conventional lead-frame-based packages except that an additional lead frame is needed to hold the silicon chip. This technology is called MF-LOC, and the manufacturing process is presented in Fig. 2.9. After dicing, the chip is mounted on the die pad frame. Then another lead frame with lead fingers is placed on the top of the mounted die pad frame with appropriate alignment. After wire bonding, the package is en-

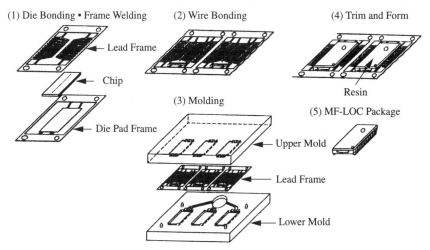

Figure 2.9   Manufacturing process for MF-LOC packages.

capsulated by transfer molding. With the final singulation and trimming, the MF-LOC package is completed.

It has been reported that the internal stress and the bottom land coplanarity of leadless packages are highly dependent upon the thickness of the package [1]. In order to optimize the encapsulation parameters, a finite-element analysis was performed to investigate the effects of package thickness. The results of this analysis are presented in Fig. 2.10. Both MF-LOC (SON type) and Tape-LOC (BLP type) were studied for comparison. It was found that MF-LOC has lower internal stress and less package warpage than Tape-LOC. Furthermore, for MF-LOC, both quantities increase when the package becomes thicker. Based on the results of this analysis, the package thickness of SON was determined to be 0.7 mm. Similar investigation was performed on SOC packages. It was found that with a package thickness of 0.75 mm, the package warpage and the lead coplanarity are less than 50 μm and 75 μm, respectively. Such figures are quite acceptable for surface-mounted components (SMC). Therefore, good mountability is expected for both SON and SOC packages.

2.5   Electrical and Thermal Performance

The electrical performance of SON and SOC has been characterized by simulation. The self-inductance L_s and mutual inductance L_m of a SON-26 package are given and benchmarked with the performance of a TSOP-26 in Table 2.2. The simulation was conducted using GREENFIELD software at a clock frequency of 100 MHz [4]. It was

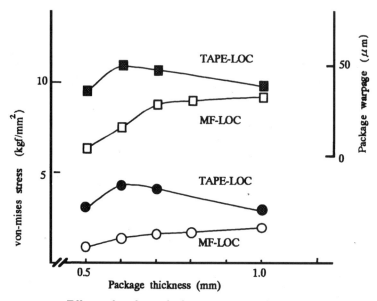

Figure 2.10 Effects of package thickness on internal stress and package warpage.

TABLE 2.2   Comparison of Electrical Performance of SON and TSOP

	Inductance (nH)	
	L_s	L_m
SON-26	2.354	1.096
TSOP-26	6.090	2.860

TABLE 2.3   Comparison of Electrical Performance of SOC and TSOP

	SOC-46		TSOP-50
	Cu	Alloy 42	Alloy 42
R (mΩ)	28	251	363
C (pF)	0.75	0.75	1.34
L (nH)	2.7	4.6	7.8

found that the inductance of SON-26 is about one-third that of TSOP-26. The R/L (resistance/inductance) and C (capacitance) of the SOC-46 package were investigated by Parasitic Parameter-3D and Maxwell-3D, respectively. The simulation was performed at a clock frequency of 200 MHz. The results are given in Table 2.3 and compared to those of a TSOP-50 package. It was found that the SOC with Cu lead frame has very good results on all measures. The superior electrical performance of SON and SOC, compared to that of TSOP, is attributed to the substantial reduction in the lead length.

The thermal performance of SON-26 and SOC-46 was characterized as well. The results are presented in Figs. 2.11 and 2.12, respectively.

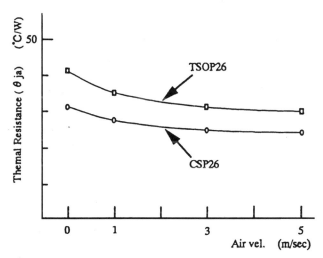

Figure 2.11   Thermal performance of SON.

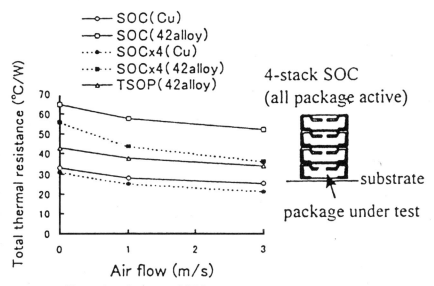

Figure 2.12   Thermal performance of SOC.

It is observed that the thermal resistance θ_{ja} of SON-26 is much lower than that of the corresponding TSOP-26. This phenomenon is expected because the die-mounting pad of SON is exposed at the top of the package, resulting in better heat dissipation capability. In Fig. 2.12, three kinds of comparison are presented. It is observed that the Cu lead frame always outperforms the alloy 42 lead frame. While TSOP has lower θ_{ja} than SOC with the same lead-frame material (alloy 42), SOC with Cu lead frame has the best performance. In addition, it is interesting to note that the four-stack 3DPM has lower thermal resistance than the single module. This improvement in thermal performance is especially obvious for SOC-46 with alloy 42 lead frame.

2.6   Qualifications and Reliability

The package reliability of SON and SOC has been investigated using comprehensive qualification tests. These tests were performed to verify the resistance of the package to soldering heat, thermal stress, and moisture absorption. The results are presented in Tables 2.4 and 2.5, respectively. Since there were no observable defects after all tests, it is concluded that SON and SOC have the same package reliability as other conventional plastic packages.

One of the concerns in introducing CSP packages is the process for mounting them on the PCB. If the assembly of CSP does not require

TABLE 2.4   Package Qualification Data for SON

Test item	Condition	Results
Package crack	85°C/85% RH · 168 h + IR (245°C max)	0/100
Temperature cycle	−65°C/150°C, 500 cycles	0/25
Pressure cooker	(Precondition: 85°C/85% RH · 168 h + IR) 121°C · 2 atm/100%, 336 h	0/25

TABLE 2.5   Package Qualification Data for SOC

Test item	Condition	Results
Package crack	Dry-up 125°C, 24 h + 85°C/85% RH, 96 h + IR 245°C max	0/10
Temperature cycle	−65 to 150°C, 1000 cycles	0/15
Pressure cooker	121°C, 2 atm, 168 h	0/10

Preconditioning: dry-up 125°C, 24 h + 85°C/85% RH, 24 h + IR 245°C max.

the modification of existing facilities, the acceptance of new packages will be highly encouraged. The mountability of SON and SOC has been proven to be acceptable. Figure 2.13 shows the pad design of PCB for SON-26. Using conventional pick and place equipment, 800 SON-26 packages were mounted without any errors. Therefore, the reliability in surface mounting is ensured.

In addition to mountability, the reliability of solder joints under thermal cycling is another concern with SMCs. Figure 2.14 shows the cross section of a solder joint from the SON-PCB assembly. For engineering estimation and benchmarking, a finite-element analysis with

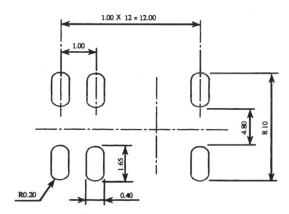

Figure 2.13   Solder pad design for SON.

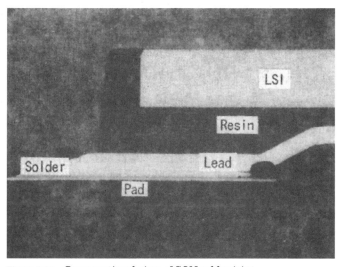

Figure 2.14   Cross-sectional view of SON solder joint.

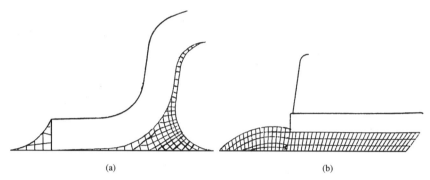

(a)   (b)

Figure 2.15   Finite-element model for solder-joint stress analysis: (*a*) TSOP, (*b*) SON.

the meshes shown in Fig. 2.15 was performed to evaluate the stress in the solder joints of TSOP and SON. The plots in Fig. 2.16 indicate that the stress in the solder joints of SON is lower than in those of TSOP. This result was confirmed by the thermal cycling test. The temperature profile used was −65 to 150°C (for 30 min at each temperature per cycle). The testing results presented in Fig. 2.17 show that the thermal fatigue life of SON solder joints is much better than that of TSOP. Further inspection of the cross section of failed solder joints was conducted as shown in Fig. 2.18. For both TSOP and SON, the crack in the solder joint was initiated at the "heel" fillet area. This observation coincides with the results of finite-element analysis.

The solder joint reliability of SOC-46 was also investigated by thermal cycling test. The adopted temperature profile was −55 to 125°C. A TSOP-50 package was also tested for benchmarking. From the results given in Table 2.6, it was observed that cracks started to appear in the TSOP solder joints after 200 cycles. On the other hand, no cracks were initiated in SOC solder joints until 400 cycles. Therefore, it may be concluded that SOC has much better solder joint reliability than TSOP.

2.7   Applications and Advantages

The SON and SOC packages were developed for low-pin-count (<50) ICs. Their main application is in memory devices. So far 16M DRAM, 16M flash memory, and 16M SDRAM have been packaged with SON and SOC [5]. The monthly production for Fujitsu's SON packages is more than 2 million. On the other hand, the stacked SOC packages find very good application in memory cards. Figure 2.19 shows a JEIDA Type II memory card with 44 two-stack SOC-3DPMs (20 and

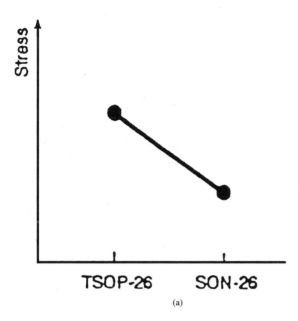

(a)

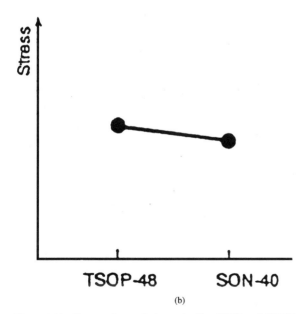

(b)

Figure 2.16   Comparison of stress in the SON and TSOP
solder joints: (a) 26 pins, (b) >40 pins.

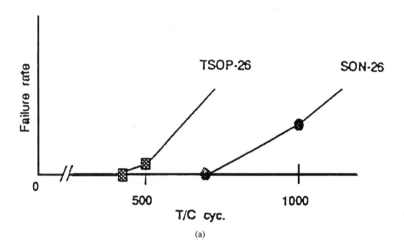

(a)

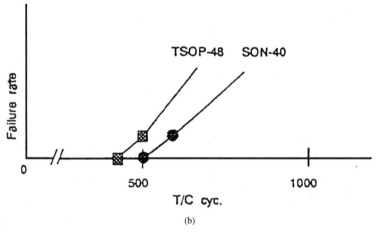

(b)

Figure 2.17 Comparison of thermal fatigue life between SON and TSOP:
(a) 2G pins, (b) ≥ 40 pins.

24 modules on the front and the rear side, respectively). The memory capacity is 176 MB. With the implementation of 64M SDRAM, the capacity of such memory cards will increase to 480 MB in the future.

The major competitor to SON/SOC is TSOP. The former two packages have much smaller form factors than the corresponding TSOP. For SON, the mounting area is 50 percent smaller and the mounting volume is 67 percent less than that of TSOP. For SOC, the reductions

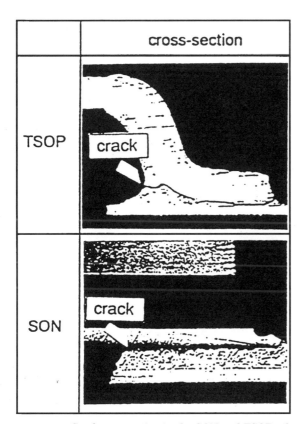

	cross-section
TSOP	crack
SON	crack

Figure 2.18 Crack propagation in the SON and TSOP solder joints.

TABLE 2.6   Comparison of Solder Joint Qualification Testing Results among SOC, TSOP, and 3DPM

	Thermal cycles				
	100	200	300	400	500
TSOP-50	Passed	Passed	3/20	6/20	13/20
SOC-46	Passed	Passed	Passed	Passed	5/30
3DPM(×2)	Passed	Passed	Passed	Passed	1/10

in these two amounts are 30 percent and 60 percent, respectively. In addition, because of the short lead length and the exposure of the die-mounting pad, the SON/SOC packages have electrical and thermal performance superior to that of TSOP. The board-level solder joint reliability is also better. Therefore, these packages have good potential to replace TSOP in the future.

Figure 2.19 Application of SOC-3DPM to JEIDA Type II memory card.

2.8 Summary and Concluding Remarks

The SON package is a leadless plastic-encapsulated CSP based on MF-LOC technology. By extending the leads of SON and wrapping them around to the back of the package, a SOC can be implemented. Both packages use wire bonds and plated flat lands as their first-level and second-level interconnects, respectively. Since the SOC has terminals at the top and bottom surfaces of the package, it is an ideal candidate for stacked modules—3DPM in Fujitsu's term.

In addition to custom-designed lead frames, the encapsulation molding compound is an essential element for the fabrication of SON/SOC packages. A highly filled molding resin was developed to ensure package reliability at elevated temperature. Because of the short lead length and the exposure of the die-mounting pad, SON and SOC have superior electrical and thermal performance. In addition, the SOC-3DPM has lower thermal resistance than the single module.

The package reliability of SON and SOC was verified by comprehensive qualification tests. They are considered to have the same reliability as conventional plastic packages. The board-level solder joint reliability was investigated by thermal cycling tests. SON/SOC had longer thermal fatigue life than TSOP.

SON and SOC have relatively small form factors and are aimed at packaging low-pin-count (<50) ICs such as memory devices. They have good potential to substitute for conventional TSOP packages. The current applications include 16M DRAM, 16M flash memory, and 16M SDRAM. Also, the 3DPM configuration can be used on JEIDA Type II memory cards. With the implementation of 64M SDRAM, a memory card of 480-MB capacity can be achieved in the future.

2.9   References

1. J. Kasai, M. Sato, and T. Fujisawa, "Low Cost Chip Scale Package for Memory Products," *Proceedings of the Surface Mount International Conference,* San Jose, Calif., September 1995, pp. 6–17.
2. T. Hamano, T. Fujisawa, M. Seki, K. Mitobe, S. Orimo, M. Sato, and J. Kasai, "SOC Package and 3D-Package Module for Memory Device," *Proceedings of the IMAPS Advanced Technology Workshop on Chip Scale Packaging,* Austin, Tex., August 1997.
3. K.-B. Cha, Y.-G. Kim, and T.-K. Kang, "Ultra-thin and Crack-free Bottom Leaded Plastic (BLP) Package Design," *Proceedings of the 45th ECTC,* Las Vegas, Nev., May 1995, pp. 224–228.
4. J. Kasai, "Low Cost Chip Size Package Solutions for ASIC and Memory Application," *Advanced IC Packaging Technologies: Array & Chip Scale,* Semicon/Europa'96, March 1996.
5. Fujitsu Ltd., "The Latest Trend in LSI Packaging Technology," *Fujitsu Electronic Device News: Find,* vol. 15, no. 3, 1997, pp. 20–29.

Chapter

3

Fujitsu's
Bump Chip Carrier (BCC)

3.1 Introduction and Overview

The bump chip carrier (BCC) package was developed by Fujitsu Ltd. in 1996 as a substitution for shrink small outline packages (SSOP) and chip-on-board (COB) in communication equipment. The original name is SON for ASSP [1]. This package is aimed at applications involving low-cost ICs with low pin count (<50). The BCC package features resin bumps with plated metal thin film at the bottom. Although this package is classified as the lead-frame type of CSP, the lead frame just acts as intermediate tooling and does not exist in the completed package. According to Fujitsu, the BCC package has performance and reliability equivalent to those of SSOP and has a form factor close to that of COB. Currently this package is in mass production for cellular phone components. It is expected that BCC will replace SSOP to a substantial extent in the future.

3.2 Design Concepts and
Package Structure

The BCC package was first released in 1996 [1]. The design concept was to develop a new packaging technology for low-cost ICs with low pin count (<50). The main objective was to replace SSOP and COB in ASSP applications. The BCC is a leadless plastic package with mounting terminals at the bottom. The early version was called SON for ASSP and had 24 dual-in-line bumps (SON-24) [1]. Currently the package with 16 perimeter bumps (BCC-16), as shown in Fig. 3.1, is in mass production. The bottom bump width is 0.4 mm. The lead

Top View

Bottom View

Figure 3.1   Top and bottom views of BCC packages.

TABLE 3.1   Fujitsu's BCC Family

Package name	Lead pitch	Number of I/O leads				
		20	40	60	80	100
BCC	0.65	16● 20◊	32◊	48◊	64◊	
	0.50		32○	48○	64○	

● Volume production.
◊ Under development.
○ R&D.

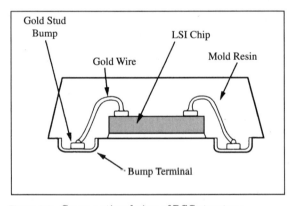

Figure 3.2   Cross-sectional view of BCC structure.

pitch is 0.65 mm and 0.8 mm along the longitudinal and the transverse sides, respectively. New generations of BCC with finer lead pitch and higher pin count are under development and evaluation, as presented in Table 3.1.

The cross section of a BCC package is shown in Fig. 3.2. It can be seen that this package has a very simple structure. The major feature of BCC is the resin bumps at the bottom of the package. These bumps are covered with metal thin film and serve as the interconnects from the package to the printed circuit board (PCB). The first-level interconnects between the die and the bump terminals are gold wire bonds. In order to improve the robustness of the manufacturing process, a proprietary wire bonding technique was developed by Fujitsu [2]. This special technique is implemented by adding an extra Au stud bump at the bottom of the resin bump. Good bonding reliability was reported [2].

The BCC package does not have a die-mounting paddle. Although it is classified as the lead-frame type of CSP, the lead frame does not exist in the completed package but just acts as intermediate tooling to form the resin bumps. Like other plastic packages, the BCC is encapsulated by epoxy molding compound. Low profile is one of the major advantages of BCC. The typical package thickness is less than 0.8 mm. Since the package size is 2 mm plus the die size, BCC is a true chip scale package.

3.3   Material Issues

In addition to the silicon die, the major constituents of the BCC package include the gold bonding wires, the die-attach adhesive, the encapsulation molding compound, and the metallization on the resin bumps. Although the BCC package belongs to the category of lead-frame-based CSP, there is no lead frame in the completed package. The function of the lead frame is to provide intermediate tooling to form the resin bumps. The lead-frame material is Cu alloy. The whole lead frame is removed by etching at the final step of the fabrication process.

The most critical part of the BCC package is the metallization on the resin bumps. This metallization consists of four layers of metal thin films. It not only provides a bondable surface for the Au wire bonds, but also serves as a terminal for the board-level interconnect. The sequence of metallization, from the inside to the outside of the resin bump, is Pd-Ni-Pd-Au. The inner Pd (500-nm) layer and outer Pd (100-nm) and Au (5-nm) layers are for wire bondability and solder wettability, respectively, while the intermediate Ni (5-μm) layer serves as a diffusion barrier [2]. These metal thin films are originally deposited on the inner surface of cavities in the Cu lead frame by electroplating. When the lead frame is etched away at a later stage, the metallization is transferred to the outer surface of the resin bumps.

3.4 Manufacturing Process

The fabrication of BCC is a proprietary process of Fujitsu Ltd. The first step is to prepare the lead frame, as shown in Fig. 3.3. Etching resist is applied on the upper surface of the Cu lead frame to form the land pattern of resin bumps, which are to be made at a later stage. The patterned lead frame is then half-etched to create craters at the top side. With the existing pattern, four layers of metal thin film are deposited on the inner surface of the created craters by electroplating. After the etching resist is stripped away, the lead frame is ready for the next step of fabrication.

Once the lead frame is prepared, the subsequent process can be carried out as shown in Fig. 3.4. The silicon chip is mounted on the top side of the lead frame with die-attach adhesive. Au wire bonding is then conducted to make the first-level interconnects between the bond pads on the die and the metallization in the craters of the lead frame. It should be noted that the bonding area on the metallization is restricted because the crater is relatively deep and has a curved surface. Therefore, the conventional wire-bonding method cannot be used directly, and a modified bonding technology was developed by Fujitsu for the BCC package. This modification is implemented by adding an Au stud bump on the metallization area first. Forming this extra stud bump increases the margin of the bonding direction. As a

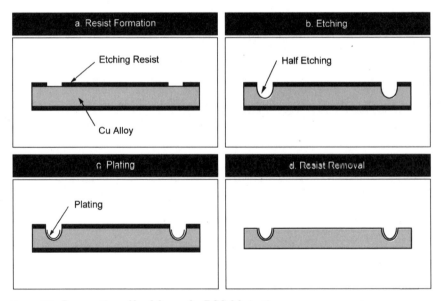

Figure 3.3 Preparation of lead frame for BCC fabrication.

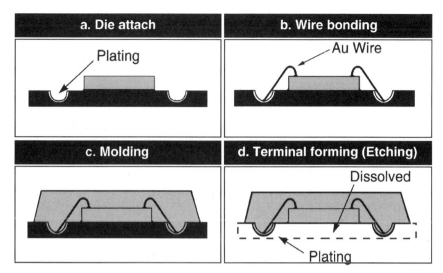

Figure 3.4 Manufacturing process for the BCC package.

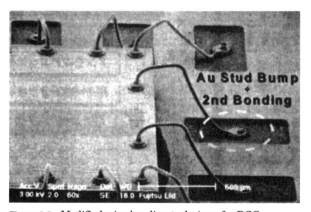

Figure 3.5 Modified wire-bonding technique for BCC.

result, narrow-area bonding becomes feasible. Following formation of the stud bump, the standard ball-stitch wire bonding is performed. The first bond lands on the bond pad of the silicon die, while the second bond is placed on the top of the existing stud bump as shown in Fig. 3.5. With this bonding technique, a wire length of 0.6 mm and a loop height of 150 μm can be implemented [2].

After wire bonding, the die and the wires are encapsulated by molding compound. The epoxy flows into the craters in the lead frame to form the resin bumps. The final procedure is to strip the lead frame by chemical etching. An ammoniac solution which dissolves only Cu is used as the etchant. Once the lead frame is removed, the resin bumps are exposed and the preexisting metallization is transferred to their surface. The typical bump height is 85 μm.

3.5 Electrical and Thermal Performance

The BCC package has been evaluated for electrical and thermal performance. Since the competitor to BCC is SSOP, the performance of SSOP with equal pin count was evaluated under the same testing conditions for benchmarking. The self-inductance and self-capacitance were measured by the time-domain reflectometry (TDR) method. The results and comparison are given in Table 3.2. Since the BCC package does not have inductance and parasitic capacitance caused by the leads, its electrical performance is much superior to that of SSOP.

The thermal performance in terms of θ_{ja} was evaluated for both BCC-16 and SSOP-16 by experimental measurement. The results are presented in Fig. 3.6. For forced convection cases, it seems that SSOP

TABLE 3.2   Electrical Performance of BCC

	BCC-16	SSOP-16
Self-inductance	3.35 nH	4.33 nH
Self-capacitance	0.055 pF	0.278 pF

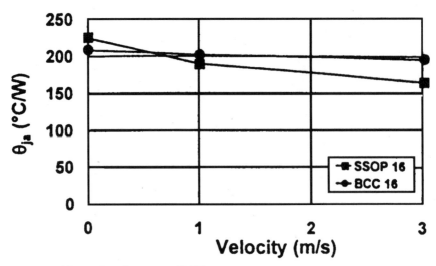

Figure 3.6   Thermal performance of BCC.

has a lower thermal resistance. However, in the actual applications of BCC packages (e.g., in the chassis of a cellular phone), the heat transfer environment is closer to natural convection. In this case, BCC outperforms SSOP in heat dissipation.

3.6 Qualifications and Reliability

Qualification and reliability testing are the essential issues for newly developed packages. The BCC package has been investigated for its wire bondability, crack resistance, and solder joint reliability. Since a modified wire-bonding technique (stitch bond on stud bump) is implemented in the BCC package, the bondability of Au wire (especially the second bond) is a major concern. A series of wire-pulling tests were performed on BCC-16 packages, and the results are shown in Fig. 3.7. While the specification is 4g, the measured wire-pulling strength is >10g on average. Besides, the failure of wire bonds was mainly in C mode (with a minor population in B or D mode). No E mode failure was observed. Therefore, it may be concluded that the modified bonding technique is by no means inferior to the conventional wire-bonding method.

The crack resistance of BCC packages was evaluated by both solder dipping and IR reflow methods, as shown in Table 3.3. The specimens were preconditioned under 85°C/85 percent RH for 48 h first. After the moisture absorption, some packages were dipped into molten solder at 260°C for 10 s while the others were subjected to the IR reflow

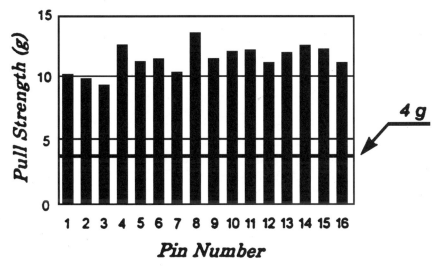

Figure 3.7   Results of wire-pulling tests.

TABLE 3.3   Evaluation of Crack Resistance for BCC

- Condition
  BCC-16
  Pretreatment: 85°C/85% RH at 48 h
  Solder dip/IR reflow

- Results
  Solder dip 260°C for 10 s OK
  IR reflow 245°C for 10 s OK

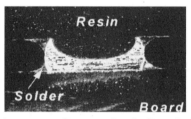

Figure 3.8 Cross-sectional view of solder joint of BCC-PCB assembly.

TABLE 3.4   Results of Solder Wettability Tests

| | Solder dip | |
Pretreatment	215°C for 3 s	260°C for 10 s
1. Steam aging (100°C/100%) for 8 h	OK	OK
2. Temperature storage (125°C) for 192 h	OK	OK
3. Temperature storage (150°C) for 16 h	OK	OK

condition at 245°C for 10 s. No package cracking or surface peeling was observed after either test. This result indicates that the BCC package has the same crack-resistance capability as SSOP. Therefore, BCC can be considered to be a package requiring no humidity control.

There are two concerns with solder joint reliability for BCC packages. One is the solder wettability of the resin bumps, and another is the joint strength after thermal cycling. Figure 3.8 shows the cross section of a solder joint from the BCC-PCB assembly. It is obvious that the solder wraps around the resin bump, indicating an excellent wettability on the surface of metallization. In order to investigate the environmental effect on the wettability, a series of solder dipping tests were performed under various conditions, as shown in Table 3.4. At first the BCC specimens were subjected to three different kinds of aging conditions. After flux dispensing, the specimens were dipped into molten solder at 215°C and 260°C for 3 s and 10 s, respectively. Cross-sectional inspection was conducted afterwards.

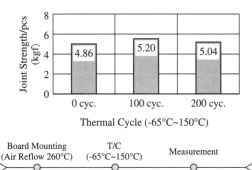

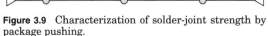

Figure 3.9 Characterization of solder-joint strength by package pushing.

For all testing conditions, no degradation in solder wettability was observed.

The strength of BCC solder joints was evaluated by the package pushing method. The specimens were reflowed on the PCB first. Through a preexisting hole in the PCB underneath the mounted specimen, a probe was used to push away the BCC package until separation from the PCB occurred. The peak load during pushing is evaluated as an index of solder joint strength. In order to investigate the effect of thermal fatigue on solder joint reliability, the specimens were subjected to temperature cycling (T/C) between −65 and 150°C. The results of solder joint strength before T/C, after 100 cycles, and after 200 cycles are presented in Fig. 3.9. No degradation in solder joint strength was observed during the tests.

Further tests with high-temperature storage were conducted to investigate the effectiveness of the diffusion barrier in the metallization of resin bumps. The specimens were aged at 150°C for 504 h. Cross-sectional inspection by SEM/EPMA was performed on the solder joint. It was found that the Pd-Au thin films had thoroughly diffused into the solder. A Sn-Ni intermetallic layer was formed at the interface. However, no Sn or Pb was identified on the other side of the Ni layer. Therefore, the effectiveness of the diffusion barrier in the metallization was verified.

3.7 Applications and Advantages

The BCC package was developed for low-cost ICs with low pin count. The main objective is to replace SSOP or COB in portable consumer electronics for telecommunication. Currently BCC is in mass production as phase-locked loop (PLL) synthesizers in cellular phones. The development of new applications with finer lead pitch and higher pin count is under way [3].

The major advantages of BCC are its simple structure and compactness. Compared to SSOP with the same pin count, the BCC package requires 40 percent less mounting area and 67 percent less mounting volume. Besides, as a surface-mount-compatible chip scale package, the BCC can avoid known-good-die (KGD) issues. Since this package has performance and reliability equivalent to those of SSOP and has a form factor close to that of COB, the BCC package has become an attractive choice for ASSP applications [4].

3.8 Summary and Concluding Remarks

The BCC is a leadless and low-profile plastic-encapsulated package. It is a true CSP because its package size is equal to the die size plus 2 mm. Although this package belongs to the category of lead-frame-based CSPs, the lead frame does not exist in the completed package; it only serves as intermediate tooling to form the bottom terminals and the associated metallization. The BCC has a simple package structure and features proprietary resin bumps. The first-level interconnects are implemented by a modified wire-bonding technique. Because of the short wire length and the leadless configuration, the BCC package has superior electrical performance. On the other hand, BCC is comparable to SSOP in heat dissipation. Through various qualification tests, the BCC package has been proven to be a reliable technology.

BCC was developed as a substitution for SSOP and COB to package low-cost ICs with low pin count (<50). From the released data, this package has performance and reliability equivalent to those of SSOP and has a form factor close to that of COB. Currently the BCC package is in mass production for applications in portable communication electronics. It is expected that BCC will replace SSOP to a substantial extent in the near future.

3.9 References

1. J. Kasai, "Low Cost Chip Size Package Solutions for ASIC and Memory Application," *Advanced IC Packaging Technologies: Array & Chip Scale,* Semicon/Europa'96, March 1996.
2. J. Kasai, "Chip Size Packaging for ASSP Application," *Proceedings of Microelectronic Packaging into the 21st Century,* Atlanta, Georgia, December 1996, pp. 1–6.
3. Fujitsu Ltd., "The Latest Trend in LSI Packaging Technology," *Fujitsu Electronic Device News: Find,* vol. 15, no. 3, 1997, pp. 20–29.
4. Y. Yoneda, "Development of BCC (Bump Chip Carrier)," *Technical Digest 3rd VLSI Packaging Workshop of Japan,* Kyoto, Japan, December 1996, pp. 71–72.

4

Fujitsu's Microbga and Quad Flat Nonleaded Package (QFN)

4.1 Introduction and Overview

In 1996, Fujitsu implemented the terminal formation etching technology to develop two new CSPs, namely, MicroBGA and quad flat nonleaded (QFN) packages [1]. These packages are classified as lead-frame-based CSPs because custom-designed lead frames are used as the interposer. The major feature of MicroBGA and QFN is that their lead frames are etched after encapsulation molding. In addition to the lead pattern, the former is distinguished from the latter by having one extra layer of buildup resin on the top of the lead frame. While both packages have the same lead pitch, they are aimed at different pin-count ranges. The structure of Fujitsu's MicroBGA is very different from that of the other well-known "micro BGA" by Tessera (see Chapter 16), which has a flexible interposer. On the other hand, QFN looks similar to other lead-frame-based CSPs and is targeted to compete with conventional SSOP. The MicroBGA and the QFN are relatively new and are still under evaluation. Some reliability data have been released, but electrical and thermal performance have not yet been reported at this time.

4.2 Design Concepts and Package Structure

The terminal formation etching method is the core technology shared by MicroBGA and QFN [2]. The design concept is to use a low-cost lead frame together with conventional wire bonding to achieve cost reduction for chip scale packaging. This technology is illustrated in Fig. 4.1. At first, the lead frame is half-etched to make the terminal pattern at both sides of the lead frame. Then the top side is molded with resin epoxy for encapsulation. Finally, the bottom side is further etched to separate the terminals. With such technology, both area-array and peripheral leads can be formed at the bottom surface of the package.

Unlike other kinds of ball grid array (BGA) packages, Fujitsu's MicroBGA uses a metal lead frame as the package substrate. The interconnection between the die and the lead frame is by Au-wire bonding. The bond pads are staggered, with a pitch of 140 μm. The board-level interconnects are eutectic solder bumps attached to short studs on the etched lead frame. The bump diameter is 0.5 mm (20 mils) with a pitch of 0.8 mm. A resin buildup layer with Cu-plated via and trace is added to the top of the lead frame in order to redistribute the peripheral wire-bond pads to the area-array solder bumps. The current lineup for MicroBGA is 256 pins (16×16, full grid). A picture of this package is shown in Fig. 4.2. It can be seen that the package size

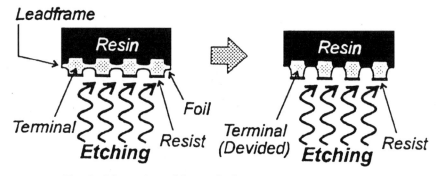

Figure 4.1 Terminal-formation etching method.

Figure 4.2 Top and bottom views of MicroBGA.

is as small as a U.S. nickel (5-cent) coin. The MicroBGA is classified as a CSP not because of its package-to-chip size ratio (currently 2 instead of 1.2) but because of its fine bump pitch (0.8 mm). The outline of the MicroBGA package is given in Fig. 4.3.

QFN was developed to replace shrink small outline packages (SSOP). It does not have the outer leads in order to reduce the mounting space. The board-level interconnects are peripheral plated flat lands at the bottom of the package, while Au-wire bonds are employed for the first-level interconnection. The land pitch is 0.8 mm. QFN is very compact in size, as shown in Fig. 4.4. The current lineup for QFN is 24 pins. The outline of the QFN package is given in Fig. 4.5.

Fujitsu's MicroBGA and QFN share a lot of common features, as

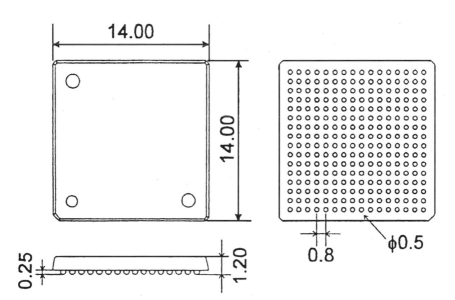

Figure 4.3   Outline of MicroBGA package.

Figure 4.4   Top and bottom views of QFN.

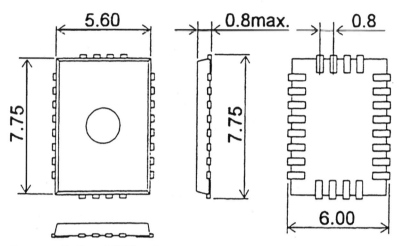

Figure 4.5 Outline of QFN package.

TABLE 4.1   Comparison between MicroBGA and QFN

	MicroBGA	QFN
Number of terminals	256 (16 × 16 full matrix)	24 (peripheral)
Terminal pitch	0.8 mm	0.8 mm
Mounting area	14.0 × 14.0	6.0 × 8.05
Seated height	1.2 mm max.	0.8 mm max.
Interposer	Lead frame + substrate	Lead frame
First bonding	Wire bonding	Wire bonding
Encapsulation	Molding	Molding
Terminal profile	Solder bump	Solder plated

presented in Table 4.1. In addition to the extra buildup layer in MicroBGA, the major difference is the number and pattern of bottom terminals. These two packages are aimed at packaging ICs with different pin-count ranges.

4.3   Material Issues

Both MicroBGA and QFN are lead-frame-based plastic packages. Besides the silicon chip, the packages are composed of gold bonding wires, die-attach adhesive, metal lead frames, and encapsulation molding compound (EMC). In addition, the MicroBGA has a buildup layer on the top of the lead frame and solder balls at the bottom of the lead frame.

The terminal formation etching method is the key technology for implementing MicroBGA and QFN. Its major feature is formation of the bottom terminals by etching after the package is molded by encapsulant. The lead frame has a thickness of 200 μm and is made of Cu alloy with Ni plating at both sides. The function of the Ni plating is to act as a surface resist during etching so that a desired terminal pattern can be created.

The base material that forms the buildup layer in the MicroBGA is epoxy resin. In order to redistribute the circuit from the peripheral bond pads to the area-array bottom terminals, Cu traces must be plated on the top of the buildup resin. The surface finishing of Cu traces is Ni/Au by electroplating.

The package interconnects of MicroBGA are eutectic solder balls. These balls are formed by dipping the bottom terminals in the solder paste followed by reflow. On the other hand, the bottom terminals of QFN are flat lands. The lands are formed by etching from the back side of the lead frame. The finishing is plated solder in order to produce a solder-wettable surface for board-level assembly.

4.4   Manufacturing Process

The general manufacturing process for Fujitsu's MicroBGA and QFN includes lead-frame preparation, die attach, wire bonding, encapsulation, terminal formation, and bottom lead finishing. In principle, there is no difference in the performance of die attach, wire bonding, and encapsulation for the captioned packages and other wire-bonded plastic packages. However, the rest of the fabrication procedures are different.

For MicroBGA, the most significant step in fabrication is lead-frame preparation, as shown in Fig. 4.6. The lead frame is a 200-μm-thick Cu alloy sheet with Ni plating on the top and bottom surfaces. The Ni layer is arranged to form the area-array pattern for the bottom terminals and acts as an etching resist. The lead frame is then half-etched to create short studs, as shown in Fig. 4.7 at both sides of the lead frame. The dimensions of these terminals are given in Table 4.2. The next step is to establish the buildup layer. First, a resin layer is coated on the top of the lead frame, as shown in Fig. 4.6b. Afterwards a photo-defined micro-via is made on the top of each terminal, as illustrated in Fig. 4.6c. On top of the buildup resin, a Cu layer is deposited by plating and then etched to form the desired pattern, as presented in Fig. 4.8. The design rule is that no more than four traces can pass through the space between any two vias. Finally, the Cu traces and lands are finished with Ni/Au by electroplating to form an Au wire-bondable surface. It should be noted that the wire-bonding pads have

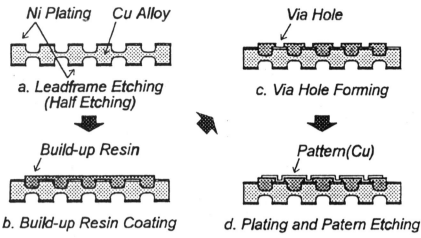

Ni Plating   Cu Alloy   Via Hole

a. Leadframe Etching   c. Via Hole Forming
   (Half Etching)

Build-up Resin   Pattern(Cu)

b. Build-up Resin Coating   d. Plating and Patern Etching

Figure 4.6   Preparation of lead frames for MicroBGA.

Figure 4.7   Short stud terminals on the lead frame of MicroBGA.

TABLE 4.2   Features of Lead Frame for MicroBGA

Terminal diameter (upper side)	400 μm
Half etching depth (upper side)	50 μm
Terminal diameter (lower side)	500 μm
Half etching depth (lower side)	100 μm

a staggered pattern and are located between the outermost row and the second row of terminals. These pads have a rectangular shape with a size of 100×200 μm. The pitch between pads is 140 μm. All other dimensions related to the buildup layer are given in Table 4.3.

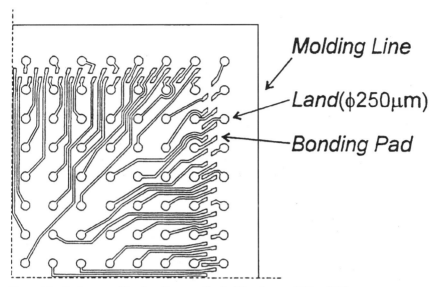

Figure 4.8   Cu trace and land pattern on the buildup layer of MicroBGA.

TABLE 4.3   Features of Buildup Layer for MicroBGA

Buildup resin thickness	40 μm
Via hole diameter	100 μm
Cu plated thickness	12 μm
Cover land diameter	250 μm
Line width/space	60 μm/60 μm
Pattern treatment	Ni (3 μm) + Au (0.3 μm) (electroplating)

Once the lead frame is completed, the subsequent packaging process of MicroBGA can be carried out as illustrated in Fig. 4.9. After die attach, wire bonding is performed to make the first-level interconnection. Since the bond pads are staggered, there are two wire loops. The maximum loop heights for the lower and the higher loop are 150 μm and 250 μm, respectively. The minimum distance between the two wire loops is 40 μm. It was reported by Fujitsu that a bonding speed of 6 wires per second can be achieved for MicroBGA.

Package encapsulation is the next step after wire bonding. The thickness of the molding epoxy is 0.8 mm. Afterwards, back etching is conducted to form the separate terminals, as shown in Fig. 4.9c. It should be noted that the Ni plating is on the bottom surface of the terminals and serves as a resist layer during back etching. Figure 4.10 presents a cross-sectional view of the lead frame with various etching duration. The shape of the terminals can be clearly observed. It was found that an

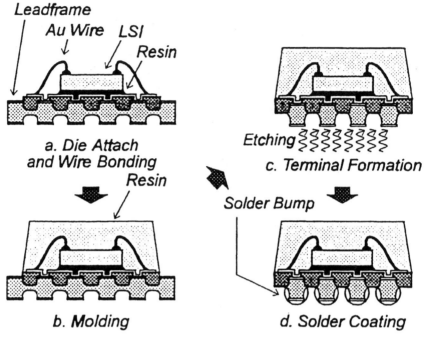

Figure 4.9 Assembly process for MicroBGA package.

etching time of 120 s gives the best result. Although there is a concern regarding the distortion of the area-array pattern due to the back etching, it has been confirmed that the maximum position shifting of bottom lands in the terminal formation etching process is 20 μm. Compared to the diameter of the terminals (500 μm), this deviation is negligible.

The last step in the fabrication of MicroBGA is the attachment of solder bumps to the bottom terminals. This procedure is illustrated in Fig. 4.11, and the associated geometries and conditions are given in Table 4.4. A tooling plate with dimples that match the area-array pattern of MicroBGA is fabricated first. Then the eutectic solder paste is printed over the tooling plate to fill the dimples on the surface. The bottom terminals of MicroBGA are dipped into the filled dimples with appropriate alignment. After reflow heating, the solder bumps are formed and attached to the bottom terminals of MicroBGA. A picture of the finished solder bumps is shown in Fig. 4.12.

Compared to the process just described, the fabrication of QFN is much simpler. A half-etched lead frame is prepared as shown in Fig. 4.13. Like that for MicroBGA, the lead frame for QFN is a Cu alloy sheet with Ni plating on both sides of the surface as an etching resist layer. After die attach, wire bonding, and encapsulation as illus-

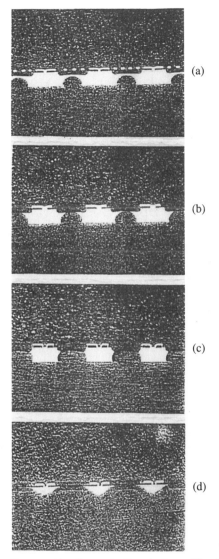

Figure 4.10 Cross-sectional view of lead frame during terminal-formation etching process: (*a*) 30 s, (*b*) 60 s, (*c*) 120 s, (*d*) 240 s.

trated in Fig. 4.14, the shaded portion of the lead frame in Fig. 4.13 is etched away to form the terminals. At last the bottom lands are finished with plated solder to make a solder-wettable surface for board assembly. It should be noted that since the QFN has peripheral terminals which are outside the die-attach area, no buildup layer is needed for circuit redistribution.

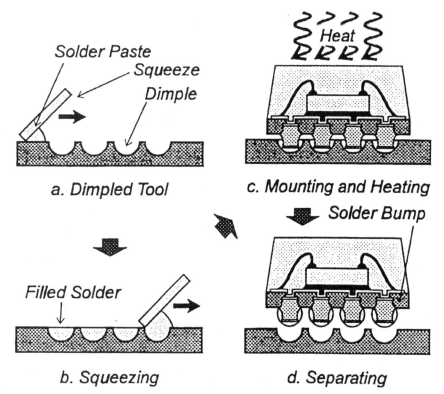

Figure 4.11 Formation of solder bumps for MicroBGA.

TABLE 4.4   Solder Bump Forming Conditions for MicroBGA

Dimple dimension	600 μm × 300 μm (depth) Shape: Hemisphere
Solder paste	Eutectic solder (RA type)
Reflow condition	225°C (N$_2$ atmosphere)
Solder bump dimension	500 μm × 250 μm (depth) Shape: Hemisphere

4.5   Qualifications and Reliability

Since Fujitsu's MicroBGA and QFN packages are relatively new, the results of electrical and thermal performance tests have not yet been released at this time. Furthermore, only limited reliability testing data are available. The specimens were preconditioned in an 85°C/85 percent RH environment for 48 h first. After moisture absorption, the

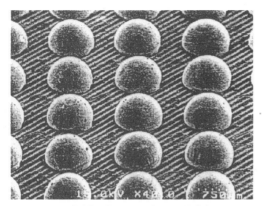

Figure 4.12 Solder bumps of MicroBGA in area-array pattern.

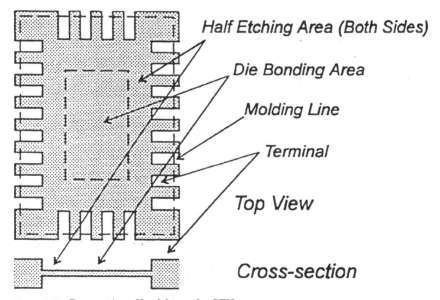

Half Etching Area (Both Sides)

Die Bonding Area

Molding Line

Terminal

Top View

Cross-section

Figure 4.13 Preparation of lead frame for QFN.

MicroBGA was heated by IR reflow at 245°C and the QFN was dipped into molten solder at 260°C. No surface cracking or internal delamination was observed after the test. Therefore, the package reliability of MicroBGA and QFN has been verified.

Since the MicroBGA has a buildup resin layer between the lead frame and the silicon chip, the reliability of die mounting is a concern. A die shear test, as shown in Fig. 4.15a, was performed to investigate the integrity of the die-attach adhesive. After testing, it

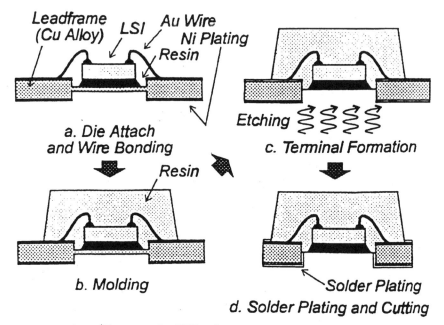

Figure 4.14   Assembly process for QFN package.

was observed that failure always occurred in the resin layer. The die shear strength at room temperature was 192 N. With elevated temperature, the shear strength was reduced to 111 N and 80 N at 150°C and 200°C, respectively. These values are considered acceptable for MicroBGA.

The solder bumps of MicroBGA are not produced in a conventional way. A ball shear test, as shown in Fig. 4.15b, was performed to characterize the shear strength of the solder bumps. After testing, it was found that failure occurred in the buildup resin, not in the solder bump. The strength for ball shear is 5.7 N. This value is much lower than the shear strength of regular BGA solder balls. In addition to the ball shear test, the MicroBGA was assembled on a PCB to investigate the solder joint reliability. A cross-sectional view of MicroBGA solder joints is shown in Fig. 4.16. The land on the PCB is solder-mask-defined and has a diameter of 0.5 mm (20 mils). The thickness of the solder mask is 150 μm. The solder was reflowed at 230°C. The strength of the solder joints was evaluated by a ball pull test, as shown in Fig. 4.15c. The result was 1.4 N per bump. Again the failure occurred in the buildup resin. Therefore, it may be concluded that the buildup layer of MicroBGA is a "weak spot" for reliability.

Shear Stress

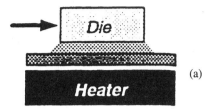

(a)

Shear Stress

(b)

Tensile Stress

(c)

Figure 4.15 Reliability tests of MicroBGA: (*a*) die shear, (*b*) ball shear, (*c*) ball pull.

4.6 Applications and Advantages

The design concept of MicroBGA and QFN is to implement the exist-ing lead-frame etching technology to develop fine-pitch (<1.0 mm) chip scale packages. The MicroBGA has a relatively large package size compared to other CSPs, but can accommodate high-pin-count ICs. However, actual applications have not yet been reported.

Figure 4.16   Cross-sectional view of MicroBGA solder joints.

TABLE 4.5   Comparison between QFN and SSOP

	QFN-24	SSOP-24
Terminal pitch (mm)	0.8	0.65
Body size (mm × mm)	5.6 × 7.75	5.6 × 7.95
Mounting area (mm × mm)	6.0 × 8.05	7.6 × 7.95
Seated height (mm)	0.8 max.	1.45 max.
Mounting volume	38.64	75.53
(Ratio)	(0.51)	(1)

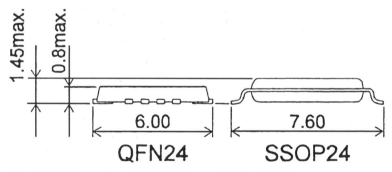

Figure 4.17   Comparison of outline between QFN and SSOP.

The QFN is very compact in size and is aimed at packaging low-pin-count ICs. Although so far there has been no release for practical applications, this package is targeted to replace SSOP for logic devices. The comparisons given in Table 4.5 and Fig. 4.17 indicate that the QFN has form factors superior to those of the SSOP. With proven reliability and performance, this package could be a potential competitor to conventional SOICs.

4.7 Summary and Concluding Remarks

Fujitsu's MicroBGA and QFN are lead-frame-based CSPs without outer leads. Both packages adopt Au-wire bonds as the first-level interconnects. For the board level, the MicroBGA uses full matrix ball grid array solder bumps, while the QFN employs peripheral solder-plated flat lands. In addition to the difference in terminal pattern, the MicroBGA has an extra buildup layer on the top of the lead frame to redistribute the circuit from the peripheral wire-bonding pads to the area-array bottom terminals. Although both packages have the same lead pitch, their pin counts are aimed at different ranges.

The terminal formation etching method is the core technology for implementing these two packages. This technology requires a Ni-plated Cu lead frame and features back etching after encapsulation molding. It has been proved that lead-frame etching can produce a precise terminal pattern. Therefore, a fine lead pitch of less than 0.8 mm may be expected.

Since the MicroBGA and the QFN are relatively new, the reports on their reliability are still limited. Also, the results of electrical and thermal performance tests have not yet been released. From the available testing data, it may be concluded that the MicroBGA and QFN have the same package reliability as other conventional plastic packages. However, the solder ball shear and ball pull tests indicate that the buildup resin could be a "weak spot" for the solder joint reliability of MicroBGA.

The MicroBGA is classified as a CSP because of its fine pitch rather than because of the package-to-chip size ratio. It has a relatively large package size compared to other CSPs, but can accommodate high-pin-count ICs. Although so far there are no reports of actual applications of MicroBGA, it is still worth introducing this CSP because of its unique package structure.

The QFN is very compact in size and is developed to package low-pin-count ICs. With the superior form factors, this package is targeted to replace SSOP for logic devices. Upon complete verification of reliability and performance, the QFN package may be a potential competitor to conventional SOICs.

4.8 References

1. *BGA Development Update Service,* TechSearch Int'l, Ltd., March 1996.
2. J. Kasai, K. Tsuji, Y. Yoneda, S. Orimo, R. Nomoto, H. Sakoda, and M. Onodera, "Development of MicroBGA and QFN by Leadframe Terminal Formation Method," *Proceedings of the 1st Pan Pacific Microelectronics Symposium,* Honolulu, Hawaii, February 1996, pp. 23–30.

Hitachi Cable's Lead-on-Chip Chip Scale Package (LOC-CSP)

5.1 Introduction and Overview

Hitachi Cable's LOC-CSP was released in 1996 based on the existing lead-on-chip (LOC) technology [1]. The main objective in developing this new chip scale package was to extend the LOC package's capacity to stacked three-dimensional (3-D) memory modules [2]. Compared to standard LOC packages, the LOC-CSP has a smaller molding area at the upper portion of the package and a wider shoulder at the top of the leads for ease of mounting stacked modules. The package is much thinner as well. The LOC-CSP is a leaded plastic package which fully complies with the common definition for CSP. The chip-to-package size ratio is close to 90 percent. This package is aimed at applications involving low-pin-count (40 to 60) devices. Good package reliability through thermal shock testing was reported. Currently Hitachi Cable produces one hundred million LOC packages per month for DRAMs. The stacked LOC-CSP is at the preproduction stage. Once fully qualified, this package may find wide application for 3-D memory modules.

5.2 Design Concepts and Package Structure

The LOC configuration is an innovation which allows die expansion without increasing the package size and footprint. This technology was developed in the early 1990s and has become a popular choice for packaging memory devices, as shown in Fig. 5.1 [3]. Since most 16-MB and above DRAMs have center bond pads, the LOC package will

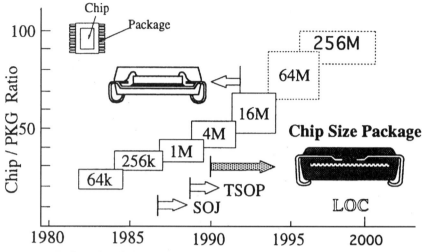

Figure 5.1   Package trends for DRAM devices.

be a major player in the arena of high-capacity memory modules. The LOC technology features overlapping of the lead frame on the top of the IC and can be implemented in two ways. One way is to attach the die to the lead fingers by double-sided adhesive tape (Tape-LOC), and the other is to use an additional lead frame to hold the silicon chip (MF-LOC) [4]. Hitachi Cable adopted the former configuration when it developed its standard LOC package, presented in Fig. 5.2 [3].

Because of the rapid growth in demand for higher-capacity DRAM, the 3-D memory module has attracted a lot of attention in the electronics industry. In view of this, Hitachi Cable modified its standard LOC package and developed a new LOC-CSP. Compared to standard LOC packages, this new CSP has a smaller molding area at the upper portion of the package and a wider shoulder at the top of the leads, as shown in Fig. 5.3. In addition, the package is much thinner. Such modifications are intended to make stacking packages more convenient, as illustrated in Fig. 5.4.

Unlike the USON type of CSP, the LOC-CSP is a leaded plastic package. The chip-level and board-level interconnects are Au wire bonds and plated flat lands, respectively. The current lineup for this package is in the range of 40 to 60 pins with lead pitches of 0.3, 0.4, and 0.5 mm. For 3-D memory modules, the present configuration is four-stack, which allows the expansion of 16-MB DRAM to 64 MB. The LOC-CSP has a relatively compact package size. For a 15×7×0.28 mm LSI chip, the package dimensions are 15.6×7.6×0.6 mm. The chip-to-package size ratio is about 90 percent. Therefore, the LOC-CSP is considered a true chip scale package.

Chip   Lead frame   Wire bond   Mold compound

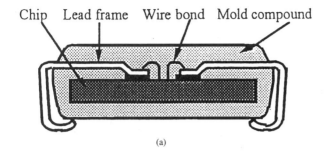

(a)

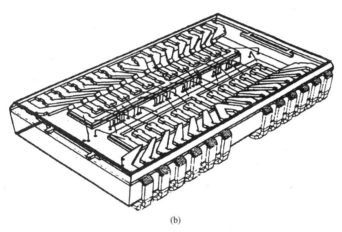

(b)

Figure 5.2   Package structure of LOC: (*a*) cross-sectional view, (*b*) three-dimensional view.

Figure 5.3   Cross-sectional view of Hitachi Cable's LOC-CSP.

5.3   Material Issues

The LOC-CSP is a lead-frame-based CSP implemented by the Tape-LOC technology. Besides the silicon chip, the major package constituents include the gold bonding wires, the die-attach adhesive tape, the custom-designed lead frames, and the encapsulation molding compound (EMC). The die-attach adhesive tape is a base film with adhesive on both sides. The thickness of the tape is about 100

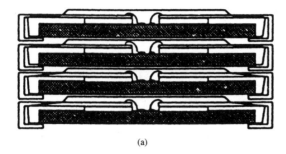

(a)

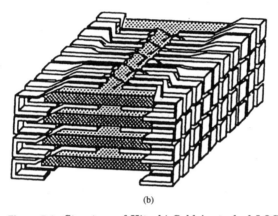

(b)

Figure 5.4 Structure of Hitachi Cable's stacked LOC-CSP: (a) cross-sectional view, (b) three-dimensional view.

μm, which contributes substantially to the package thickness. In order to reduce the thickness, instead of using adhesive tape, Hitachi Cable developed a technique to coat liquid adhesive at the bottom of the inner leads. After die mounting, the adhesive is only 10 μm thick, resulting in considerable reduction in the package thickness.

From the cross section shown in Fig. 5.3, it can be seen that the LOC-CSP is a highly "unbalanced" (in the thickness direction) package. Therefore, package warpage under thermal loading is a major concern. Since it is difficult to change the die and the lead frame, tuning the EMC becomes the way to minimize the warpage. Hitachi Cable has investigated various EMCs; their material properties are given in Table 5.1. The experimental results are presented in Fig. 5.5. It was decided that 50 μm is the maximum allowable package warpage. After analysis, it was concluded that an EMC with low coef-

TABLE 5.1   Properties of Various Molding Compounds for LOC-CSP

Item	Unit	A	B	C	D	E	F	G	H
Viscosity	poise	—	116	170	400	100	100	130	130
α_1	ppm	10	7	12	13	9	9	8	8
α_2	ppm	46	30	45	40	35	35	32	32
T_g	°C	150	160	125	135	130	130	130	130

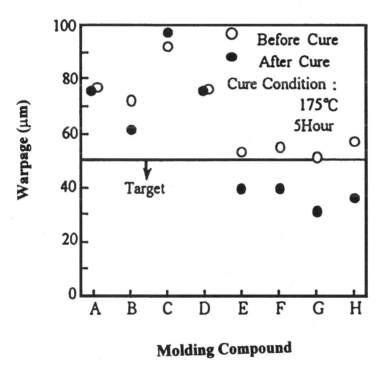

Molding Compound

Figure 5.5   Warpage of LOC-CSP package with various molding compounds.

ficient of thermal expansion (CTE) and strong adhesion to the lead frame can reduce the warpage and, consequently, improve the package reliability.

5.4   Manufacturing Process

The manufacturing process for Hitachi Cable's LOC-CSP is given in Fig. 5.6. This process is very similar to that for the fabrication of standard LOC packages, but with minor modifications. Therefore, the introduction of LOC-CSP has a minimum cost impact on the present packaging infrastructure.

Lead Frame with Adhesive

Chip mounting ; Wire Bonding

Molding

Trimming / Forming , Test

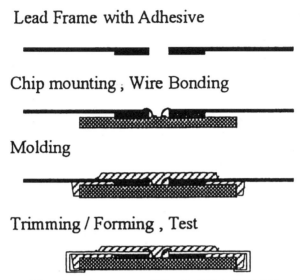

Figure 5.6 Manufacturing process of Hitachi Cable's LOC-CSP.

Since the LOC-CSP is a lead-frame-based package, the first step in fabrication is lead-frame preparation. In particular, the deposition of die-mounting adhesive at the bottom of the inner leads is a critical issue for LOC packages. There are two ways to implement this. The conventional way is to install an adhesive tape which consists of a base film and adhesive on both sides. The major advantage of this method is the thickness uniformity of the adhesive layer. However, the drawbacks are considerably high material cost and relatively large thickness. More recently, Hitachi Cable has developed a new technique for depositing the die-mounting adhesive for LOC-CSP. A liquid adhesive is coated on the lead frame with a thickness of 20 μm. After die mounting and curing, the thickness of the adhesive is reduced to 10 μm, which is substantially thinner than the adhesive tape. The material cost is much lower as well. The next step is wire bonding. For a single LOC-CSP package, the wire bonds are exactly the same as those in the standard LOC. However, for stacked modules, the wire bonds need to be arranged in a specific pattern, as shown in Fig. 5.7, so that the expansion of capacity can be implemented.

After die mounting and wire bonding, the IC is encapsulated by molding compound. The major differences between the standard LOC and the LOC-CSP occur in this step. In Fig. 5.6, the molding area at the upper portion of the LOC-CSP is much smaller than that at the lower portion, exposing a wider shoulder area at the top of the leads. In addition, the encapsulation does not cover the bottom of the die, resulting in a thinner package. Such modifications are intended to ease

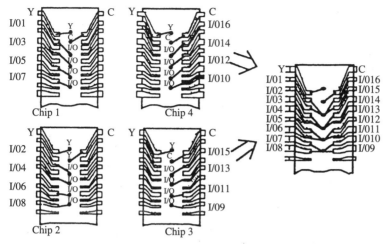

Figure 5.7 Wire-bonding pattern for Hitachi Cable's four-stack LOC-CSP 3-D module.

the stacking of 3-D modules. Following the encapsulation is the trimming of leads. The leads are then bent down to the bottom of the package, forming flat lands as the second-level interconnects. It should be noted that a solder plating is necessary to make a wettable surface on the leads for board-level assembly.

5.5 Electrical Performance and Package Reliability

Since the lead length of LOC-CSP is rather short, the lead inductance and capacitance are not the major concerns for electrical performance. Instead, because the distance between the die and the lead frame is very small (10 to 100 μm), the electrostatic capacitance needs to be considered [1]. An experimental study was conducted to investigate the effect of thickness of die-mounting adhesive on this property, and the results are given in Fig. 5.8. In this figure, the two thinner samples were produced by the liquid adhesive coating method, while the thicker ones were by the double-sided adhesive tape. It was found that the maximum deviation in electrostatic capacitance due to the variation in adhesive thickness is 1 pF. Since this value is very small, it may be concluded that the effect of adhesive thickness on the electrical performance of LOC-CSP is negligible. The thermal performance of Hitachi Cable's LOC-CSP has not yet been released. However, since the package is rather thin and the bottom of the silicon chip is not covered by encapsulant, good capability for heat dissipation should be expected.

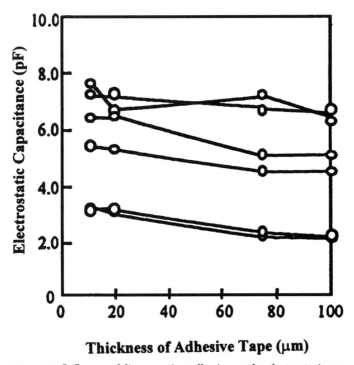

Thickness of Adhesive Tape (μm)

Figure 5.8 Influence of die-mounting adhesive on the electrostatic capacitance.

In addition to the electrical performance, the package reliability of LOC-CSP has been evaluated by thermal shock [5]. The testing specification is −55 to 150°C with 30 min of dwell time per cycle at each temperature. After 600 cycles, the specimens were inspected by scanning acoustic microscope (SAM). No package cracking or internal delamination could be detected. Therefore, the package quality was verified.

5.6   Applications and Advantages

The LOC-CSP was developed for packaging low-pin-count (40 to 60) ICs. The main application aimed at is DRAMs, especially stacked memory modules. According to Hitachi Cable, because of its compact size and stackable structure, the LOC-CSP is very suitable for memory modules used in image data processing or in workstations.

In addition to the stackable configuration, the LOC-CSP has all the advantages of the standard LOC packages. Among them are improved moisture resistance, easy lead pattern design, low lead inductance and noise, high current density, and good package reliability. Once fully qualified, this package will find wide application in three-dimensional memory modules.

5.7 Summary and Concluding Remarks

The LOC-CSP is a leaded plastic package based on the existing LOC technology. The main objective in developing this new chip scale package is to extend the LOC package's capacity to stacked 3-D memory modules. This package uses wire bonds and plated flat lands as the first-level and second-level interconnects, respectively. The manufacturing process for the LOC-CSP is very similar to that for the standard LOC except for certain modifications in encapsulation molding and lead forming. These adjustments are necessary so that package stacking can be easily implemented. It should be noted that although adhesive tape is a conventional method of die mounting in LOC packages, Hitachi Cable has developed a liquid adhesive coating technique to replace the adhesive tape. As a result, both material cost and package thickness are reduced. Since the LOC-CSP has a highly unbalanced structure, package warpage could be a concern for reliability. It was found that tuning of EMC properties could alleviate this problem. By selecting EMCs with low CTE and high bonding capability, the package warpage can be controlled to under 50 μm, which is quite acceptable for surface-mounted components.

The electrical performance of the LOC-CSP was evaluated for the electrostatic capacitance between the lead frame and the silicon chip. It was found that the adhesive thickness has little influence on this characteristic. Although data on thermal performance are not found in the literature, good heat dissipation capability is expected because the package is rather thin and the bottom of the silicon chip is exposed to the outside. In addition, the package reliability of LOC-CSP has been verified by thermal shock testing. However, data on the board-level solder joint reliability are not yet available.

The LOC-CSP has a compact package size which fully conforms to the common definition for CSP. The chip-to-package size ratio is about 90 percent. This package is aimed at applications for low-pin-count devices such as DRAM. Currently the standard LOC is in mass production. The LOC-CSP is still at the preproduction stage. Because of its compact size and stackable structure, this package may be suitable for applications such as image processing which require massive stacked memory modules.

5.8 References

1. N. Taketani, K. Hatano, H. Sugimoto, O. Yoshioka, and G. Murakami, "CSP with LOC Technology," *Proceedings of the ISHM International Symposium on Microelectronics,* Minneapolis, Minn., October 1996, pp. 594–599.
2. N. Taketani, K. Hatano, O. Yoshioka, and G. Murakami, "Hitachi Cable's Most Advanced CSPs," *Proceedings of the Chip Scale Packages Symposium,* Singapore, February 1998, pp. 1–9.
3. G. Murakami, "Rationale for Chip Scale Packaging (CSP) Rather than Multichip Module (MCM)," *Proceedings of the Surface Mount International Conference,* San Jose, Calif., September 1995, pp. 1–5.
4. J. Kasai, M. Sato, and T. Fujisawa, "Low Cost Chip Scale Package for Memory Products," *Proceedings of the Surface Mount International Conference,* San Jose, Calif., September 1995, pp. 6–17.
5. K. Hatano, "Reliability of CSP Manufactured by Using LOC Package Technology," *Proceedings of the IMAPS Advanced Technology Workshop on Chip Scale Packaging,* Austin, Tex., August 1997.

6

Hitachi Cable's
Micro Stud Array Package (MSA)

6.1 Introduction and Overview

The micro stud array (MSA) package is a new CSP developed by
Hitachi Cable in 1995 for high-pin-count applications [1]. This package is also known as metal stud array or μ-stud BGA [2]. The MSA is
a lead-frame-based package. Two sets of stud arrays are fabricated at
both sides of the lead frame by photochemical etching. Each array
consists of four processions of tiny studs. The connection from the die
to the PCB is via Au wire bonds and the studs only. There are no
leads or routing on the lead frame. The package size of the MSA is not
very compact, so that it does not conform to the common definition for
CSP. However, because of the fine-pitch stud array, the MSA is still
recognized as a chip scale package. The main application of MSA is
expected to be ASICs because of its high-pin-count capacity. Currently
(1998) Hitachi Cable is producing MSA prototypes for customers' evaluation. The reliability data and electrical/thermal performance data
are not yet available in the literature.

6.2 Design Concepts and
Package Structure

Most chip scale packages are designed for applications with a low to
medium range of pin counts. Unlike the others, Hitachi Cable's MSA
was developed to accommodate ICs with a large number of I/Os
(>500), as shown in Fig. 6.1. This package features fine-pitch interconnects by lead-frame etching. There are four processions of studs at
each side of the lead frame, as shown in Fig. 6.2. Each stud is basically a cylindrical column which is made from the lead frame by photo-

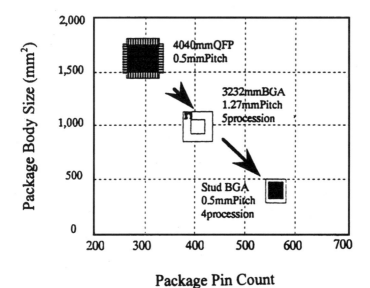

Package Pin Count

Figure 6.1 Package size reduction from QFP to BGA and to MSA.

Figure 6.2 Top stud array of a 576-pin MSA.

chemical etching. Both end faces of the stud are Au-plated, providing wire-bondable and solder-wettable surfaces for interconnection. All studs have the same pitch of 0.5 mm. It should be noted that although the bottom studs are aligned as a uniform-perimeter array, the studs at the top side are arranged in a staggered pattern to allow fine-pitch wire bonding.

The package structure of MSA is illustrated in Fig. 6.3. The first-level interconnects are Au wire bonds, while the second-level interconnects are Au-plated stud bumps. If necessary, solder balls may be attached to those bottom studs to form conventional BGA. It should be noted that it is impossible to perform wire bonding on the studs at the four corners of the package. As a result, those studs are not really the chip I/Os but are reserved pins for mounting passive components for noise reduction and impedance matching. Therefore, for instance, a 576-pin MSA actually has only 456 I/Os for the LSI inside. From the cross-sectional view of MSA structure, it can be seen that the connection from the die to the PCB is via Au wire bonds and the studs only. There are no leads or routing on the lead frame. At the bottom of the MSA package, there is a heat spreader, as shown by the cross-sectional view in Fig. 6.3 and the bottom view in Fig. 6.4. This is actually a remaining portion of the etched lead frame, and the die is directly attached to the other side of this heat spreader.

Compared to other CSPs, the package size of MSA is relatively large. For a silicon chip of $15 \times 15 \times 0.3$ mm, the package size is $21 \times 21 \times 1.0$ mm. Therefore, MSA does not comply with the common definition of CSP. However, because of the fine-pitch stud array, MSA is still considered a chip scale package. Table 6.1 presents the current lineup of MSA packages. It should be noted that the actual package

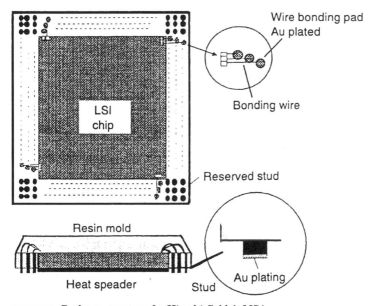

Figure 6.3 Package structure for Hitachi Cable's MSA.

Figure 6.4 Bottom view of a 576-pin MSA (with solder balls attached to the bottom studs).

TABLE 6.1 Hitachi Cable's MSA Package Family

Chip pin count	Wire bonding pitch (μm)	Overall package size (mm)	Overall package thickness (mm)
456	125	21 × 21	1.0
568	100	24 × 24	1.0
712	83	26 × 26	1.3
812	71	27 × 27	1.5
948	62	30 × 30	1.75

pin count is the sum of the chip I/O pin count and the number of re-served studs at the corners. Because of its high-pin-count capacity, the main application of MSA is expected to be the domain of ASICs.

6.3   Material Issues and Manufacturing Process

Hitachi Cable's MSA employs the conventional lead-frame-etching and wire-bonding technology. Therefore, the fabrication of this package has a minimum cost impact on the present manufacturing infrastructure.

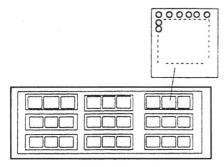

Au point plating on both side of metal plate

Photo-chemical etching

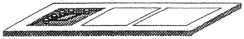

Epoxy resin injection

Photo-chemical etching of back side

Die attach , wire bonding and mold

Figure 6.5   Manufacturing process for Hitachi Cable's MSA.

The manufacturing process for MSA is presented in Fig. 6.5. Since this is a lead-frame-based package, the first step is lead-frame preparation. It is reported that the lead-frame material can be either alloy 42 or Cu. The thickness is 300 μm. A layer of 0.5-μm Au is electroplated on both sides of the lead frame to form the desired stud land pattern. The lead frame is then subjected to photochemical etching to form the stud array at the top side. A square cavity at the die-attach area is also made at this stage for subsequent die mounting. Afterwards, by an injection-molding process, a thin layer (150 μm) of epoxy resin is deposited at the top side of lead frame to contain the existing studs, as shown in Fig. 6.6. The next step is to etch the back side of the lead

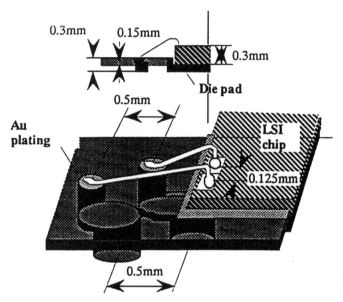

Figure 6.6 Cross-sectional and three-dimensional views of the internal configuration of Hitachi Cable's MSA.

frame to form the bottom stud array and separate the adjacent stud columns. It should be noted that although all studs have the same pitch of 0.5 mm, the standard diameters of top studs and bottom studs are 0.1 mm and 0.2 mm, respectively. Once the lead frame is prepared, the die is mounted in the aforementioned cavity by silver paste. The next steps are wire bonding and encapsulation molding, which are the standard procedures used for most other packages. In order to reduce the thermal mismatch, the molding encapsulant is the same epoxy resin used to contain the studs in the earlier step.

MSA is a wire-bonded package for devices with a large number of I/Os. To accommodate the numerous interconnects, fine-pitch wire bonding with a staggered configuration is necessary. In Fig. 6.7, the pattern of wire bonding for the MSA package is illustrated. Pictures of actual wire bonds are presented in Fig. 6.8. The typical bond-pad pitch on the die is 125 μm. For better wire bonding, the top studs in different processions are offset by the same distance, as shown in Fig. 6.7. However, the bottom studs must be aligned in a uniform perimeter array in order to match the solder pads on the PCB. In general, the board-level interconnects of MSA are Au-plated stud bumps. The package-to-board assembly follows the conventional SMT process. If necessary, solder balls may be attached to the bottom

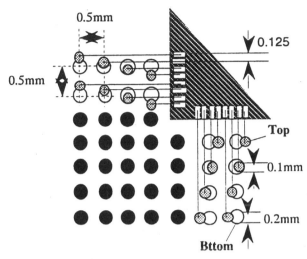

Figure 6.7   Staggered wire-bonding pattern of MSA.

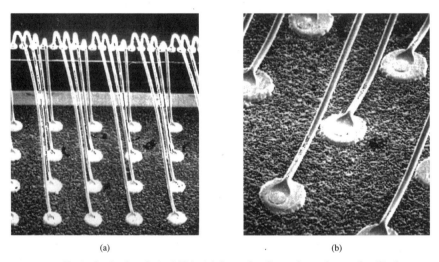

(a)	(b)

Figure 6.8   Typical wire bonds in MSA: (a) from the die pads to the studs, (b) close-up view of the wire bonds on studs.

studs by solder paste printing and reflow to form conventional BGA, as shown in Fig. 6.9.

6.4   Performance and Reliability

MSA is a relatively new CSP and is not yet in mass production. Currently Hitachi Cable is making MSA prototypes for customers' evaluation. The reliability data and electrical/thermal performance

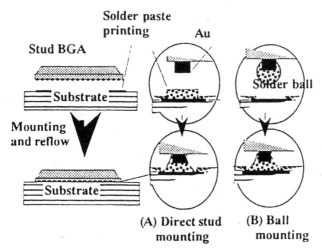

Figure 6.9   Surface-mount assembly of Hitachi Cable's MSA.

data are not yet available in the literature. However, from the design and structure of MSA, certain performance characteristics may be anticipated.

One of the features of the MSA package is the rather short electrical passage length from the die to the PCB. The connection is through wire bonds and studs only. There are no leads or routing on the lead frame. Therefore, the inductance and capacitance should be very low. At the bottom of the MSA package there is a metal heat spreader to which the die is directly attached. Since this is the back portion of the original lead frame, the heat spreader and the bottom stud array share the same level. After surface-mount assembly, the heat spreader is very close to the PCB. Therefore, the heat-dissipation capability of MSA should be relatively good. Hitachi Cable also suggests the use of underfill, which will further improve the thermal performance.

In addition to the silicon chip, the bulk of the MSA package is molding encapsulant. Hitachi Cable claims that the epoxy resin can be tuned to minimize the thermal mismatch. As a result, reasonable package reliability should be achieved. A qualification test such as −55 to 150°C thermal cycling is planned. However, the test data have not yet been released at this time.

6.5   Applications and Advantages

MSA was developed for high-pin-count applications. The current lineup ranges from 456 to 948 pins (real chip I/Os). Therefore, the MSA is suitable for packaging ASIC devices. The major competitor to MSA is the PBGA package for the similar pin-count range. However, the former has much smaller form factors than the latter. For instance, a conventional PBGA (1.27-mm ball pitch) with 484 pins has an outline of 33×33 mm. The package size of MSA (0.5-mm stud pitch) with 456 pins (real chip I/Os) is 21×21 mm. The package is thinner as well. In addition to the form factors, another advantage of MSA is the cost factor. Because MSA adopts the conventional lead-frame-etching and wire-bonding technology, the fabrication of this package has minimum cost impact on the existing infrastructure.

6.6 Summary and Concluding Remarks

MSA is a leadless plastic-encapsulated package based on the conventional lead-frame-etching and wire-bonding technology. The bulk of the lead frame is etched to form stud arrays and a bottom heat spreader. The wire bonds and studs are the only interconnects between the die and the PCB. There are no leads or routing on the lead frame. Since the passage is rather short, MSA should have relatively good electrical performance. Also, because of the existence of the heat spreader, the heat generated by the chip should be dissipated effectively.

Judging from the package size, MSA does not fit the conventional definition for CSP. However, because of the fine-pitch stud array, MSA is still recognized as a chip scale package. MSA was developed for packaging high-pin-count ICs. The current lineup ranges from 456 to 948 chip I/Os. The major competitor to MSA is the PBGA package for the similar pin-count range. Currently (1998) MSA is still under prototyping and evaluation. Once the package and the solder joint reliability are qualified, this package should find certain applications for ASIC devices.

6.7 References

1. M. Mita, G. Murakami, T. Kumakura, N. Okabe, and S. Shinzawa, "Advanced Interconnect and Low Cost μ Stud BGA," *Proceedings of the International Electronics Manufacturing Technology Symposium,* Austin, Tex., September 1995, pp. 428–433.
2. N. Taketani, K. Hatano, O. Yoshioka, and G. Murakami, "Hitachi Cable's Most Advanced CSPs," *Proceedings of the Chip Scale Packages Symposium,* Singapore, February 1998, pp. 1–9.

7

LG Semicon's Bottom-Leaded Plastic Package (BLP)

7.1 Introduction and Overview

The bottom-leaded plastic (BLP) package is a lead-on-chip type of CSP. It was developed by LG Semicon Co., Ltd. (Korea), and can be considered as an extension of the thin small outline package (TSOP). This packaging technology is implemented with custom-designed lead frames. The interconnects at the chip level and the package level are wire bonds and plated flat lands, respectively. Depending upon the bond pad location, there are two types of configurations. Because of the paddleless design, adhesive tape is used for die mounting. Low profile is the major feature of BLP packages. The package outline complies with the JEDEC standard for USON. This package is designed for low-I/O (<100) applications. The main application of BLP packages is for memory devices. Currently BLP is in mass production and is available for license to other companies. The use of BLP for stacked memory modules is under evaluation.

7.2 Design Concepts and Package Structure

The design of the BLP package first appeared in 1995 [1]. The basic idea was to implement an ultra-thin leadless plastic package based on the conventional TSOP structure. In order to achieve this objective, a custom-designed lead frame is needed. The lead-on-chip scheme is adopted so that the package-to-die ratio can satisfy the requirement for CSPs. Wire bonding is used for the chip-level interconnects. There are two variations of the configuration; they differ in the location of bond pads. The S-BLP is for packaging ICs with peripheral bond pads, while the C-BLP is for center bond pads. The cross sections of both packages are shown in Fig. 7.1.

Like other plastic packages, BLP is encapsulated by epoxy molding compound. Low profile is the major feature of BLP. Typical package thickness is within the range of 0.48 to 0.82 mm. The packaged die is usually prepared for a thickness of 0.28 mm or less. In order to achieve this thinness, the die paddle is omitted and the loop height of wire bonding must be well controlled. Because of the paddleless design, the die is mounted on the lead fingers by thermoplastic adhesive tape. A comparison of the dimensions of various surface-mount plastic packages is given in Table 7.1 and Fig. 7.2.

BLP is a leadless package. The package-level interconnects are plated flat lands. The package outline conforms to the JEDEC MO-196 standard. The standard land configuration is 20 pins for S-BLP

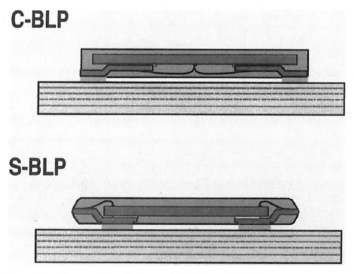

Figure 7.1   Cross-sectional views of C-BLP and S-BLP.

TABLE 7.1   Comparison of Geometry among SOJ, TSOP, and BLP

	SOJ	TSOP	S-BLP
Height (mm)	3.76	1.27	0.48–0.82
Width (mm)	8.50	9.22	5.21
Package area (mm^2)	149	160	107
Package volume (mm^3)	548	200	90
Package weight (g)	0.82	0.3	0.2

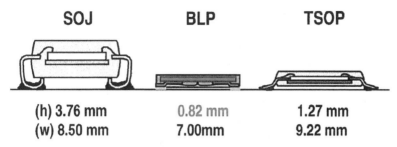

Figure 7.2 Comparison of package dimensions for SOJ, TSOP, and BLP.

Figure 7.3 Top and bottom views of BLP packages: (*a*) C-BLP, (*b*) S-BLP.

and 28 pins for C-BLP, as shown in Fig. 7.3. The typical land width
and pitch are 0.4 mm and 0.8 mm, respectively. Currently the BLP
package is under evaluation for three-dimensional memory modules.
The stacked structure of BLP and a comparison with other packages
are presented in Fig. 7.4.

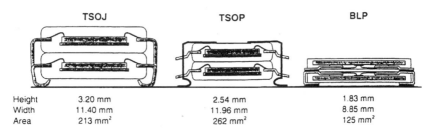

	TSOJ	TSOP	BLP
Height	3.20 mm	2.54 mm	1.83 mm
Width	11.40 mm	11.96 mm	8.85 mm
Area	213 mm^2	262 mm^2	125 mm^2

Figure 7.4 Comparison of form factors for various stacked memory modules.

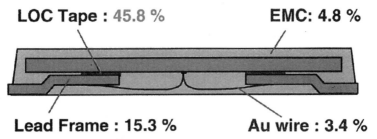

LOC Tape : 45.8 %   EMC: 4.8 %

Lead Frame : 15.3 %   Au wire : 3.4 %

Figure 7.5 Material cost breakdown for the major constituents of the BLP package.

7.3 Material Issues

BLP is a lead-frame-based plastic package. In addition to the silicon chip, the major constituents of BLP include the bonding wires, the custom-designed lead frame, the die-mounting adhesive, and the encapsulation molding compound. The material cost breakdown is shown in Fig. 7.5.

The chip-level interconnects of BLP are Au wire bonds. The typical diameter of the Au wire is 30 to 33 μm. The wire connects the bond pads on the IC chip and the lead frame. Because of the ultra-thin package profile, the loop height of the bonding wire must be minimized and well controlled. For the present BLP packages, the lead frame is made of alloy 42 with a thickness of 200 μm. A copper lead frame is under evaluation for the purpose of solder joint reliability enhancement. Because of its compact size, one single lead-frame strip can contain up to 18 BLP units. The comparison of lead-frame layout for S-BLP and SOJ is presented in Fig. 7.6. For better performance, certain notches, as shown in Fig. 7.7, should be considered during the lead-frame design of S-BLP. The notch at location a eases the trimming for singulation, while the one at b assists the down-set process. In addition, the notch at c improves the hermeticity of the package by detouring the interfacial path between the lead frame and the molding encapsulant.

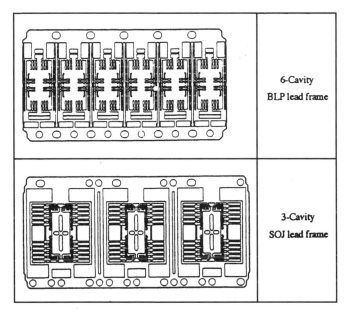

6-Cavity BLP lead frame
3-Cavity SOJ lead frame

Figure 7.6   Comparison of lead-frame design.

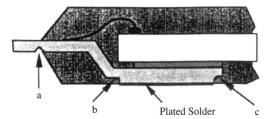

a

b   Plated Solder   c

Figure 7.7   Detailed view of notches in the S-BLP lead frame.

The BLP package has a paddleless design. It is similar to the LOC type of packages. The die is mounted on the inner pads by thermoplastic adhesive tape. The tape is a polyimide-based film. At both sides of the base film, there are adhesives for the bonding to the die and the lead frame, respectively. The total thickness of the adhesive tape is 100 μm. Good bonding strength and low contamination on the wire bond pads are the major advantages of adhesive tape. In addition, since the base polyimide film can provide uniform thickness, which leads to good coplanarity of the inner lead bond pads, the wire bondability is rather reliable. However, the material cost of adhesive tape is very high. Therefore, an effort is being made to replace the ad-

TABLE 7.2   Material Properties of Modified Epoxy for Die-Mounting
Adhesive Paste

Material properties	Epoxy I	Epoxy II
Viscosity, CPS@25°C	80,000	154,000
Thixotropic index	4.0	3.0
T_g (°C)	205	113
Elastic modulus (kg/mm²)	300	250
Moisture absorption (%)	0.12	0.19
Dielectric constant	3.5	2.8

hesive tape with low-cost adhesive paste. Two types of modified epoxy
are shown in Table 7.2. The major material properties of the developed adhesive paste that are of concern include the high-temperature
adhesion strength, the curing time duration, and the resistance to
moisture absorption. In addition, since the width of the inner leads is
rather narrow (0.3 mm), high viscosity is essential for the adhesive
paste to prevent contamination at the wire-bonding areas. Once the
feasibility is proven, it is expected that the new adhesive paste could
reduce the cost of BLP packages by as much as 42 percent!

Molding is one of the most important procedures for plastic packages. Because of the ultra-thin package profile, the thickness of the
encapsulant above and underneath the die is less than 270 μm.
Although the low volume of encapsulation molding compound (EMC)
makes the raw material cost relatively low (4.8 percent of the cost of
the whole package), the rather small EMC thickness results in high
molding flow resistance. Consequently, there are a lot of challenges
involved in the design of the mold gate and the tuning of flow pressure and transfer time.

7.4   Manufacturing Process

The manufacturing process of BLP is similar to that for conventional
plastic packages with certain modifications. Because of the bottom
lead configuration, the lead-frame design is a unique feature of BLP
packages. After wafer sawing, there are seven procedures, as shown
in Fig. 7.8, before the package is completed [1]. The first step is die
mounting. Because of the paddleless design, the die is bonded to the
inner leads by a double-sided adhesive tape. This process can be implemented with a conventional LOC die bonder. After prebaking the
lead frame together with the adhesive tape, the IC chip is mounted.
The bonding condition is 400°C for 2 to 4 s with a pressure of 5
kg/cm².

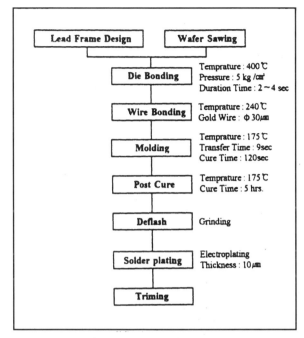

Figure 7.8 Manufacturing process for BLP package.

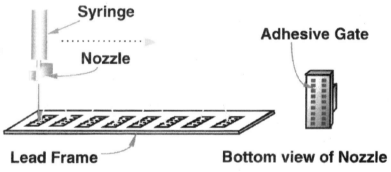

Figure 7.9 Dotting mechanism for dispensing die-mounting adhesive paste.

Since the material cost of adhesive tape is very high (almost one-half of the whole package cost), recently LG Semicon has developed a low-cost epoxy-based paste to replace the adhesive tape [2]. The adhesive paste is dispensed on the lead frame by a dotting process, as shown in Fig. 7.9. Because uniform adhesive thickness is critical to the yield of wire bonding, the amount of adhesive paste must be precisely controlled during dispensing. This objective may be achieved by optimizing the three major processing parameters, namely, the air pressure in the dispensing nozzle, the viscosity of the adhesive paste, and the dotting speed. A picture of well-controlled adhesive paste on the inner leads is presented in Fig. 7.10. After the dispensing of adhesive paste, the die is mounted to the lead frame. The die-mounting process may be performed with a conventional LOC die bonder at a speed of one unit per second. The typical bonding condition is 230°C for 1 to 2 s. The thickness of the adhesive paste after die mounting is 45 μm. Currently LG Semicon has completed the LOC adhesive paste development project. It is expected the new process will be in mass production and will replace the existing adhesive tape in the near future.

The process right after die mounting is wire bonding. Currently the BLP package uses Au wire with a diameter of 30 to 33 μm. The wire bond is conducted at 240°C. Ball bonding and wedge bonding are used at the chip side and the lead-frame side, respectively. Since BLP packages are very thin, the loop height of the bonding wires must be minimized and precisely controlled. The two steps after wire bonding are encapsulation and postcure. The molding is performed at 175°C with 9 s of transfer time and 120 s of curing time. The total thickness of EMC is less than 540 μm. With such a small dimension, the resistance of the molding flow is relatively high. Therefore, the molding design parameters such as the position of the mold gate, the lead-frame geometry, and the transfer pressure must be carefully determined. After encapsulation, the packages require a postcure treatment at 175°C for 5 h.

The thin flash on the bottom lands is fatal to the board-level solder joint reliability. Therefore, a deflash process is needed after the package is encapsulated. For C-BLP packages, the deflash is done by con-

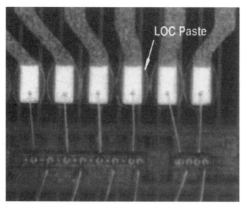

Figure 7.10 Die-mounting adhesive paste on the bond pads of the BLP lead frame.

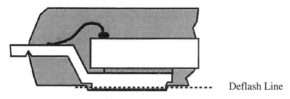

Deflash Line

Figure 7.11   Deflash line of S-BLP.

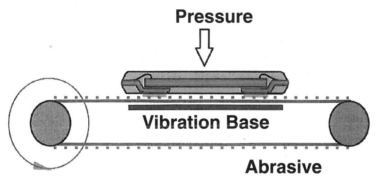

Figure 7.12   Deflashing by mechanical grinding.

ventional chemical wetting and water jet. It is reported that this process is very stable. However, for S-BLP packages, a mechanical grinding process is necessary [3]. The bottom lands of the lead frame have been designed with a 30-μm extrusion, as shown in Fig. 7.11. A mechanical grinder, illustrated in Fig. 7.12, has been developed to deflash the bottom of S-BLP packages. By optimizing the applied pressure, the abrasive surface, and the base vibration mode, it is reported that the deflash thickness can be precisely controlled.

In order to improve the solderability for surface mount, after the deflash process, the bottom leads of BLP packages are coated with 90Pb/10Sn solder by electroplating. The thickness of the plated solder is 10 μm. Since there are no external leads, a forming process is not required. The last procedure in fabricating BLP is trimming for package singulation.

7.5   Electrical and Thermal Performance

The BLP packages have been tested extensively for their electrical and thermal performance [4]. Since the package has a very short internal I/O path, good electrical performance is expected. For the purpose of characterization, a test package without an IC chip was fabricated and mounted on a test board with a known ground reference.

High-frequency S-parameter data were measured using a network analyzer such as HP8510C. The lead inductance and capacitance were calculated by the MDS program. The package was tested within the frequency range between 50 MHz and 3 GHz. The results for an S-BLP package are listed in Table 7.3 and benchmarked with a commercial PQFP. In addition, with further characterization, the internal noise level in the package was found to be less than 4 percent of the driving voltage. With such electrical performance, it could be concluded that the BLP package is applicable for high-speed devices with clock frequency of 1 GHz or higher.

Heat dissipation capability is another critical index for evaluating the performance of an electronic package. A three-dimensional finite-element analysis was conducted to evaluate the thermal performance of BLP packages. The thermal conductivity of the package constituents is listed in Table 7.4. For a typical simulation under the natural convection (NC) condition, with a 0.3-W power supply, which is a typical value for memory devices, it was found that the temperature difference between the junction and ambient temperatures (ΔT_{ja}) was 26.23°C, which corresponds to a thermal resistance θ_{ja} of 87.43°C/W. Since this value is acceptable for practical applications, BLP is suitable for packaging devices which consume 1 W or less. Table 7.5 shows the comparison between various packages under the NC condi-

TABLE 7.3   Electrical Performance of BLP

	BLP	PQFP
Test conditions	20 pins, 50 MHz–3 GHz	196 pins, 50 MHz–10 GHz
Self-inductance (nH)	1.7–2.7	5.4–5.8
Mutual inductance (nH)	0.6–0.8	2.1–2.3
Self-capacitance (pF)	0.15–0.25	1.48–1.74
Mutual capacitance (pF)	0.04–0.10	0.62–0.71

TABLE 7.4   Thermal Conductivity of BLP Package Constituents

Package materials	Thermal conductivity (W/m · K)
Lead frame (alloy 42)	14.70
Epoxy molding compound	0.88
Adhesive tape	0.17
Silicon die	139
Printed circuit board	0.21
Air gap (between BLP and PCB)	0.02

TABLE 7.5   Comparison of Thermal Resistance among Various Packages

	SOJ	TSOP	BLP
θ_{ja} with 4M DRAM (°C/W)	66.5	90.2	87.4
θ_{ja} with 16M DRAM (°C/W)	62.4	61.8	51.4

TABLE 7.6   Film Coefficient of Materials

	Film coefficient (W/m^2 · K)		
	Natural convection	Air flow: 1 m/s	Air flow: 5 m/s
Package	15.55	27.52	58.07
PCB	7.80	13.80	29.13

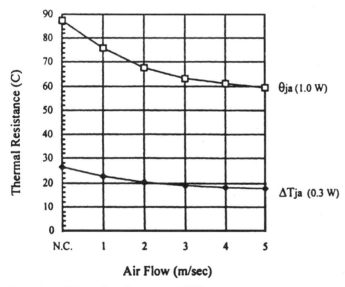

Figure 7.13   Thermal performance of BLP.

tion. It may be concluded that the BLP package has better thermal performance than the conventional TSOP.

In addition to natural convection, several forced convection cases were investigated as well, using different film coefficients (given in Table 7.6). The results of thermal resistance vs. air flow are presented in Fig. 7.13. It is observed that the thermal resistance decreases with respect to the increase in air flow speed. Further parametric studies were conducted to investigate the effects of underfill, lead-frame material, and PCB thermal conductivity on the thermal performance. The comparison of thermal resistance for various cases is presented

TABLE 7.7   Effects of Underfill, Lead Frame, and PCB on BLP's Thermal Conductivity

	Original conductivity (W/m · K)	Modified conductivity (W/m · K)	Change in θ_{ja}
Underfill	0.02 (air)	4.00 (epoxy)	−1.41%
Lead frame	14.70 (alloy 42)	390.10 (Cu)	−0.53%
PCB	0.21 (normal)	0.25 (high Cu content)	−0.50%

in Table 7.7. From these results, it may be deduced that the major heat dissipation of BLP is from the top side of the package instead of the bottom side.

7.6   Qualifications and Reliability

The reliability of plastic packages is a major concern for electronic devices. Every new package design must be subjected to a series of qualification tests before being adopted for real applications. The BLP packages have been qualified for package reliability and solder joint reliability [5]. For the former, the IR reflow test method was employed to evaluate the package resistance to moisture-induced damage. The first step was prebaking. The specimens were placed in a chamber at 125°C for 16 h. This procedure is necessary to ensure that the specimens contain the minimum amount of moisture before testing. After prebaking, the specimens were subjected to an environment of 85°C and 85 percent relative humidity (RH) for 168 h. Once this moisture exposure period was completed, the specimens were cooled down to room temperature with 100 percent RH. Subsequently, an IR reflow condition with the temperature profile specified in Fig. 7.14 was applied to the specimens three times. To investigate the crack resistance of BLP packages, visual inspection was performed after the reflow procedure. No observable external cracks could be identified. Further inspection was conducted using scanning acoustic microscopy (SAM). No internal delamination could be detected, as shown in Fig. 7.15. Finally, the specimens were cross-sectioned for microscopic investigation. As shown in Fig. 7.16, there is no separation at any of the interfaces inside the package. As a result, the BLP package was qualified for JEDEC level 1 IR crack resistance. In addition to the crack resistance, further reliability tests were performed as presented in Table 7.8. The results show that the BLP package passed all qualification tests.

The BLP package is a low-profile surface-mount component (SMC) without lateral leads. For such a leadless SMC, the reliability of board-level interconnects is always a major concern. To characterize the solder joint reliability of BLP packages, a series of accelerated

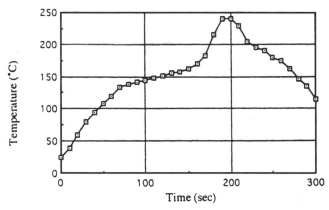

Figure 7.14   IR reflow temperature profile for the assembly of BLP on PCB.

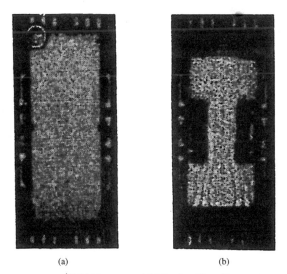

(a)   (b)

Figure 7.15   C-SAM image of BLP after IR reflow: (a) top view, (b) bottom view.

thermal cycling (ATC) tests were conducted [6]. The specimens were C-BLP with a specially designed daisy chain connection, as shown in Fig. 7.17. The packages were surface-mounted on FR-4 test boards with a thickness of 1 mm. In order to control the volume of solder paste, various sizes of solder pads on the PCB were designed (0.8 to 1.2×0.45 mm). The composition of the solder used for ATC tests was 62Sn/2Ag/36Pb, and no-clean flux was employed. The solder paste was printed on the solder pads through a 150-μm-thick stencil. After

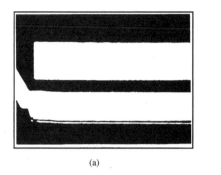
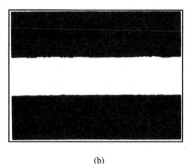

(a)   (b)

Figure 7.16 SEM inspection of the cross section of BLP: (*a*) chip–tape–lead frame interface, (*b*) chip-EMC interface.

TABLE 7.8   Package Qualification Testing Results of BLP

Test item	Test conditions	Test duration	Results
IR reflow	85°C/85% RH, 168 h, 240°C	3 times	0/40
PCT	121°C,100% RH, 2 atm	168 h	0/96
T/C	−65 to 150°C, air to air	1000 cycles	0/135
T/S	−65 to 150°C, liquid to liquid	300 cycles	0/22
HTST	150°C storage	1000 h	0/55
HTOL	125°C, 4.6V	1000 h	0/152
THB	85°C/85% RH, 3.6 V	1000 h	0/228

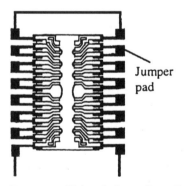

Jumper
pad

Figure 7.17 Daisy chain configuration
for solder joint reliability tests.

reflow, the solder joint height was about 110 μm. The cross section of
a typical C-BLP solder joint is presented in Fig. 7.18. It should be
noted that, since the trimmed edge of the lead frame is not plated, a
proper solder fillet cannot be formed at the toe area of the bottom
leads. In order to improve this situation, a modified lead-frame design

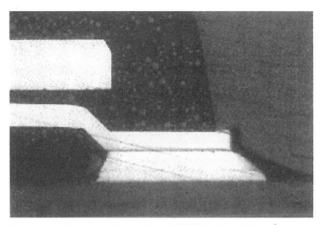

Figure 7.18   Cross-sectional view of BLP-I solder joint.

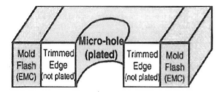

Figure 7.19   Schematic diagram for the bottom lead tip of BLP-II.

called BLP-II was proposed (the original is termed BLP-I) [7]. For the BLP-II design, additional microholes are made on the trim line of the bottom leads during the fabrication of the lead frame. A layer of 90Pb/10Sn will be formed on the inner wall of such microholes when the lead frame is subjected to solder plating. After the final trimming process, half of the microhole will remain at the edge of the bottom lead, as shown in Fig. 7.19, offering an ideal solder-wettable surface for fillet forming. A typical cross section of a BLP-II solder joint is presented in Fig. 7.20. A clear solder fillet can be observed at the toe of the bottom lead.

Both BLP-I and BLP-II were tested under thermal cycling for solder joint reliability. The testing temperature profile followed Condition B of Mil. Std. 883C, for which the temperature range is −55 to 125°C and the time period per cycle is 26 min, as shown in Fig. 7.21. The resistance of the daisy chain circuit of each specimen was monitored during the thermal cycling test. Failure of a solder joint was considered to be a 20 percent or more increase in electrical resistance. The accumulated percentage of failures for every 100 cycles in ATC test is summarized in Table 7.9. It is observed that the

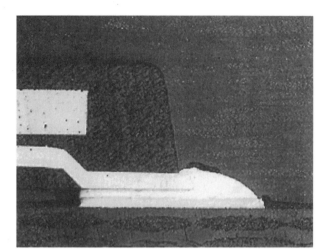

Figure 7.20 Cross-sectional view of BLP-II solder joint.

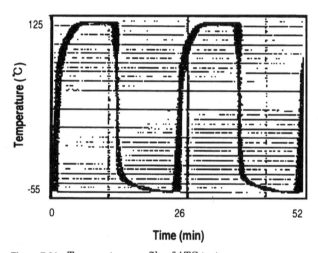

Time (min)

Figure 7.21 Temperature profile of ATC tests.

TABLE 7.9 Accumulated Percentage of Failure in ATC Tests

		Cycles							
Type	PCB pad size	600	700	800	900	1000	1100	1200	1300
BLP-I	0.8 × 0.45	10.6%	17.0%	40.4%	51.1%	68.1%	91.5%	—	—
	1.0 × 0.45	—	8.7%	32.6%	54.3%	63.0%	87.0%	—	—
	0.8 × 0.45	—	7.3%	26.8%	58.5%	82.9%	95.1%	100%	—
BLP-II	1.0 × 0.45	1.4%	2.8%	9.7%	19.4%	40.3%	52.8%	70.8%	87.6%
	1.2 × 0.45	—	—	—	4.8%	7.1%	14.3%	40.5%	61.9%

TABLE 7.10   Statistical Results of Solder Joint Reliability

Type	PCB pad size	Failure-free life	Weibull life with 63% of failures	Weibull shape parameter
BLP-I	0.8 × 0.45	500	991	6.0
	1.0 × 0.45	600	1002	8.0
	0.8 × 0.45	600	991	8.2
BLP-II	1.0 × 0.45	500	1198	6.5
	1.2 × 0.45	800	1358	9.2

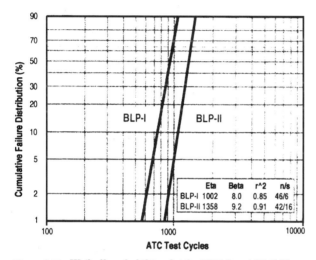

Figure 7.22   Weibull probability plot for BLP-I and BLP-II.

first solder joint failure does not occur until 600 cycles. For shorter solder pad dimensions, it seems that the difference between BLP-I and BLP-II is not very significant. However, substantial improvement in solder joint reliability can be identified for BLP-II with a longer solder pad. The Weibull distribution parameters of the ATC test results are calculated and listed in Table 7.10. The corresponding Weibull probability plot is given in Fig. 7.22. It can be seen that the solder joint reliability of BLP-II is indeed much better than that of BLP-I [8].

It should be noted that the aforementioned ATC test was for C-BLP only. In a previous experimental study by moiré interferometry, as shown in Fig. 7.23, it was found that the longitudinal effective coefficients of thermal expansion (CTE) of S-BLP and C-BLP are 3.9 and 6.25 ppm/°C, respectively. Since the CTE of FR-4 is 18 ppm/°C, it is obvious that the former package has a much larger thermal mismatch

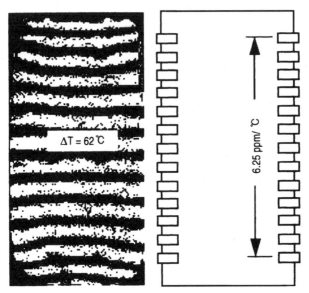

Figure 7.23 Moiré fringes and corresponding effective CTE (C-BLP).

with the PCB and, consequently, has poorer reliability for solder joints at the board level.

7.7 Applications and Advantages

Memory devices are the major application for BLP packages. The early version of the bottom-leaded plastic package was fabricated for a 4M DRAM module using the S-BLP configuration. However, since all 16M and 64M DRAM have center bond pads, LG Semicon decided not to produce S-BLP for the moment. Currently, the C-BLP is in mass production for 16M DRAM. As shown in Fig. 7.24, the application of BLP to notebook PCs has been implemented. Application for 256M SDRAM is planned for 1999. Several companies have licensed or have been evaluating this package, as shown in Table 7.11.

The major advantages of BLP packages are small footprint and low profile. The former can be illustrated by the PCMCIA cards shown in Fig. 7.25. The original configuration has 12 TSOP-II modules. With BLP packaging of the same IC, 24 modules can be mounted on the same size card, giving a 100 percent increase in capacity. The advantage of low profile is presented in Fig. 7.26. The thickness of the shown 8 MB SIMM is 9.0 mm if SOJ packages are used. With the application of BLP, the thickness can be reduced to 3.2 mm. In addition, only 63 percent of the original mounting area is needed. The smaller

Figure 7.24 Application of BLP packages to the notebook PC.

TABLE 7.11   Current Status of BLP Packages

Company	Country	Application	Status
Daewoo Computer	Korea	On-board and SO-DIMM for notebook	In production
Sambo Computer	Korea	On-board memory for notebook	In production
LG Electronics, Ltd.	Korea	On-board and SO-DIMM for notebook	Under evaluation
Siemens	Germany	Package technology license	Evaluation finished
TEPC	Taiwan	Memory module (SO-DIMM)	Evaluation for system applications
Transend	Taiwan	Memory module (SO-DIMM)	Evaluation finished
King Max	Taiwan	Memory module (SO-DIMM)	Evaluation for system applications

form factor has made BLP a very attractive package for portable device applications.

7.8   Summary and Concluding Remarks

BLP is a leadless plastic-encapsulated package. Its outline complies with the JEDEC standard for USON. The basic structure of BLP has a LOC configuration. This packaging technology is implemented

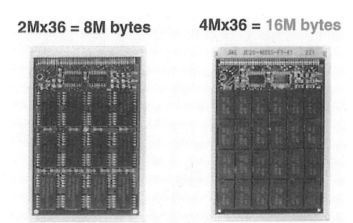

2Mx36 = 8M bytes   4Mx36 = 16M bytes

Figure 7.25 Comparison of mounting density for PCMCIA application (TSOP versus BLP).

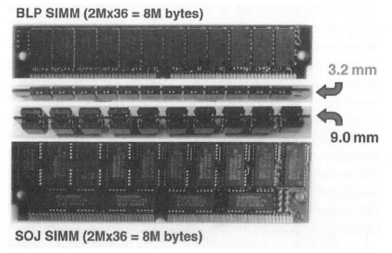

BLP SIMM (2Mx36 = 8M bytes)

3.2 mm

9.0 mm

SOJ SIMM (2Mx36 = 8M bytes)

Figure 7.26 Comparison of form factors for SIMM application (SOJ versus BLP).

using custom-designed lead frames. Because of the paddleless design, adhesive tape is needed for die mounting. A new adhesive paste has been developed in order to reduce the relatively high material cost. The first-level and second-level interconnects are wire bonds and plated flat lands, respectively. Depending upon the bond pad location, there are two configurations, namely, S-BLP and C-BLP. Because of the short distance of the interconnects and the compact size, the BLP has relatively good electrical and thermal performance. It is expected that this package can be used for high-speed devices with clock frequency of 1 GHz or even higher. The package reliability of BLP has been investigated via certain qualification tests. Good crack resistance was reported. Because of the leadless configuration, the reliability of solder joints at the board level is a major concern for BLP packages. ATC tests have been conducted to evaluate the solder joint reliability of the BLP-PCB assembly. The first failure of solder joints does not occur until 600 cycles. With the modification in lead-frame design and the change in solder pad dimension, the first failure can be postponed to 900 cycles. Further study of solder joint reliability is under way.

The BLP package is designed for low-I/O (<100) applications. Small footprint and low profile are the major advantages of BLP. These features make BLP a good candidate to replace or to expand the capacity of TSOP-II and SOJ. The main devices packaged by BLP technology are memory modules. Currently C-BLP is in mass production for 16M DRAM. Applications involving notebook PC and PCMCIA cards have been reported. The use of BLP for stacked memory modules is under evaluation. Application for 256M SDRAM is planned for 1999.

7.9 References

1. K.-B. Cha, Y.-G. Kim, and T.-K. Kang, "Ultra-thin and Crack-free Bottom Leaded Plastic (BLP) Package Design," *Proceedings of the 45th ECTC,* Las Vegas, Nev., May 1995, pp. 224–228.
2. T.-G. Kang, Y.-G. Kim, and H.-S. Chun, "Bottom-Leaded Plastic (BLP) Package: A New Design with Low Cost Adhesive Material," *Proceedings of NEPCON-West,* Anaheim, Calif., February 1997, pp. 63–73.
3. Y.-G. Kim, J.-S. Kim, C.-J. Song, and T.-K. Kang, "New Infrastructure for the BLP Package and Solder Joint Reliability," *Proceedings of Semicon-West,* San Francisco, Calif., July 1996, pp. 39–45.
4. Y.-G. Kim, T.-K. Kang, and B.-S. Seol, "Electrical and Thermal Performance Characterization for the Bottom Leaded Plastic (BLP) Package," *Proceedings of the IEPS Conference,* International Electronic Packaging Society, San Diego, Calif., September 1995, pp. 63–73.
5. Y.-G. Kim, B. Han, S. Choi, and M.-K. Kim, "Bottom-Leaded Plastic (BLP) Package: A New Design with Enhanced Solder Joint Reliability," *Proceedings of the 46th ECTC,* Orlando, Fla., May 1996, pp. 448–452.
6. K.-S. Choi, S. Choi, Y.-G. Kim, and K.-Y. Cho, "Solder Joint Reliability of the BLP (Bottom-Leaded Plastic) Package," *Proceedings of Semicon Korea,* Seoul, Korea, February 1997, pp. 63–73.
7. Y.-G. Kim, K.-S. Choi, and S. Choi, "Solder Reliability of the Leaded and Leadless Packages: New BLP Design," *Proceedings of Semicon-West,* San Jose, Calif., July 1997, pp. 39–45.
8. Y.-G. Kim, K.-S. Choi, and S. Choi, "Solder Joint Reliability of Leadless CSP: New BLP Design," *Chip Scale Review,* 1997, pp. 63–73.

8

TI Japan's
Memory Chip Scale Package
with LOC (MCSP)

8.1 Introduction and Overview

The memory chip scale package (MCSP) was released by Texas
Instruments (TI) Japan in 1997 [1]. There are two configurations of
MCSP. One is based on the tapeless lead-on-chip (LOC) technology,
and the other uses a flexible substrate as the interposer. This chapter
focuses on the lead-frame-based MCSP. The MCSP with flexible substrate is introduced in Chap. 18.

TI Japan's MCSP is a low-pin-count but large-lead-pitch package.
The first- and second-level interconnects are wire bonds and BGA, re-
spectively. The current lineup is 52 pins with a pitch of 1.27 mm. The
MCSP uses a unique process for wafer preparation with polyimide
coating. This thin layer is for the purpose of protection and adhesion.
Good electrical performance and package reliability were reported. TI
Japan has just finished its internal qualifications for the MCSP. This
package is aimed at replacing SOJ and TSOP for DRAM applications.

8.2 Design Concepts and
Package Structure

The LOC technology is widely used in the IC packaging industry for
memory devices. Conventional LOC packages include SOJ and TSOP.
In general, there are two ways to implement the LOC structure. One
is to use an extra lead frame at the bottom to hold the silicon chip
(MF-LOC), and the other is to mount the die on the inner leads using

a double-sided adhesive tape (Tape-LOC) [2]. In 1994, TI Japan developed a new technology called tapeless-LOC which does not require die-mounting adhesive tape [3]. Instead, the lead frame is placed directly on the chip, on the surface of which there is a thin layer of polyimide. This polyimide film is deposited by spin coating; its purposes are protection and adhesion. TI Japan has employed this tapeless-LOC technology to implement a lead-frame-based chip scale package, namely, MCSP [1].

The MCSP is a plastic-encapsulated package. The first- and second-level interconnects are Au wire bonds and solder ball grid array, respectively, as shown in Fig. 8.1. Because of the LOC configuration, this package is for center pad devices only. For a silicon chip of 18.4×8.77 mm, the package size of the MCSP is approximately 18.4×10.2 mm (725×400 mils). Therefore, the dimensions of the MCSP comply with the conventional definition of CSP. It should be noted that in order to achieve the minimum size, the bottom of the die is not covered by the molding encapsulant. The package thickness is 0.8 mm. The current lineup for the MCSP is 52 pins. The solder ball pitch and diameter are 1.27 mm and 1 mm, respectively. The ball population is arranged in the array pattern shown in Fig. 8.2. Compared to a TSOP package with a similar pin count, the outline of the MCSP is smaller.

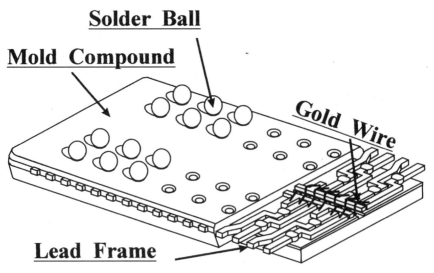

Solder Ball

Mold Compound

Gold Wire

Lead Frame

Figure 8.1 Package structure of lead-frame-based MCSP.

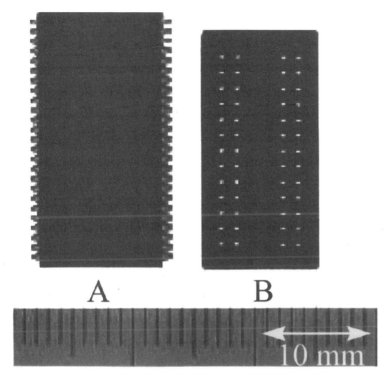

Figure 8.2   Bottom view of (a) TSOP and (b) MCSP.

8.3   Material Issues

The MCSP is a lead-frame-based CSP implemented by tapeless-LOC technology. The materials used to package the silicon chip include the polyimide coating, the gold bonding wires, the custom-designed lead frame, the encapsulation molding compound (EMC), and the eutectic solder balls. According to TI Japan, both Cu and alloy 42 can be used for the lead-frame material.

The most critical material for the tapeless-LOC technology is the I-line polyimide coating on the active surface of the die [4]. The polyimide used is a photosensitive material so that it can be patterned to open windows on the center pads of the IC for wire bonding. The functions of this thin layer are protection of the chip and adhesion to the lead frame [5]. The polyimide is originally in a liquid form and is deposited on the surface of the wafer by spin coating. After curing, the polyimide becomes a part of the wafer and serves as a protective layer. From postcure surface analysis by x-ray photoemission spectroscopy (XPS), it is observed that there are a considerable number of carbonyl and carboxyl groups, which make the polyimide layer more hydrophilic. This surface feature can enhance the wettability and adhesion strength between the lead frame and the polyimide layer on the chip. From the results of further analysis by Fourier transform infrared spectroscopy (FTIR), it is suggested that the bonding between the lead frame and the polyimide is mainly by the dipole moment, as illustrated in Fig. 8.3. The material properties of this polyimide are given in Table 8.1. The related reliability issue will be discussed in detail later in this chapter.

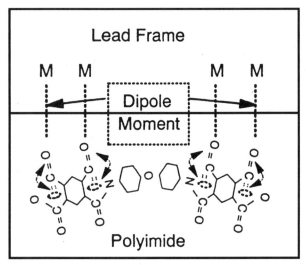

Figure 8.3 Adhesion bonding between the lead frame and the polyimide layer.

TABLE 8.1   Material Properties of Polyimide Coated on Wafer Surface

Property	Value
Thickness (μm)	18
Tensile strength (MPa)	120
Elongation (%)	50
T_g (°C)	250
Wafer internal stress (MPa)	30
5% weight loss (°C)	500
Relative dielectric constant	3.3
Dielectric tangent	0.003
Adhesion strength to Si (MPa)	70

TABLE 8.2   Material Properties of Various Encapsulation Molding
Compounds

	A	B	C	D
Spiral flow (in)	56	36	33	28
Viscosity (poise)	110	210	240	360
Gel times (s)	21	17	22	20
Hardness	73	81	79	78
CTE (ppm)	13	13	13	9
Strength (kg/mm^2)	12	12	12	12
Modulus (kg/mm^2)	1320	1270	1340	1680
Gravity	1.89	1.89	1.90	1.96

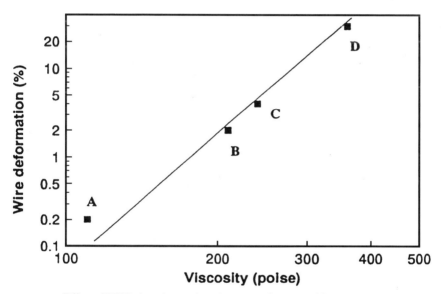

Figure 8.4   Effect of EMC viscosity on the wire sweep during molding.

It was reported that the wire sweep during the encapsulation molding could be a concern for the MCSP [1]. It is believed that the degree
of wire deformation mainly depends on the viscosity of the molding
resin. A study was conducted to investigate this issue using four
EMCs with the material properties given in Table 8.2. The results are
presented in Fig. 8.4. In order to balance the wire sweep with other
considerations such as thermal mismatch and material cost, the EMC
with 240-poise viscosity was chosen for the MCSP.

8.4 Manufacturing Process

The fabrication of the MCSP is quite different from that of other lead-frame-based CSPs. This package features a unique wafer-preparation process, involving a polyimide coating which is the key technology for tapeless-LOC. The procedures for applying the thin polyimide layer are illustrated in Fig. 8.5. The regular finish on the wafer is a thin film of PO nitride/oxide. To prepare the wafer for the tapeless-LOC process, a photosensitive polyimide is deposited on top of the PO film by spin coating (1600 rpm for 15 s) followed by soft baking (100°C for 400 s). The next step is to pattern the polyimide layer by I-line ultraviolet (UV) exposure (230 to 330 mJ/cm²). After patterning and cleaning, windows are opened on the top of the wire-bonding pads as shown in Fig. 8.6. The polyimide is further cured at 350°C for 2 to 3 h in the nitrogen environment. Because of the dehydration and chemical reaction, the thickness of the polyimide layer shrinks from 34 μm to 15 ~ 18 μm. The PO film is then removed by CF_4/O_2 plasma etching. Consequently, the bonding pads are exposed and ready for wire bonding.

Once the wafer preparation is completed, the process presented in Fig. 8.7 will be performed to package the MCSP. The lead frame is directly attached to the aforementioned polyimide layer on the silicon chip. No additional adhesive tape is required. The lead frame is in a strip form with the pattern shown in Fig. 8.8. The lead-frame materi-

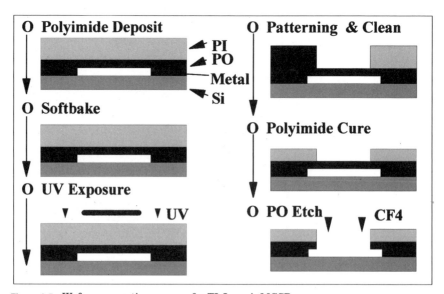

Figure 8.5 Wafer preparation process for TI Japan's MCSP.

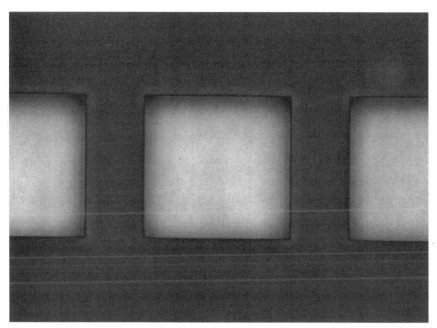

Figure 8.6   Windows opened on the polyimide layer for wire-bonding pads.

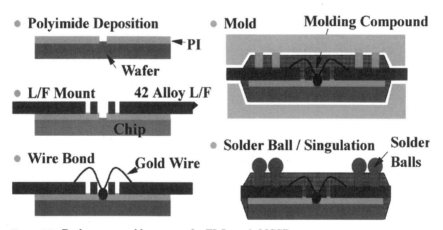

Figure 8.7   Package assembly process for TI Japan's MCSP.

al can be either alloy 42 or Cu. The next steps are wire bonding and encapsulation. It should be noted that the bottom of the die is not molded in order to minimize the package size. Besides, special tooling is arranged in the mold so that a designated land array on the lead frame can be exposed for the subsequent solder ball deposition. Prior

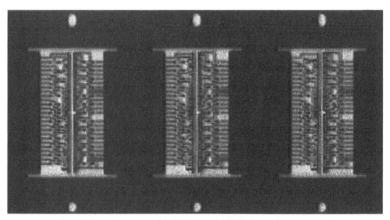

Figure 8.8   Lead-frame pattern for TI Japan's MCSP.

to the placement of solder balls, a Ni layer is plated on the exposed solder lands as a diffusion barrier. Finally, the 63Sn/37Pb solder balls are attached to the lead frame by reflow at 230 to 240°C.

8.5   Electrical and Thermal Performance

Since the MCSP employs the tapeless-LOC technology and the lead frame sits directly on the surface of LSI, the alpha particle attack from lead frame and EMC to IC is a major concern. The only barrier between the packaging materials and the silicon chip is the polyimide layer. Therefore, the thickness of this protective barrier is an important factor to be considered. In order to identify the valid thickness, an accelerated soft error rate (ASER) test and an SER simulation (SSER) were performed [1]. From the results shown in Fig. 8.9, the accelerated testing and numerical simulation correlate with each other reasonably well. It was determined that an 18-μm polyimide layer is thick enough to protect the IC from alpha particle attack.

The main application of the MCSP is to package DRAM. One of the representative performance indices for memory devices is the access time. This quantity is closely related to the package capacitance, which depends on the distance between the leads and the IC. Therefore, this issue may be a concern for the MCSP, since the lead frame sits directly on the die. A series of measurements were made to compare the access time for the conventional LOC package and the MCSP. The former uses a 75-μm-thick tape for die mounting, while the latter has an 18-μm-thick polyimide layer. From the results presented in Table 8.3, it can be concluded that there is no significant degradation in access time caused by the design of the MCSP.

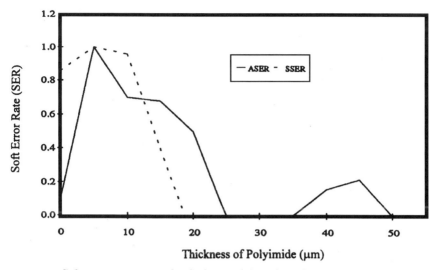

Figure 8.9   Soft error rate versus the thickness of the polyimide layer.

TABLE 8.3   Comparison of Access Times for Conventional and Tapeless LOC

Access time (ns)	Conventional LOC	Tapeless LOC
Maximum	85	84
Minimum	71	71
Average	75.95	76.86

The thermal performance of the MCSP has not yet been reported in the literature at this time. However, since the bottom of the die is not molded, good heat dissipation capability should be expected.

8.6   Qualifications and Reliability

The package reliability of MCSP has been evaluated by wire pulling, finite-element analysis, and qualification tests. These tasks were performed to verify the resistance of the package to wire-bond failure, delamination, reflow heating, and moisture absorption. The results of the wire pull strength test are given in Table 8.4. The average value is higher than the common industrial requirement, and the deviation seems to be within control. Furthermore, all failures occur in the wire instead of the bonding sites. This indicates that the wire-bonding process for the MCSP is very reliable.

TABLE 8.4   Wire Pulling Test Results

Wire pull strength (g)			Failure (%)	
Minimum	Maximum	Average	Ball off	Stitch off
7.8	9.2	8.5	0/20	0/20

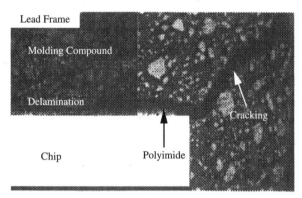

Figure 8.10   Delamination and cracking in the MCSP.

The MCSP is a plastic package. It was reported that internal damage such as delamination and cracking, shown in Fig. 8.10, may appear in the package as a result of hygrothermal loading [5]. These failure modes are closely related to the bonding strength between the polyimide coating and the EMC. A computational analysis was performed during the development of the MCSP to investigate this issue [6]. Since the failure is usually initiated at the area with stress concentration, a finite-element model was established to simulate the neighborhood of a corner of the die, as shown in Fig. 8.11. The peel test shown in Fig. 8.12 was conducted as well to evaluate the adhesion strength between the polyimide and the EMC. With the stress analysis and the peel test, the interface toughness can be calculated [7]. The results for eight different polyimide materials are given in Table 8.5. By the criterion of mixed-mode fracture, the possibility of creating delamination under thermal cyclic loading can be estimated. From the results shown in Fig. 8.13, it can be seen that three kinds of materials are tough enough to prevent delamination. Accordingly, the polyimide with the highest toughness was used for the fabrication of the MCSP. Through the qualification tests listed in Table 8.6, no delamination and crack were observed in the package. Therefore, the package reliability of the MCSP has been verified.

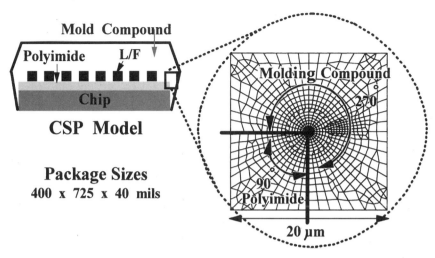

Figure 8.11   Finite-element model for the stress analysis of TI Japan's MCSP.

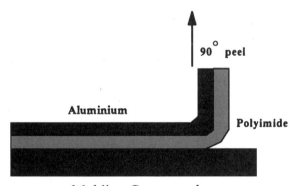

Figure 8.12   Peel test for adhesion strength characterization.

TABLE 8.5   Stress Intensity Factors for Various Polyimide Coatings for TI Japan's MCSP

Adhesion (g/cm)	$K\sigma_{\theta c}$ (kgf · mm$^{-3/2}$)	$K\tau_{r\theta c}$ (kgf · mm$^{-3/2}$)
20	0.026	0.024
30	0.028	0.024
60	0.033	0.024
70	0.050	0.024
190	0.056	0.024
280	0.072	0.024
300	0.076	0.024
440	0.100	0.024

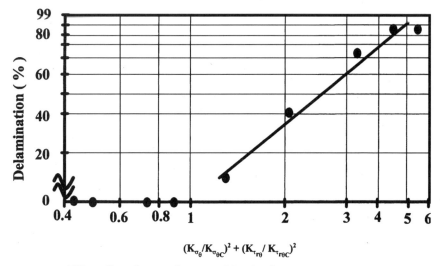

$$(K_{\sigma_\theta}/K_{\sigma_{\theta C}})^2 + (K_{\tau_{r\theta}}/ K_{\tau_{r\theta C}})^2$$

Figure 8.13   Effect of interface toughness on delamination.

TABLE 8.6   Package Qualification Test Results for TI Japan's MCSP

Test	TSOP with LOC	MCSP with tapeless LOC
85/85 + IR	0/10	0/20
Temperature cycle	0/10	0/20
PCT	0/10	0/20
High-temperature storage	0/10	0/20

8.7   Applications and Advantages

The MCSP was developed for low-pin-count ICs. The current lineup is 52 pins with a ball pitch of 1.27 mm. Because of the LOC configuration, the MCSP can package only LSI with center pads. The targeted application is DRAM devices. The outline of the MCSP fully conforms to the common definition of CSP. The package size is slightly smaller than that of a TSOP with a similar number of pins. The objective of TI Japan is to use MCSP to replace SOJ and TSOP in the future.

The major advantage of the MCSP is its cost-effectiveness. Because of the tapeless-LOC technology, there is no need to use die-mounting adhesive tape, which is the most expensive packaging material. According to TI Japan, the MCSP can substantially reduce the material cost for packaging. The other advantages of the MCSP include good electrical performance and package reliability. The BGA interconnects also provide self-alignment capability during surface-mount assembly. However, the BGA pattern of the MCSP seems to be different from the conventional ones. Therefore, custom solder pad design on the PCB may be required.

8.8   Summary and Concluding Remarks

TI Japan's MCSP is a lead-frame-based plastic package. It features a unique wafer-preparation process using a polyimide coating. The first- and second-level interconnects are wire bonds and BGA, respectively. The MCSP was developed to package low-pin-count ICs. The current lineup is 52 pins with a ball pitch of 1.27 mm. The outline of the MCSP conforms to the definition of CSP. Good electrical performance and package reliability have been reported. Data on thermal performance have not yet been released at this time. However, because of the design of the package structure, good heat dissipation capability should be expected.

The tapeless-LOC technology is employed for the fabrication of MCSP. This approach can substantially reduce the material cost of packaging. TI Japan has just finished its internal qualifications for MCSP. This package is aimed at replacing SOJ and TSOP for DRAM applications.

8.9   References

1. M. Amagai, H. Sano, T. Maeda, T. Imura, and T. Saitoh, "Development of Chip Scale Packages (CSP) for Center Pad Devices," *Proceedings of the 47th ECTC,* San Jose, Calif., May 1997, pp. 343–352.
2. J. Kasai, M. Sato, and T. Fujisawa, "Low Cost Chip Scale Package for Memory Products," *Proceedings of the Surface Mount International Conference,* San Jose, Calif., September 1995, pp. 6–17.
3. M. Amagai, R. Baumann, S. Kamei, M. Ohsumi, E. Kawasaki, and H. Kitagawa, "Development of a Tapeless Lead-on-Chip (LOC) Package," *Proceedings of the 44th ECTC,* Washington, D.C., May 1994, pp. 506–512.
4. M. Amagai, T. Saitoh, M. Ohsumi, E. Kawasaki, C. K. Yew, L. T. Chye, J. Toh, and S. Y. Khim, "Development of Tapeless Lead-on-Chip (LOC) Packaging Process with I-line Photosensitive Polyimide," *Proceedings of the IEEE/CPMT International Electronics Manufacturing Technology Symposium,* Austin, Tex., October 1997, pp. 237–244.
5. M. Amagai, "The Effect of Adhesive Surface Chemistry and Morphology on Package Cracking in Tapeless Lead-on-Chip (LOC) Packages," *Proceedings of the 45th ECTC,* Las Vegas, Nev., May 1995, pp. 719–727.
6. M. Amagai, H. Seno, K. Ebe, R. Baumann, and H. Kitagawa, "Cracking Failures in Lead-on-Chip Packages Induced by Chip Backside Contamination," *Proceedings of the 44th ECTC,* May 1994, pp. 171–176.
7. M. Amagai, "Investigation of Stress Singularity Fields and Stress Intensity Factors for Interfacial Delamination," *Proceedings of the 46th ECTC,* Orlando, Fla., May 1996, pp. 414–429.

9

3M's
Enhanced Flex CSP

9.1 Introduction and Overview

This small enhanced flex array package (Enhanced Flex CSP) was developed by 3M in 1997 based on the company's Microflex TBGA technology [1]. This package belongs to the category of flex-based CSP because a polyimide tape is used as the interposer. Furthermore, an additional copper lead frame is employed in order to reduce the thermal mismatch between the package and the PCB. The first- and second-level interconnects of this package are Au wire bonds and eutectic BGA solder balls, respectively. 3M's enhanced flex CSP has both cavity-up and cavity-down configurations. The former may have full-grid BGA and is designed for high-performance ICs with peripheral bond pads, such as digital signal processors (DSPs) and microcontrollers. The latter can have only perimeter BGA and is developed for low-pin-count devices with center bond pads, such as DRAM. Qualification and reliability testing data for this package are not yet available at this time. From the results of computational modeling, relatively good solder joint reliability is anticipated.

9.2 Design Concepts and Package Structure

There are quite a few CSPs with flexible interposers in the commercial market nowadays. Typical examples include Tessera's μBGA (Chapter 16) and TI Japan's μStar BGA (Chapter 17), as shown in Fig. 9.1. The former is probably the most recognized flex-based CSP so far. It has an elastomer layer to relax thermal stresses, resulting in outstanding

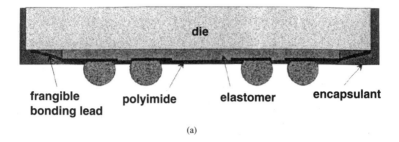

frangible   polyimide   elastomer   encapsulant
bonding lead

(a)

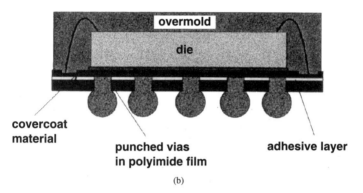

covercoat
material
                punched vias   adhesive layer
                in polyimide film

(b)

Figure 9.1   Various flex-based CSPs for comparison: (*a*) Tessera's μBGA; (*b*) TI Japan's μStar BGA.

board-level reliability. Nevertheless, the μBGA requires a new manufacturing infrastructure and is not flexible for various die-shrink configurations. Therefore, it is more suitable for niche applications such as flash memory. On the other hand, the μStar BGA employs wire bonds as the chip-level interconnects, leading to a more adaptive package structure for configuration changes. However, this package is not equipped with a buffering layer for thermal mismatch and uses punched vias for solder-ball attachment. Therefore, solder joint reliability is a critical concern for the μStar BGA. In view of these CSPs, with certain modifications to its existing technology, 3M developed a wire-bondable flex-based CSP with CTE matching to the PCB [1].

Microflex circuits are a technology of 3M for tape ball grid array (TBGA) packages [2]. This technology materialized wire-bondable metallization on the circuitry of a flexible polyimide tape so that the chip-level interconnects of TBGA can be implemented by wire bonding. In 1997, 3M modified its Microflex TBGA to Enhanced Flex CSP by introducing an additional Cu lead frame between the silicon chip and the flex interposer. This metallic carrier does not provide any electrical interconnection in the package. Instead, because of the relatively high CTE (16.6 ppm/°C), it reduces the thermal mismatch between the package and the printed circuit board.

3M's enhanced flex CSP is considered a flex-based chip scale package because the interposer is a thin (50-μm) polyimide tape with Cu traces (35-μm line/space width). However, unlike that of the Microflex TBGA, the whole substrate of this package is not a single layer structure but a laminate consisting of the polyimide tape and a 127-μm-thick (5-mil) Cu lead frame. These two elements are bonded together by a 25-μm-thick thermoplastic adhesive film. In order to minimize the stress due to thermal mismatch, the silicon chip is mounted on the Cu lead frame using a low-modulus die-attach adhesive. The first-level interconnects of 3M's CSP are Au wire bonds, while the interconnects at the board level are eutectic BGA solder balls. The standard ball pitch is 0.8 mm. The nominal ball diameter is 0.5 mm (20 mil). It should be noted that, since the Cu lead frame is between the die and the polyimide tape, the Au wires are connected to the bond pads on the flex through the slots on the lead frame and the bonding adhesive film.

The enhanced flex CSP has two configurations for various applications. The cavity-up CSP is for ICs with peripheral bond pads. The package is overmolded, as shown in Fig. 9.2*a*. This version may have full-grid BGA and, hence, can accommodate more I/Os (100 to 300). The targeted applications include high-performance DSPs and microcontrollers. The cavity-down CSP is for ICs with center bond pads. The package is encapsulated at the wire bonding area, as shown in Fig. 9.2*b*. This configuration can have perimeter BGA only. Therefore, the cavity-down CSP is aimed at low-pin-count (50 to 100) applications such as DRAMs.

9.3 Material Issues

3M's enhanced flex CSP is a plastic-encapsulated flex-based package with wire bonds and BGA as the interconnects. Besides the silicon chip, the package materials include the gold bonding wires, the die-attach adhesive, the laminated substrate, the encapsulation molding compound (EMC), and the eutectic solder balls. The most important element of this CSP is the substrate, as shown in Fig. 9.3. Although this package belongs to the category of flex-based CSP, the interposer is actually a laminate consisting of a polyimide tape and a copper lead frame. The former is a flex circuit with Cu traces on one side. The thicknesses of the polyimide tape and the Cu trace are 50 μm and 25 μm, respectively. The current design rule for the line/space width of

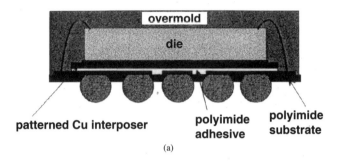

patterned Cu interposer   polyimide   polyimide
                                 adhesive   substrate

(a)

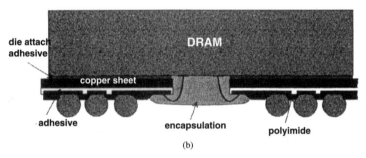

die attach
adhesive

copper sheet

·adhesive   encapsulation
                                    polyimide

(b)

Figure 9.2 Cross-sectional view of 3M's enhanced flex CSP: (a) cavity-up configuration; (b) cavity-down configuration.

the Cu traces is 35 μm. The development of a 25-μm-width design rule is underway.

It should be noted that 3M's microflex circuits can accommodate thermosonic Au wire bonding. In order to achieve a wire-bondable surface, appropriate metallization must be implemented on the bond pads. The current configuration is 1.5-μm Ni plating on top of the Cu trace, followed by 0.76-μm Au plating. Another major feature of 3M's flex is the tapered solder-pad via formed by chemical etching. An area array of such pads is shown in Fig. 9.4. The nominal solder-pad diameter and pitch are 375 μm and 800 μm, respectively. It is reported that the tapered via wall can avoid stress concentration in the deposited solder ball.

The additional lead frame is a 127-μm-thick (5-mil) Cu strip. This metallic carrier does not provide any electrical interconnection in the package. Instead, because of the relatively high CTE (16.6 ppm/°C), it reduces the thermal mismatch between the CSP and the PCB. The

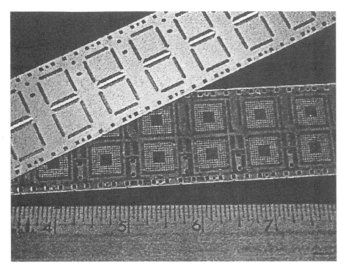

Figure 9.3   Top and bottom views of 3M's enhanced flex substrate.

Figure 9.4   Solder-pad via formed by chemical etching with tapered edge on the flex.

flex circuit and the lead frame are bonded together by a 25-μm-thick adhesive film. This material is a thermoplastic polyimide with a glass transition temperature T_g close to 220°C. This property combined with the excellent adhesion provides this enhanced flex substrate with very high moisture resistance. It should be noted that, in order to access the bond pads on the flex for wire bonding, slots are opened at the corresponding locations on the lead frame and the adhesive film, as shown in Fig. 9.5.

Since the enhanced flex CSP has two configurations for the chip

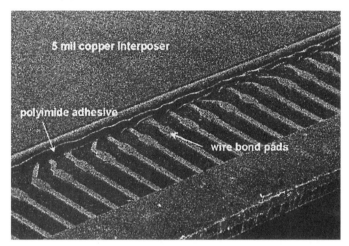

Figure 9.5 Wire bond pads on the flex located in the slot of the Cu lead frame.

orientation, different die-attach adhesives are used. For the cavity-up version, the die-attach adhesive is a low-modulus adhesive in order to minimize the stresses due to thermal mismatch. For the cavity-down CSP, the standard adhesive used in the lead-on-chip (LOC) technology is adopted. The encapsulant of this package is a regular molding compound. Also, the BGA balls are eutectic solder, 63Sn/37Pb. The standard ball diameter is 0.5 mm (20 mil). The BGA balls are attached to the solder pads on the flex using a conventional reflow process. The geometry of the solder pad is defined by the polyimide tape instead of by the solder mask. Since the via is formed by chemical etching and has a tapered side wall, the solder ball should have a very smooth profile, as shown in Fig. 9.6a. Another solder ball with a solder-mask-defined bond pad is shown in Fig. 9.6b for comparison. A right-angle corner at the root of solder ball resulting from the existence of the solder mask can be observed. This geometry will introduce substantial stress concentration to the solder ball.

9.4   Manufacturing Process

The manufacturing process for 3M's enhanced flex CSP is very straightforward. The major procedures are summarized in Fig. 9.7. The key issue is the making of the package interposer. The manufacturing process begins with a prefabricated polyimide tape with patterned Cu traces and solder bond pads. Then the flex circuits and a Cu lead frame

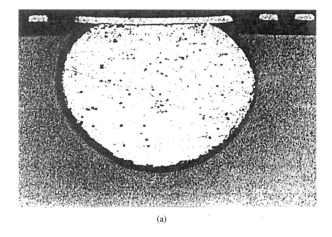

(a)

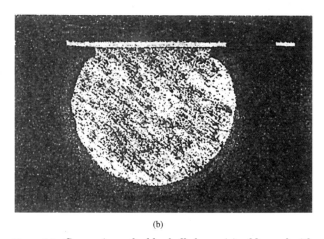

(b)

Figure 9.6 Comparison of solder-ball shape: (*a*) solder pad with chemical etching via; (*b*) solder pad defined by solder mask.

are bonded together by a thermoplastic adhesive to form a laminated substrate. In order to access the bond pads on the flex for subsequent wire bonding, slots should be opened at the corresponding locations on the lead frame and the adhesive film, as shown in Fig. 9.7.

Once the laminated substrate is ready, the subsequent steps are

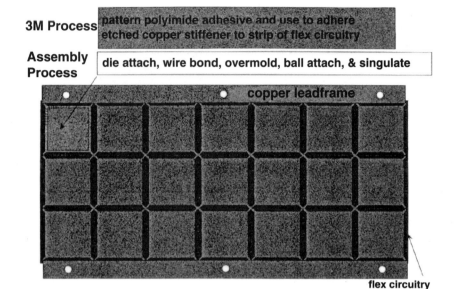

3M Process — pattern polyimide adhesive and use to adhere etched copper stiffener to strip of flex circuitry

Assembly Process — die attach, wire bond, overmold, ball attach, & singulate

copper leadframe

flex circuitry

Figure 9.7   Manufacturing process for 3M's enhanced flex CSP.

conventional die attachment, wire bonding, encapsulation, and ball attachment. Since the Cu lead-frame carrier is in strip form and has sufficient rigidity, the handling of this CSP during manufacturing is quite easy. The last procedure in fabrication is to singulate the package from the strip. This singulation can be performed by mechanical punching or by dicing with a saw. No undesirable silicone materials are present anywhere in the process.

3M claimed that the manufacturing process for enhanced-flex CSP fits into the existing infrastructure of the electronic packaging industry. Therefore, the production cost should be relatively low. In addition, the wire-bonding interconnects are flexible and can allow various die-shrink configurations with the same substrate. Therefore, the lead time for packaging new ICs can be minimized. Nevertheless, because of the introduction of the additional Cu lead frame, the material cost is substantially increased. Therefore, this CSP will not be cost-effective unless the production volume is quite high.

9.5 Performance and Reliability

The experimental results for the electrical and thermal performance of enhanced flex CSP have not yet been reported in the literature at this time. From engineering considerations, 3M claimed that the existence of the lead-frame carrier might improve both electrical and thermal performance as follows [1]: Since the Cu lead frame is only 25 µm away from the signal traces on the flex, it provides an excellent reference plane which can reduce the self- and mutual inductance of the traces. Future high-frequency applications may also employ wire bonding directly to the Cu lead frame to create an active ground plane [3]. In addition, with the assistance of the Cu carrier, the enhanced-flex CSP should be able to distribute the heat efficiently via solder balls to the PCB.

The qualification data for 3M's enhanced flex CSP are not available so far. However, the laminated substrate alone has been tested for moisture sensitivity and has passed the JEDEC level 1 requirements. This outstanding performance is attributed to the high T_g and high modulus of the polyimide adhesive between the flex and the Cu lead frame. Based on previous experience [2], 3M projected that the CSP should achieve JEDEC level 2 moisture resistance. Furthermore, the laminated substrate has been qualified for 240 h under the pressure cooker testing (PCT) condition. Since the enhanced flex CSP does not have a solder mask, which is most susceptible to the PCT condition, the whole package is expected to have no problem with this qualification test.

The main selling point of 3M's enhanced flex CSP is the CTE matching to the PCB. The temperature cycling test for thermal fatigue life is under way, but the experimental data are not yet available. However, because of the existence of the high-CTE Cu lead-frame carrier, the tapered-wall solder-pad vias, and the low-modulus die-attach adhesive, good solder joint reliability is anticipated. In order to verify these design concepts, a computational modeling using the finite-element method was performed to compare the board-level reliability among two different packages, namely, µStar BGA, and enhanced flex CSP [1]. The modeling was a two-dimensional elasto-plastic analysis on the diagonal cross section of the CSP-PCB assembly. The finite-element model was constrained to suppress out-of-plane bending. The major dimensions and material properties are given in Table 9.1. The simulation began from 20 to 125°C, and then went down to −55°C. A plastic strain contour at −55°C for both packages is presented in Fig. 9.8 for comparison. A rather high strain concentration can be observed at the sharp corner of µStar BGA's solder joint. In contrast, the enhanced flex CSP with tapered-wall solder-pad vias, have a relatively smooth contour distribution and much lower plastic strain. Further comparison for normal strain at both −55 and 125°C is presented in Fig. 9.9. It can be seen that the enhanced flex CSP always has the smallest strains. Therefore, the concept of CTE matching by introducing an additional Cu lead frame is verified.

TABLE 9.1 Dimensions and Material Properties Used in the Finite-Element Analysis

μStar

Item	Material	Dimensions	CTE (ppm/°C)	Thickness	Pitch	Poisson's ratio	Modulus (MPa)
Base	UPILEX-S	12 mm × 12 mm	8	75 μm		0.34	8,240
Copper	VLP Cu	12 mm × 12 mm	16.6	25 μm		NA	118,590
Adhesive	TOMOEGAWA X Type	12 mm × 12 mm	60°@25°C 160°@104°C	12 μm		0.467	1,961 118,590
Capture pad	Au over Ni over Cu	478 μm × 478 μm	16.6	0.5 μm Au/Ni, 1 mil Cu	800 μm	NA	NA
Resist	CCR 240GS	NA	100	25 μm		0.3	4,903
Array	NA	13 × 13 4 row pop.	NA	NA	800 μm	NA	NA
Via (punched)	NA	375 μm	NA	NA	800 μm	NA	NA
Solder ball	Eutectic 63/37	500 μm	24.7	NA	800 μm	NA	See [1–3]
Test board	FR4		15	1.6 mm		NA	36,124
Die attach	NA	NA	40	25.4 μm		0.3	NA
Die	Silicon	10 mm × 10 mm	2.1	279 μm		NA	130,312
Overmold		12 mm × 12 mm	17.1	800 μm		NA	25,924

3M CSP

Base	Polyimide E-film	12 mm × 12 mm	13	50 μm	NA	0.34	5,515
Copper	EHS 194 alloy	12 mm × 12 mm	16.6	25 μm & 125 μm		NA	118,590
Adhesive	KJ	12 mm × 12 mm	60	50 μm		0.34	2,758
Capture pad	Au over Ni over Cu	478 μm × 478 μm	16.6	0.5 μm Au/Ni, 1 mil Cu	800 μm	NA	118,590
Array	NA	13 × 13 4 row pop.	NA	NA	800 μm	NA	NA
Via	NA	375 μm	NA	NA	800 μm	NA	NA
Solder ball	Eutectic 63/37	500 μm	24.7	NA	800 μm	NA	NA
Test board	FR4		15	1.6 mm		NA	36,124
Die attach		NA	40	25.4 μm		0.3	NA
Die	Silicon	10 mm × 10 mm	2.1	279 μm		NA	130,312
Overmold		12 mm × 12 mm	17.1	800 μm		NA	25,924

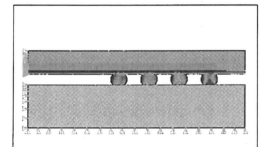

Finite Element Analysis for Plastic Strain
(most critical solder ball at outermost location)

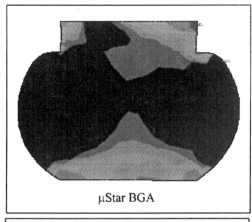

μStar BGA

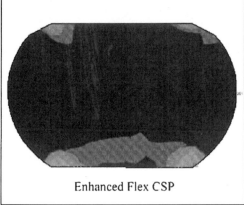

Enhanced Flex CSP

Figure 9.8   Comparison of plastic strain contours in
the most critical solder joint.

Mechanical Modeling of Solder Joints

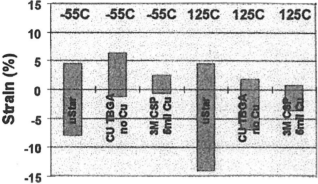

Figure 9.9   Comparison of tensile strain values in the most critical solder joint.

9.6   Applications and Advantages

3M's enhanced flex CSP is a flex-based plastic-encapsulated package which can be used to accommodate ICs with either peripheral or center wire bond pads. There are two configurations for this CSP. The cavity-down version has fewer I/Os and is designed for low-pin-count (50 to 100) devices with center pads, such as 16M and 64M DRAMs. The cavity-up version was developed for devices with peripheral bond pads and medium pin counts (100 to 300). Potential applications include high-performance DSPs and microcontrollers.

The major feature of 3M's enhanced flex CSP is the CTE matching to the PCB. From the results of computational modeling, relatively good solder joint reliability should be anticipated. In addition, with the assistance of the Cu lead frame, the heat generated by the IC can be distributed efficiently via solder balls to the PCB. This package uses conventional wire bonds for chip-level interconnection. Compared to that of Tessera's µBGA, the wire bonding of 3M's CSP is more flexible and can allow various die-shrink configurations with the same substrate. Therefore, the lead time for packaging new ICs can be minimized. In addition, the fine-pitch capability of the flex circuit can push the ring of wire bond pads as close as possible to the edge of the die. Consequently, a smaller package size can be achieved.

3M claims that the manufacturing process for the enhanced flex CSP fits in the current packaging infrastructure. As a result, the production cost should be relatively low. However, because of the additional Cu lead frame, the material cost is considerably increased. Therefore, this CSP may not be cost-effective unless the production volume is quite high.

9.7   Summary and Concluding Remarks

The enhanced flex CSP is a chip scale package with flexible substrate based on 3M's Microflex TBGA technology. The first-level and second-level interconnects of this package are Au wire bonds and eutectic BGA solder balls, respectively. A polyimide tape with wire-bondable metallization is employed as the interposer. An additional copper lead frame is bonded to the flex in order to reduce the thermal mismatch between the package and the PCB. In addition, this substrate is equipped with tapered-wall solder-pad vias formed by chemical etching. These two features together with the low-modulus die-attach adhesive should result in relatively good board-level reliability.

3M's enhanced flex CSP has both cavity-up and cavity-down configurations. The former is overmolded and was developed for high-performance ICs with peripheral bond pads, such as DSPs and microcontrollers. The latter is partially encapsulated and is designed for low-pin-count devices with center bond pads, such as DRAMs. Although the manufacturing process of this new CSP fits into the existing infrastructure of the packaging industry, because of the addition of the Cu lead frame, the fabrication may not be cost-effective unless high-volume production is carried out.

9.8   References

1. R. D. Schueller, "New Chip Scale Package with CTE Matching to the Board," *Proceedings of the International Electronics Manufacturing Technology Symposium,* Austin, Tex., October 1997, pp. 205–215.
2. R. D. Schueller, "Reliability Results for a Wire Bondable TBGA Package," *Proceedings of the Surface Mount International Conference,* San Jose, Calif., September 1996, pp. 12–26.
3. R. D. Schueller, P. Harvey, R. Heidick, and A. Kinningham, "Electrical Performance and Reliability of a TBGA with Grounded Stiffener," *Proceedings of the Surface Mount International Conference,* San Jose, Calif., September 1997, pp. 16–27.

General Electric's Chip-on-Flex Chip Scale Package (COF-CSP)

10.1 Introduction and Overview

The chip-on-flex chip scale package (COF-CSP) was developed by General Electric (GE) based on its multichip-on-flex (MCM-F) technology [1]. This package uses a thin polyimide (PI) film with either a single-side or double-side Cu trace as the interposer. The unique feature of GE's COF-CSP is the laser-drilled vias on the flexible substrate. The interconnects between the chip and the interposer are the metallization in the said vias by a sputtering/electroplating process. BGA solder balls are the regular interconnects at the board level. The standard ball pitch is 0.5 mm. In addition, other types of interconnection can be implemented without difficulty. This package was designed for ICs with low to medium pin counts. The package size is very compact and conforms to the definition of CSP. With a thinning process on the back side of the silicon chip, the package thickness can reach a minimum of 0.25 mm. Currently this package is under qualification and customers' evaluation. The potential application of COF-CSP is the chip sets for cellular phones.

10.2 Design Concepts and Package Structure

The COF technology was an offshoot of the overlay high-density interconnect (HDI) originally developed by GE and Lockheed-Martin for thin-film multichip modules (MCMs) [2]. The HDI technology is mainly aimed at high-performance applications such as aerospace and high-end computers. The bare dies are mounted in the cavities of a rigid substrate and directly interconnected by a flexible film with metal traces. No wire bonds, TABs, or solder bumps are required.

In the COF technology, a prefabricated thin polyimide tape with copper traces on both sides is used as the substrate. The silicon chip is mounted on the substrate with its active face toward the flex circuit by an adhesive layer coated on the polyimide tape. This side of the tape is then encapsulated by epoxy for ease of subsequent handling and for chip protection. Via holes are formed through the flexible substrate and the adhesive layer by laser drilling. The vias and the outer surface of the polyimide tape are metallized by sputtering/electroplating followed by photoimage patterning to establish the interconnect structure. The COF technology can be easily implemented to make MCM-F, as illustrated in Fig. 10.1. Since the whole substrate is in panel form during the fabrication process and all chips on the same panel can be interconnected as shown in Fig. 10.2, it is very easy to test the function of chips or to burn in for known good die (KGD). In addition to MCM-F, the COF technology can be used to make a single-chip module (SCM-F) as well. A thin-zero outline package (TZOP), as presented in Fig. 10.3, was developed by GE in 1995 using this technology [2]. It can be seen that, through the metallization on the flexible substrate, the fine-pitch peripheral bond pads on the chip have been reconfigured to an area array with much coarser pitch. This redistribution not only can change the location and geometry of the bond pads, but also can modify the conventional die-pad metallurgy to Au, Ni, Cu, or solder. With this advantage, the finished COF modules are compatible with wire-bonding, TAB, and flip-chip technologies and can be placed on a lead frame, a flexible/rigid substrate, or BGA solder balls for the subsequent interconnection and assembly.

Based on the aforementioned technology, GE implemented two types of COF-CSP with BGA configuration. The first type is a true chip size package with fan-in circuit and pads underneath the die, as

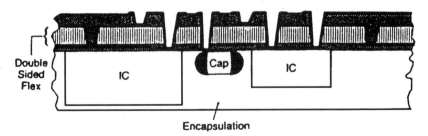

Figure 10.1   Schematic diagram for cross section of GE's MCM-F.

Figure 10.2   Twenty-five-chip-array COF test evaluation modules.

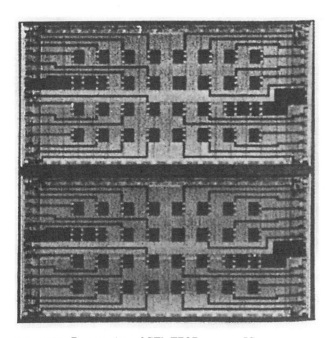

Figure 10.3   Bottom view of GE's TZOP memory IC.

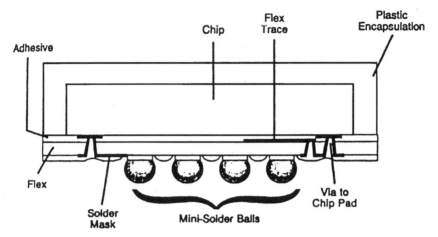

Figure 10.4   Cross-sectional view of COF-CSP.

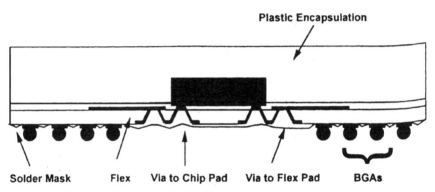

Figure 10.5   Cross-sectional view of COF-CSP with pads external to the chip area.

shown in Fig. 10.4, while the second type is a nearly chip scale package with fan-out circuit and pads external to the chip area, as illustrated in Fig. 10.5 [3]. In GE's terminology, the former is called a mini-BGA COF package and the latter is called a BGA COF package [4]. The detailed local structures of the two packages are given in Figs. 10.6 and 10.7, respectively.

In general, the COF-CSP employs all fabrication procedures from the COF technology except that the flexible substrate may have either single-side or double-side Cu traces, depending on the number of I/Os and the complexity of the circuit pattern. Although the line/space width of the Cu trace on the polyimide tape can be made even smaller, currently the COF-CSP uses a 2/2 (50 μm/50 μm) design rule for better yield and lower cost. The encapsulation of COF-CSP may have

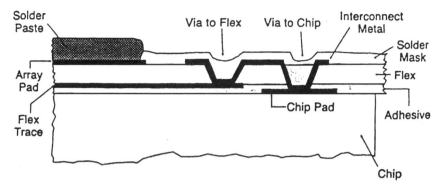

Figure 10.6   Local structure of mini-BGA COF package.

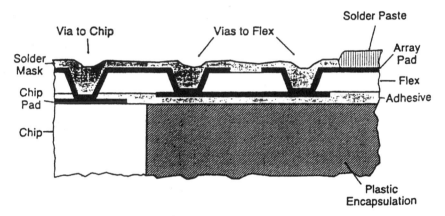

Figure 10.7   Local structure of BGA COF package.

three kinds of configuration, as presented in Fig. 10.8. The die can be encapsulated only around the periphery or overmolded. Also, a heat sink may be installed at the back of the silicon chip and then encapsulated at the four sides. In order to achieve the low-profile feature, the die of COF-CSP can be thinned by mechanical grinding either before or after the encapsulation. The minimum package thickness may reach 0.25 mm. The package size can be from 0.5 to 10 mm larger than the chip size. In most configurations, the COF package can comply with the dimension requirement for chip scale packages.

The COF-CSP package was developed for low- to medium-pin-count ICs. The potential applications include memory devices and ASICs for portable equipment. The standard ball diameter and pitch are 0.25 mm and 0.5 mm, respectively. Other kinds of configuration are available as well. The BGA pattern may be either full-grid or perimeter, as

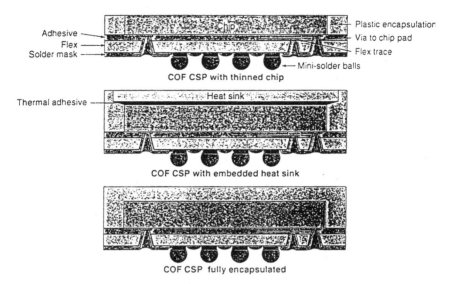

Figure 10.8   Various encapsulation configurations for COF-CSP.

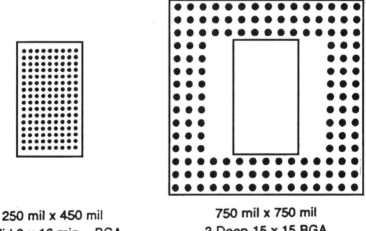

250 mil x 450 mil
Solid 9 x 16 min. - BGA
25 mil Grid, 144 Pads
Chip Scale

750 mil x 750 mil
3-Deep 15 x 15 BGA
50 mil Grid, 144 Pads
Standard BGA

Figure 10.9   Typical bottom layout of mini-BGA and BGA COF packages.

TABLE 10.1   Design Rules for GE's COF-CSP

Feature	Specification
Trace line width	50 µm (2 mils)
Line space	50 µm (2 mils)
Solder-pad diameter	0.25 mm (10 mils)
Traces between BGA pads	One (top layer)
Traces below BGA pads	Four or five (non-top layer)
Chip pad to solder pad minimum	0.5 mm (20 mils)
Package edge to chip edge minimum	0.25 mm (10 mils)
Package edge to chip edge maximum	>5 mm (>200 mils)

shown in Fig. 10.9. The former is for mini-BGA COF only, while the latter is mainly for BGA COF packages. According to GE's recommendation, for the flexible substrate with single-side prepatterned Cu trace, the solder-ball population should not exceed six rows deep if the standard ball pitch of 0.5 mm is adopted. The design rules for such a package configuration are summarized in Table 10.1.

10.3   Material Issues

GE's COF-CSP is a plastic-encapsulated package with flexible interposer and BGA interconnects. The major package constituents in addition to the silicon chip include the flexible substrate with metal trace, the die-attach adhesive, the encapsulant, the metallization for bond pads, and BGA solder balls. The substrate for COF-CSP is a prefabricated polyimide tape such as DuPont's Kapton, Allied Signal's Apical, or GE's Ultem [1]. The substrate thickness is 25 µm. Depending on the number of I/Os and the complexity of the circuit, this flexible tape may have a patterned Cu trace on one side or both sides. The nominal thickness of this Cu trace is 4 to 10 µm. However, for higher-power applications, this thickness may be increased to carry a higher electric current. For better yield and lower cost, the Cu trace line/space width follows a 2/2 (50 µm/50 µm) design rule, although smaller dimensions can be achieved with the present technology. The die-attach adhesive is a thermoplastic material. This adhesive is deposited on the surface of polyimide tape by spray or spin coating. The thickness is 10 to 15 µm. The encapsulant for COF-CSP is an industrial standard epoxy with 70 percent silica filler. The main reason for the high percentage of filler is to relax the internal stress. In addition to protecting the IC, the other function of this encapsulation is to serve as a carrier for ease of subsequent handling because the flexible substrate is in the form of a large panel.

The metallization to interconnect the bond pads on the chip and the interposer is deposited by sputtering/electroplating. The final surface finishing of the bond pads on the polyimide tape can be Au, Ni, Cu, or solder. It should be noted that the original bond pads on the die may maintain the conventional metallurgy, such as Al or Au. No modification of the die pads is needed for the aforementioned metallization process. The BGA solder balls of COF-CSP are conventional eutectic solder. The balls are not deposited on the flexible substrate in a sphere form. Instead, the solder paste is printed on the bond pads and then reflowed to form the solder balls by surface tension. The solder-ball diameter and ball height are 0.25 to 0.3 mm and 0.15 to 0.18 mm, respectively.

10.4 Manufacturing Process

The manufacturing process for COF-CSP is a simplification of the HDI technology for MCMs developed by GE and Lockheed-Martin. The flowchart of this process and the graphical illustrations for major steps are presented in Figs. 10.10 and 10.11, respectively. The fabrication begins with a single layer (25 μm thick) of polyimide tape. This kind of flex tape can be procured from commercial suppliers such as DuPont and Applied Signal or fabricated in house by GE. This tape is usually in a roll form with a width of 12 or 24 in and a length of 25 to 250 ft. Currently GE cuts the roll into 12×12 in sheets for the subsequent process. Since the polyimide tape is a flexible substrate, it needs to be bonded by adhesive or clamped by mechanical fixture to a rigid frame in order to improve the handling and dimensional stability during processing. GE's present configuration is an 8-in-diameter round frame, as shown in Fig. 10.12. However, the actual processing area for COF-CSP is a 6×6 in square inside the frame.

Once the tape is mounted to the frame, patterning of the metal traces and pads on the flex is performed. The metallization is a 4- to 10-μm-thick Cu layer. For high-power applications, this thickness can be increased to carry a higher electric current. The present design rule for the Cu trace pattern is 2/2 (50-μm line width/50-μm space width). Depending on the number of I/Os and the complexity of the circuit pattern, the flex may have Cu traces on one side or both sides. If the double-sided flex circuit is used, a thin polymer passivation coating should be placed on the outer side of the tape to serve as an insulation layer for the subsequent metallization on the surface.

The next step is to deposit the die-attach adhesive on the inner side of the flex by spray or spin coating. The thickness of this adhesive layer is 10 to 15 μm. Then the die is mounted to the prepatterned flex by a flip-chip pick-and-place machine. The placement accuracy should

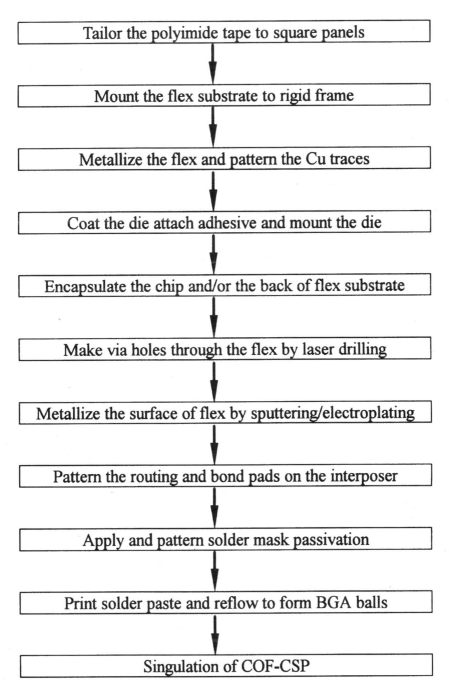

Figure 10.10 Flowchart for the fabrication of COF-CSP.

Bond chip to pre-patterned flex

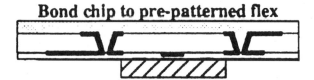

Plastic Encapsulate

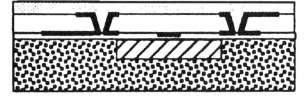

Drill vias to chip, Metal 1, and Metal 2. Sputter, electroplate

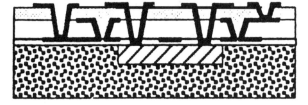

Passivate

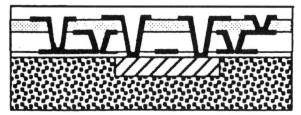

Figure 10.11   Manufacturing process for COF-CSP.

be within 10 to 20 μm. The chip-bonding process is subjected to elevated temperature and pressure. Afterwards, the chip, together with the back side of the flex substrate, is encapsulated by highly filled epoxy and cured at 100 to 150°C. The encapsulation serves two functions. One is to provide a protective enclosure to the encapsulated IC, and the other is to form a processing carrier with dimensional stability and uniform planarity.

Figure 10.12   Flexible substrate mounted on a circular frame.

Following the curing of the encapsulant, via holes are formed on the interposer using a point-and-shoot or mask-based laser ablation system. It should be noted that the vias include the holes through the flex and the die-attach adhesive to the bond pads on the chip and the passage through the flex (or the passivation layer, if a double-sided flex circuit is used) to the metallized pads on the polyimide tape. After a plasma cleaning, an additional metallization is deposited on the exposed bond pads, the via hole side walls, and the surface of the flex by a sputtering/electroplating process. Using photolithography techniques, this new metal layer is then etched to form the desired routing pattern to connect the bond pads on the die and the area-array bond pads on the interposer. The current standard pitch for the bond-pad area array is 0.5 mm. Other configurations can be implemented without difficulty. It should be noted that the metallization on the interposer may be Au, Ni, Cu, or solder. For COF-CSP with BGA configuration, in order to provide a wettable surface, the standard finishing on the solder bond pads is Cu/Ni/Au metallization. According to GE's recommendation, if a single-sided flex circuit is used, the solder-ball population of mini-BGA COF should not be more than six rows deep. The inner four rows of solder bond pads should be routed through the prepatterned Cu trace on the polyimide tape, while the outer two rows are connected by the subsequent surface metallization with fan-

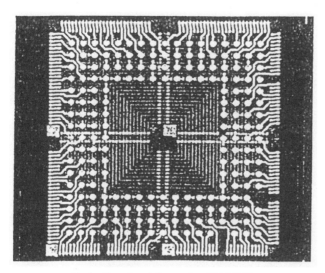

Figure 10.13   Typical bond-pad layout for COF-CSP.

in traces from the laser-drilled vias. A typical pattern of such routing is presented in Fig. 10.13.

After the patterning, a photoimageable solder mask is applied to the top surface of the substrate to passivate the metal traces and vias. This passivation layer is then patterned to expose the solder bond pads. The bond pads may be either solder-mask-defined or non-solder-mask-defined [5]. For the former, the solder pad has a diameter of 0.355 mm but is partially covered by the solder mask with an opening of 0.25 mm in diameter. For the latter, the diameters of the bond pad and the solder-mask opening are 0.25 mm and 0.355 mm, respectively. Once the solder-mask passivation has been patterned, solder paste is printed on the bond pads through a 0.15- to 0.2-mm-thick stencil and then reflowed to form the BGA solder balls. The diameter and height of the solder balls are 0.25 to 0.3 mm and 0.15 to 0.18 mm, respectively. The final step of the manufacturing process is to singulate the COF-CSP from the carrier. Before singulation, a panel with an area of 6×6 in containing all packages is cut from the circular frame, as shown in Fig. 10.14. The carrier is then sawed to the final package outline.

10.5   Performance and Reliability

The electrical and thermal performance results of COF-CSP have not yet been reported in the literature. Since the electrical passage of this package is rather short, relatively low package parasitics are expect-

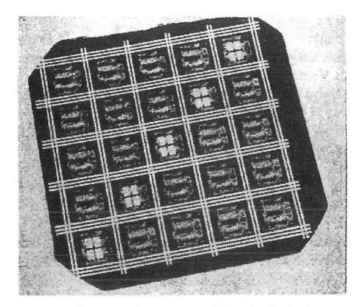

Figure 10.14   Encapsulated carrier removed from the circular frame.

ed. Also, the silicon chip can be thinned to a very low profile and a heat sink can be attached to the die, as shown in Fig. 10.8. Therefore, reasonable heat dissipation capability is anticipated for this package.

The solder joint reliability of COF-CSP is currently under investigation. The board-level assembly of this package follows the standard SMT procedures. It was reported that a solder joint height of 0.125 mm can be achieved [3]. Since the solder joints are relatively small and there is no stress relaxation elastomer between the flexible substrate and the silicon chip, underfill adhesive may be needed for COF-CSP with large die and full-grid BGA ball population.

10.6 Applications and Advantages

The COF-CSP package was developed for ICs with low to medium pin counts. Data on GE's mini-BGA COF and BGA COF families are given in Tables 10.2 and 10.3, respectively [4]. Because of its low profile and compact size, the COF-CSP package may be used for memory devices and ASICs for portable equipment. Currently this package is being evaluated by some telecom companies for chip sets in digital cellular phones. A recent design approach indicated that a common BGA bond-pad layout and footprint of COF-CSP may be shared by various devices. Four telecommunication chips with die size from 6.4×5.9 mm to 8.1×7.0 mm were involved. The chip I/Os ranged from

TABLE 10.2   GE's Mini-BGA COF Family (25-Mil Pitch)

Array area (mil × mil)	Array size	Solder balls			
		Full grid	3 deep	4 deep	5 deep
200 × 200	7 × 7	49	48	—	—
300 × 300	11 × 11	121	96	112	120
400 × 400	15 × 15	225	144	176	200
500 × 500	19 × 19	361	192	240	280
600 × 600	23 × 23	529	240	304	360
700 × 700	27 × 27	729	288	368	440
800 × 800	31 × 31	961	336	432	520

TABLE 10.3   GE's BGA COF Family (50-Mil Pitch)

Array area (mil × mil)	Array size	Solder balls		
		3 deep	4 deep	5 deep
400 × 400	7 × 7	48	—	—
500 × 500	9 × 9	72	80	81
600 × 600	11 × 11	96	112	120
700 × 700	13 × 13	120	144	160
800 × 800	15 × 15	144	176	200

Figure 10.15   Four telecom chip sets packaged by COF-CSP.

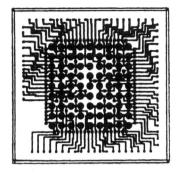
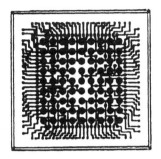
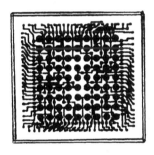
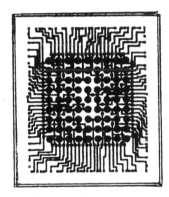

Figure 10.16   Common footprint and layout of bond pads shared by different chip sets.

72 to 80 pins. A 9×9 full-grid (81 pins) mini-BGA COF with a standard ball pitch of 0.5 mm was selected to package the aforementioned devices, as shown in Fig. 10.15. The common BGA pattern is presented in Fig. 10.16. It should be noted that, although the bond-pad layout and footprint are the same, the routing varies from one device to another. This design approach reduced the CAD layout and routing time by 50 percent and minimized the probability of layout errors.

In addition to the aforementioned benefit for package design, the COF-CSP has several other major advantages. The silicon chip can be thinned by mechanical grinding either before or after encapsulation. Therefore, this package can have a very low profile; the minimum package thickness is 0.25 mm. For mini-BGA COF, the package size is only 0.5 mm larger than the die size. Together with the fine ball pitch (0.5 mm standard), these superior form factors make COF-CSP very suitable for high-density assembly in portable electronics. Besides, the COF technology can be highly automated using the existing infrastructure for electronic packaging. Therefore, the production cost of COF-CSP can be relatively low. Most of all, the COF packages can be tested for electrical functions and burn-in for KGD while they are still on the panel carrier. Therefore, the COF-CSP becomes a very attractive option for the electronics industry.

10.7 Summary and Concluding Remarks

The COF-CSP is a chip scale BGA package with flexible interposer. The development of this package was based on the MCM-F technology originated by GE and Lockheed-Martin. The unique feature of COF-CSP is the laser-drilled vias on the flexible substrate. The interconnects between the chip and the interposer are established through the metallization by a sputtering/electroplating process. The regular board-level interconnects are eutectic BGA solder balls. The standard ball pitch is 0.5 mm.

The COF-CSP was developed to package ICs with low to medium pin counts. The potential applications include memory devices and ASICs for portable electronics. The package size is very compact and conforms to the definition of CSP. With a mechanically thinned silicon chip, the minimum package thickness can reach 0.25 mm. Currently this package is under qualification and reliability tests. Reasonable electrical and thermal performance results are expected. Some customers in the telecommunication industry are currently evaluating the COF-CSP for packaging various chip sets with a common bond-pad layout and footprint in digital cellular phones.

10.8 References

1. D. R. Kuk, R. Fillion, G. Forman, and B. Burdick, "Single Chip-on-Flex: Chip Scale Packaging Technology," *Proceedings of the Surface Mount International Conference,* San Jose, Calif., August 1995, pp. 29–35.
2. G. A. Forman, R. A. Fillion, R. F. Kole, R. J. Wojnarowski, and J. W. Rose, "Development of GE's Plastic Thin-Zero Outline Package (TZOP) Technology," *Proceedings of the 45th ECTC,* Las Vegas, Nev., May 1995, pp. 664–668.
3. R. Fillion, B. Burdick, D. Shaddock, and P. Piacente, "Chip Scale Packaging Using Chip-on-Flex Technology," *Proceedings of the 47th ECTC,* San Jose, Calif., May 1997, pp. 638–642.
4. D. Kuk, R. Fillion, G. Forman, and B. Burdick, "Demonstration of a Chip Scale Chip-on-Flex Technology," *Proceedings of the ICEMCM Conference,* 1996, pp. 351–356.
5. J. Lau, *Ball Grid Array Technology,* McGraw-Hill, New York, 1995.

11

Hitachi's Chip Scale Package for Memory Devices

11.1 Introduction and Overview

Hitachi developed a new chip scale package (CSP) for memory devices in 1997 based on the µBGA licensed from Tessera [1]. This package uses a flexible tape carrier as the interposer. The first-level interconnects are made by TAB inner lead bonding (ILB). The chip is partially encapsulated at the ILB area for protection. The board-level interconnects are BGA. The current lineup is 48 pins with a ball pitch of 0.75 mm. This package is designed for low-pin-count memory devices. Unlike Tessera's µBGA, this CSP can be applied to ICs with either peripheral (SRAM and flash) or center (DRAM) pads. Other major modifications include the improvement of material properties for elastomer and encapsulant. Hitachi's CSP is a low-cost and high-reliability package. It uses most of the existing infrastructure for IC packaging and can be fabricated with a reel-to-reel operation. The moisture sensitivity of this package is qualified for JEDEC level 1. The package reliability of Hitachi's CSP is outstanding.

11.2 Design Concepts and Package Structure

Hitachi's CSP for memory devices is a tape carrier package (TCP) with elastomer and encapsulant. The fundamental configuration is based on µBGA, licensed from Tessera. This package uses a 0.05-mm-thick polyimide (PI) tape as the substrate with a Cu trace routed on one surface. The die is mounted on the flexible substrate using an elastomer layer with adhesive on both sides. In addition to its die-mounting function, this elastomer also serves as a relaxation layer to relieve stresses due to thermal mismatch.

The chip-level interconnects of Hitachi's CSP are made by single-point TAB ILB. The inner leads are 0.018-mm-thick Cu core with double-sided Au plating on the surfaces. Note that there is no Ni layer between the Cu and Au. Unlike Tessera's µBGA, this package can be applied to ICs with either peripheral or center pads, as shown in Fig. 11.1. The chip is partially encapsulated at the ILB area for protection. A eutectic solder-ball grid array provides the board-level interconnects.

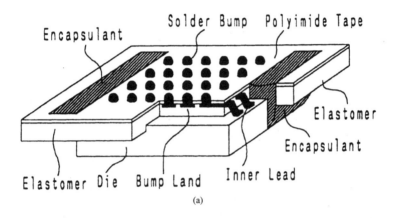

(a)

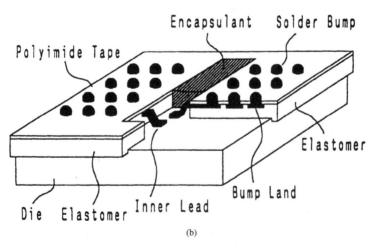

(b)

Figure 11.1 Package structure of Hitachi's CSP for dies (a) with peripheral pads and (b) with center pads.

The solder-ball diameter and pitch are 0.3 mm and 0.75 mm, respectively. This package is designed for low-pin-count memory devices such as DRAMs, SRAMs, and flash memory. The current lineup is 48 pins. The specifications of the package structure are summarized in Table 11.1.

The outline of Hitachi's CSP conforms to the standards of JEDEC and EIAJ. For CSP-48 with a die measuring 7.43×4.43 mm, the nominal package size is 8.24×5.03 mm. Therefore, this package complies with the dimensional requirements of CSP. The detailed outline of Hitachi's CSP is summarized in Fig. 11.2 and Table 11.2.

TABLE 11.1   Specifications for Hitachi's CSP Structure

		Materials Specification
Base tape	Polyimide	0.05 mm (thickness)
Solder bump	63Sn/37Pb	0.3 mm (diameter)
Inner lead core	Copper	0.018 mm (thickness)
Inner lead plating	Double-sided gold (nickelless)	
Elastomer	No-silicone double-sided epoxy adhesive–coated polymer film	
Encapsulation	No-silicone epoxy potting resin	

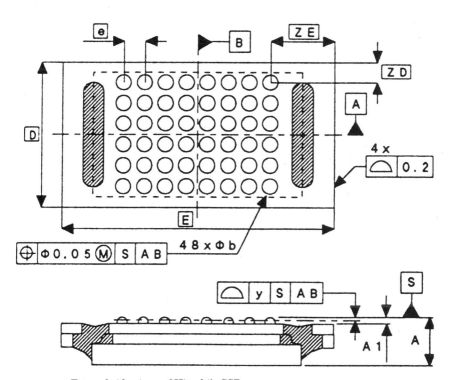

Figure 11.2   Top and side views of Hitachi's CSP.

TABLE 11.2   Package Outline for Hitachi's CSP

Reference symbol	Specification			Actual data			
	Minimum	Nominal	Maximum	Minimum	Mean	Maximum	3 σ
A	—	—	1.00	0.66	0.69	0.72	0.07
A1	0.15	0.20	0.25	0.17	0.19	0.20	0.02
D	4.83	5.03	5.23	4.94	5.02	5.03	0.06
D1	3.70	3.75	3.80	3.71	3.74	3.75	0.03
E	8.04	8.24	8.44	8.23	8.24	8.26	0.02
E1	5.20	5.25	5.30	5.25	5.25	5.26	0.01
ZD	0.54	0.64	0.74	0.59	0.64	0.66	0.03
ZE	1.40	1.50	1.60	1.45	1.49	1.54	0.05
y	—	—	0.08	0.00	0.02	0.03	0.02
b	0.25	0.30	0.35	0.34	0.35	0.36	0.01
e	0.70	0.75	0.80	0.72	0.75	0.78	0.02

11.3   Material Issues

Hitachi's CSP is a TAB bonded tape carrier package based on Tessera's μBGA. Besides the silicon chip, the package materials include the die-mounting elastomer, the flexible polyimide substrate with Au-plated Cu traces, the epoxy resin encapsulant, and the eutectic solder balls. In addition to the extended capability to package ICs with center pads, Hitachi made certain modifications to the elastomer and the encapsulant to improve the performance of the CSP.

The elastomer in Hitachi's CSP is a polymer film coated with epoxy adhesive on both sides. For the purpose of mass production, a no-flow base film was selected. Furthermore, in order to protect the inner leads and bond pads from contamination by outgassing of silicone, a no-silicone type of polymer was used. The elastic modulus of this kind of material is slightly higher than that of silicon-based elastomer. The other characteristics of this new material include low moisture absorption and short curing period. Because the surface adhesive is made from a B-stage epoxy resin, the die-attach process can be completed within 10 s under 150 to 180°C. This feature substantially improves the productivity of Hitachi's CSP. The essential material properties of the newly developed elastomer are given in Table 11.3.

The encapsulant is also a no-silicon type of material. It is made from epoxy potting resin. This resin is solventless and has low viscosity. The outgassing is minimized, and the curing period is short. The other features include strong adhesion, low moisture absorption, and

TABLE 11.3   Material Properties of Modified Elastomer for Hitachi's CSP

Item		Unit	Properties
Elastic modulus at 25°C		MPa	546
Coefficient of thermal expansion		ppm/°C	86
Glass transition temperature		°C	179
Ionic impurities	Cl⁻	ppm	0.39
	Br⁻		0.33
	Na⁺		0.10
Solvent resistance	IPA (isopropanol) 6 min immersion		No change
	HCFC 10 min ultrasonic and 60 min immersion		No change
Flammability (UL-94)			V-0

TABLE 11.4   Material Properties of Modified Encapsulant for Hitachi's CSP

Item		Unit	Properties
Viscosity at 25°C		Pa · s	7.6
Elastic modulus at 25°C		MPa	8500
Coefficient of thermal expansion		ppm/°C	38
Glass transition temperature		°C	102
Ionic impurities	Cl⁻	ppm	6.1
	Na⁺		0.1
Flammability (UL-94)			V-0

high purity. In addition, the cured resin has a good balance among glass transition temperature T_g, elastic modulus, and coefficient of thermal expansion (CTE). The material properties of the encapsulant are presented in Table 11.4. Since Hitachi's CSP was developed for DRAM applications, the requirement for flammability by UL-94 is V-0. Both the elastomer and the encapsulant fulfill this regulation.

11.4   Manufacturing Process

The manufacturing process for Hitachi's CSP is shown in Fig. 11.3. The first step is to mount the elastomer layer to the Cu trace side of the polyimide tape. Then the active face of the silicon chip is attached

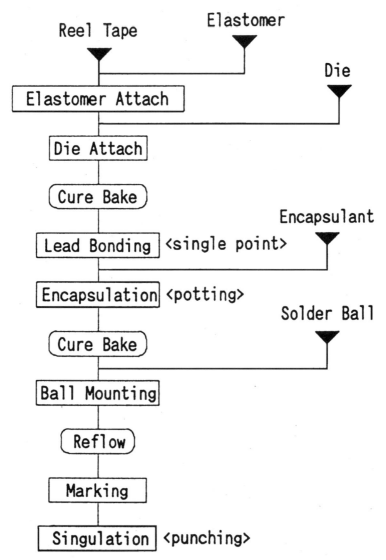

Figure 11.3   Package assembly process for Hitachi's CSP.

to the other side of the elastomer layer. Since the elastomer layer has adhesive on both sides, this die-mounting process is very similar to that for the Tape-LOC technology [2].

After the epoxy adhesive is cured by baking, single-point ILB is performed to make the chip-level interconnection [3]. It should be noted that, on the flexible polyimide tape, there are preformed open slots above the bond pads of the die for the ILB process. Because of this design, Hitachi's CSP can be applied to ICs with either peripheral or center pads. The next step is to pour the epoxy resin in the slots of the ILB for encapsulation. After ball placement followed by reflow attachment, the package is marked and then singulated from the tape by punching.

It should be noted that, just like most other TCPs, Hitachi's CSP can be fabricated with a reel-to-reel operation. Also, this package employs most of the existing infrastructure for IC packaging. Therefore, Hitachi's CSP should be a very cost-effective new chip scale package with high throughput.

11.5   Qualifications and Reliability

The package reliability of Hitachi's CSP has been verified by various qualification tests. From the results shown in Table 11.5, it can be concluded that this package has the same reliability as other conventional plastic packages, such as TSOP and PQFP.

Moisture sensitivity is always a critical issue for microelectronic packages. In order to investigate the characteristics of Hitachi's CSP in this area, a moisture absorption test was performed. From the results presented in Fig. 11.4, it can be seen that the moisture content in the package reaches saturation within a relatively short period. This is attributed to the relative thinness of the packaging materials. Furthermore, the short saturation period indicates that the conventional moistureproof bag system may not be effective for storing Hitachi's CSP. Therefore, attention should be given to investigating this issue. However, further testing data given in Table 11.6 show that the moisture sensitivity of this package is qualified for JEDEC level 1. As a result, the fast moisture saturation and bagging are no longer critical issues for Hitachi's CSP.

TABLE 11.5   Package Qualification Testing Results for Hitachi's CSP

	Condition	Result	
Temperature cycle	−55 to 125°C on board	1500 cycles	0/30
PCT	121°C, 2 atm	200 h	0/20
Moisture	65°C/95% RH	200 h	. 0/84
High temperature and humidity bias	85°C/85% RH, 7 V bias	1000 h	0/40

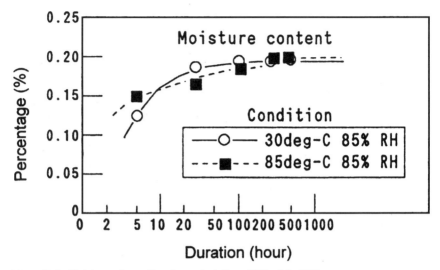

Figure 11.4 Moisture absorption characteristics of Hitachi's CSP.

TABLE 11.6   Moisture Sensitivity Test Results for
Hitachi's CSP

Precondition	Result
30°C/85% RH, 48 h	0/30
85°C/85% RH, 48 h	0/20
85°C/85% RH, 168 h	0/84
85°C/85% RH, 336 h	0/363
IR reflow condition: 240°C, 10 s, 3 times	

11.6   Applications and Advantages

Hitachi's CSP was developed for packaging memory devices. It is designed for low-pin-count applications. The current lineup is 48 pins
with a ball pitch of 0.75 mm. Based on the location of preformed slots
on the tape carrier, this package can be applied to ICs with either peripheral bond pads or center bond pads. The former represent SRAMs
and flash memory, while the latter are typical for DRAMs. Therefore,
Hitachi's CSP can be used for almost all memory devices.

In addition to this wide spectrum of applications, another major advantage of Hitachi's CSP is the feasibility of reel-to-reel operation for
fabrication. Also, since the manufacturing process adopts most of the
current IC packaging infrastructure, this package is very cost-effective and has high throughput. Most of all, the package reliability of

Hitachi's CSP is outstanding. This package can sustain temperature cycling between −55 and 125°C for 1500 cycles, and its moisture sensitivity is qualified for JEDEC level 1. All these factors make Hitachi's CSP an attractive choice for memory devices.

11.7 Summary and Concluding Remarks

Hitachi's CSP is a TAB bonded TCP based on the configuration of μBGA, licensed from Tessera. This package uses a flexible polyimide tape as the interposer. The first-level interconnects are made by single-point ILB, while the second-level interconnects are BGA. This package was designed for low-pin-count ICs such as memory devices. The current lineup is 48 pins with a ball pitch of 0.75 mm. The outline of Hitachi's CSP is relatively compact and conforms to the standards of JEDEC and EIAJ. The small outline certainly increases the density for board-level assembly. Compared to Tessera's μBGA, Hitachi's CSP has extended capability to package ICs with either peripheral (SRAMs and flash memory) or center (DRAMs) bond pads. Other major modifications include the improvement of material properties for the elastomer and encapsulant. All these developments make this new CSP more versatile and reliable.

Hitachi's CSP is a cost-effective package with outstanding reliability. It can sustain temperature cycling between −55 and 125°C for 1500 cycles, and its moisture sensitivity is qualified for JEDEC level 1. Because of the TCP configuration, this package can be fabricated with a reel-to-reel operation. Also, the manufacturing process adopts most of the existing infrastructure. As a result, Hitachi's CSP is an attractive choice for packaging memory devices.

11.8 References

1. Y. Akiyama, A. Nishimura, and I. Anjoh, "Chip Scale Packaging for Memory Devices," *Proceedings of the 48th ECTC,* Bellevue, Wash., May 1998, pp. 477–481.
2. N. Taketani, K. Hatano, H. Sugimoto, O. Yoshioka, and G. Murakami, "CSP with LOC Technology," *Proceedings of the ISHM International Symposium on Microelectronics,* Minneapolis, Minn., October 1996, pp. 594–599.
3. J. Lau, *Handbook of Tape Automated Bonding,* VNR, New York, 1992.

12

IZM's
*flex*PAC

12.1 Introduction and Overview

The *flex*PAC is a chip scale package based on several technologies developed by the Fraunhofer Institute (IZM) at the Technical University of Berlin (TU-Berlin) [1]. Pac Tech Gmbh is a spinoff company from IZM, which shares these packaging technologies and acts as a commercial carrier for equipment supply, package fabrication, and wafer bumping service. As indicated by the name, the *flex*PAC is a CSP with a flex substrate. The interposer is a carrier tape with a single layer of copper traces. The chip-level interconnects are meniscus solder bumps. This technology consists of electroless Ni bumping, eutectic Au/Sn soldering, and laser fiber-push connection (FPC). The board-level interconnects are BGA solder balls. The ball attachment is implemented by a high-speed solder-ball bumper (SBB). The ball diameter and ball pitch are 0.3 mm and 0.8 mm, respectively. Including the solder balls, the total thickness of this package is less than 1.0 mm. Because of the low-cost bumping metallurgy and the automated manufacturing process, this package could be relatively cost-effective. Certain qualification tests have been performed for the *flex*PAC. Good package reliability was reported in the literature.

12.2 Design Concepts and Package Structure

The *flex*PAC is a flex-based CSP using several unconventional interconnect technologies developed by IZM at TU-Berlin. The cross section of the package structure is illustrated in Fig. 12.1. The typical configuration of the silicon chip is 10 × 10 mm with a thickness of 0.55 mm.

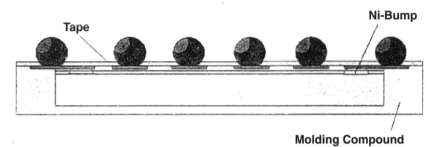

Figure 12.1   Cross section of *flex*PAC.

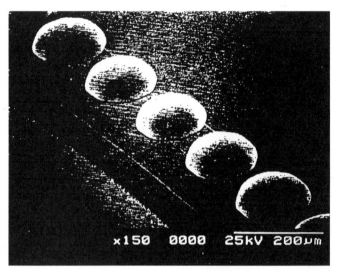

Figure 12.2   Meniscus solder bumps deposited on the Al bond pads of the die.

The die has aluminum bond pads along the periphery. A special bumping process is implemented at the wafer level to form the chip interconnects. At first, Ni bumps of 5 μm height are deposited on the Al bond pads by electroless plating [2]. Subsequently, a proprietary process is used to apply a layer of eutectic Au/Sn to the top of the electroless Ni to form meniscus solder bumps [3]. The thickness of this Au/Sn cap is 15 ± 5 μm. A picture of the meniscus solder bumps is presented in Fig. 12.2. It has been reported that if the Ni bump is 15 μm high, a total meniscus bump height of 32 ± 5 μm can be achieved [1].

The interposer of *flex*PAC is a three-layer flex circuit. A polymer carrier tape and a rolled Cu layer are laminated together by a thin adhesive film. The total thickness of this flex substrate is 85 μm. The pattern of Cu traces is formed by photolithography and etching. The

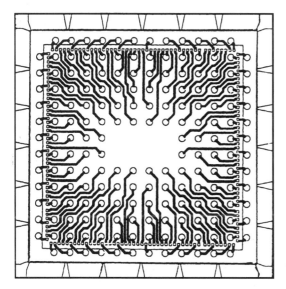

Figure 12.3 Combined fan-in and fan-out pattern of Cu traces on the flex interposer.

line/space width is 100 μm for a redistribution from 200-μm-pitch peripheral bond pads on the die to 800-μm-pitch area-array pads on the substrate. A typical combined fan-in and fan-out pattern for an interposer with 184 I/Os is shown in Fig. 12.3. In this case, the interposer is about 1.0 mm larger than the silicon chip.

The interconnection between the die and the interposer is formed by a special process called laser fiber-push connection [4]. The meniscus solder bumps are melted by optical fiber–guided laser pulses and then attached to the designated Cu pads on the interposer. The cross section of such an interconnect is presented in Fig. 12.4. In addition to these solder joints, the silicon chip and the flex substrate may be bonded by underfill adhesive as well to enhance the package reliability. After the jointing of the die and the interposer, the package thickness is about 0.65 mm. It should be noted that this dimension does not include the package encapsulation, which is an option for the *flex*PAC. The board-level interconnects are BGA. The solder balls are attached to the area-array Cu pads through windows at the bottom of the flex interposer by a proprietary process called high-speed solderball bumping (SBB) [5]. The diameter of the solder balls is 0.3 mm, and the ball pitch is 0.8 mm. A schematic diagram highlighting the aforementioned two levels of interconnects is presented in Fig. 12.5. The corresponding cross-sectional pictures of the real package are given in Fig. 12.6.

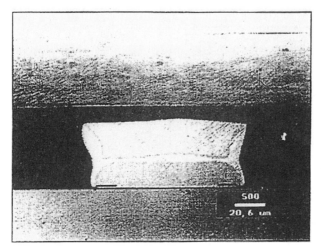

Figure 12.4 Cross section of chip-level interconnect of *flex*PAC (chip side at the bottom).

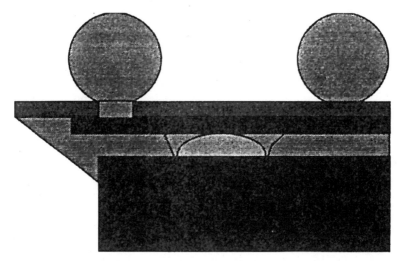

Figure 12.5 Schematic diagram highlighting the local structure of *flex*PAC.

12.3   Material Issues

The *flex*PAC is a CSP with a flexible interposer. This package contains the silicon chip, the meniscus solder bumps, the flex substrate, the die-attach adhesive, the BGA solder balls, and the encapsulation molding compound (EMC). Unlike most other flex-based CSPs, the *flex*PAC requires a special bumping process at the wafer level. To implement the so-called meniscus solder bumps, an electroless Ni layer

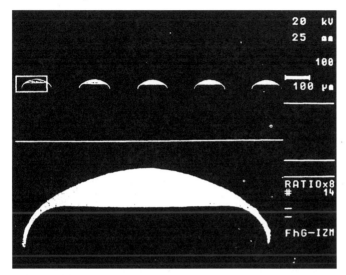

Figure 12.6 Cross section of meniscus solder bump (*bright* Au/Sn cap on *dark* Ni bump).

needs to be plated on the Al bond pads of the die to serve as the under-bump metallization (UBM) [2]. The nominal thickness of this Ni layer is 5 μm. Electroless Ni was chosen because of such advantages as low cost, high uniformity, and near-hermetic sealing on the Al pads. To protect the surface from oxidation and to improve solder wettability, the electroless Ni is covered with a 0.05-μm-thick gold flash. Afterwards, a layer of eutectic solder (80Au/20Sn) is applied to the top of the electroless Ni UBM by a well-controlled dipping process. The nominal thickness of this solder cap is 15 ± 5 μm. The selection of Au/Sn is based on its good reliability and high melting point. The cross-sectional picture of meniscus solder bumps is presented in Fig. 12.6.

It was reported that the metallurgy to form the meniscus solder bumps is Au-Sn-Ni, which has never been investigated before. From surface analysis of the cross section given in Fig. 12.6, the intermetallic Ni_3Sn_2 can be identified at the interface between the Ni UBM and the Au/Sn solder cap. The growth of this intermetallic has been measured during high-temperature storage at 175°C. The results are shown in Fig. 12.7 [1]. The intermetallic growth is obviously proportional to the square root of time, indicating that the diffusion law is the governing mechanism for this phenomenon.

The interposer of *flex*PAC is a three-layer flex circuit. A 20-μm-thick Cu foil is laminated to a 50-μm-thick polymer carrier tape by a thin adhesive film. The total thickness of this substrate is 85 μm. The

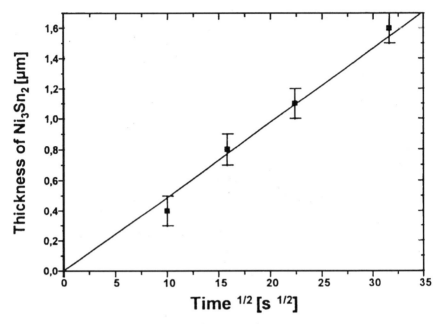

Figure 12.7 Growth of Ni_3Sn_2 intermetallic at 175°C.

pattern of Cu traces is formed by photolithography and chemical etching. The line/space width is 100 μm. To prevent oxidation, the surface of the Cu traces is plated with a thin Au (0.2 to 0.5 μm) or electroless Ni (0.5 μm) layer. For the attachment of BGA solder balls, area-array windows are opened on the polymer carrier tape for corresponding Cu bond pads.

The die-attach material of *flex*PAC is a low-viscosity and low-stress adhesive. The curing temperature is 90°C. This adhesive is not only a bonding layer between the chip and the substrate to ensure mechanical stability, but also an underfill material to encapsulate the meniscus solder joints. The BGA balls of *flex*PAC are conventional 63Sn/37Pb eutectic solder. The ball diameter is 0.3 mm. The package encapsulation is optional for *flex*PAC. If it is selected, the conventional molding material and process will be adopted to encapsulate the chip for better protection and dimensional stability.

12.4 Manufacturing Process

The fabrication of *flex*PAC employs several proprietary technologies developed by IZM and TU-Berlin. The major steps of the manufacturing process are given in Fig. 12.8. At first the meniscus solder bumps

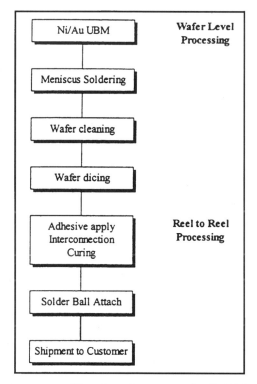

Figure 12.8 Manufacturing process for *flex*PAC package.

are formed at the wafer level. Before the eutectic Au/Sn solder is deposited, an electroless Ni plating must be performed to establish the UBM. The detailed procedures for this wafer-bumping process are given in Fig. 12.9 [6]. In order to avoid unnecessary Ni plating, the backside of the wafer must be covered by a resist layer. Prior to this chemical plating process, the Al bond pads should be cleaned by microetching to remove the oxide layer. In addition, a zincate treatment is necessary to activate the Al surface. For electroless Ni plating, a bath based on sodium hypophosphite is used. The wafer is dipped into the solution to form Ni bumps. It should be noted that the electroless Ni plating is a maskless process, which implies low-cost production. Another feature is the uniformity in bump thickness. Also, this process can achieve a near-hermetic sealing on the surface of Al bond pads. As a result, high reliability should be expected. After the Ni UBM is established, the wafer is immersed in an Au solution to coat the Ni bumps with gold flash. The function of this thin Au layer is to avoid oxidation and to improve solder wettability. Subsequently, the

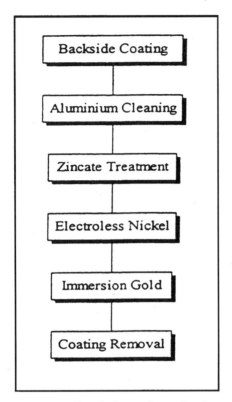

Figure 12.9 Detailed procedures for electroless Ni UBM.

backside resist coating is removed. Following the electroless Ni plating and the immersion Au coating, a well-controlled dipping process is performed to apply eutectic Au/Sn solder to the top of the Ni/Au UBM to form meniscus solder bumps on the bond pads of the wafer, as shown in Fig. 12.10.

The next two steps after meniscus soldering are wafer cleaning and dicing. Then the silicon chip is ready for the connection to the flex interposer. To ensure mechanical stability in the subsequent manufacturing process, a die-attach adhesive is used to bond the silicon chip and the flexible substrate together. The adhesive is dispensed to the active face of the chip as shown in Fig. 12.11, and then the interposer is laid over the die with the Cu trace side facing down. The curing temperature of this adhesive is 90°C, and it remains in liquid form before the chip-substrate interconnects are established.

The following procedure is laser fiber-push connection, as illustrated in Fig. 12.12. A highly stable glass fiber with a diameter of 220 μm

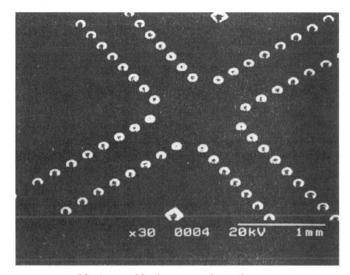

Figure 12.10   Meniscus solder bumps on the wafer.

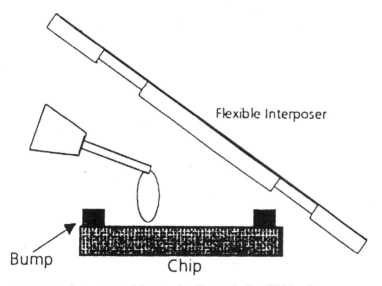

Figure 12.11   Dispensing of die-attach adhesive before FPC bonding.

is used as the mechanical clamp and the optical waveguide. The fiber is positioned above a meniscus solder bump and the corresponding Cu pad on the interposer and is pushed from a nozzle to the back of the carrier tape to apply the required bonding force. In the meanwhile, an Nd:YAG laser pulse is transmitted through the glass fiber

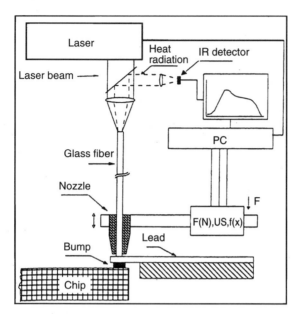

Figure 12.12   Schematic diagram for laser FPC bonding process.

TABLE 12.1   Bonding Conditions for Fiber-Push Connection

Laser power	5 W
Pulse width	10 ms
Bond force	40 cN

and heats up the contact zone between the meniscus bump and the Cu bond pad. Consequently, the Au/Sn solder is melted to form the interconnect between the bond pad on the die and the circuitry on the interposer. The bonding conditions of FPC are given in Table 12.1. It should be noted that, since the liquid adhesive has been squeezed away from the contact zone by the glass fiber, the forming of the joint is not affected by the die-attach adhesive. Furthermore, the FPC bonding method does not require an open window on the carrier tape because the polymer film has very low energy absorption at the wavelength ($\lambda = 1064$ nm) of the selected Nd:YAG laser. However, in order to prevent overheating of the flex substrate as a result of mechanical heat transfer, the IR emission during FPC is fed back to the controller for in situ monitoring. The FPC bonding is a single-shot process. It was reported that a speed of 15 bonds per second could be achieved. For each bonding, one solder joint is formed by a corre-

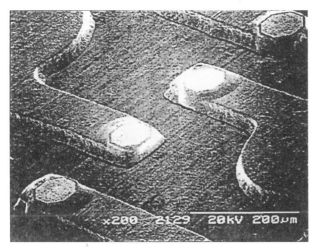

Figure 12.13 Solder joints on the inner leads of the flex interposer by FPC bonding (silicon chip has been etched away).

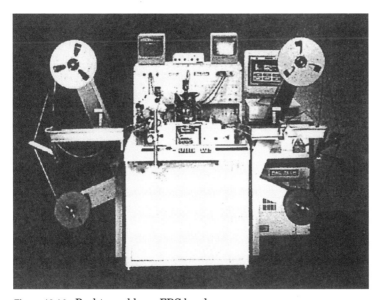

Figure 12.14 Reel-to-reel laser FPC bonder.

sponding Cu inner lead of the interposer, as presented in Fig. 12.13. While the FPC is going on, the die-attach adhesive is simultaneously cured by a hot stage. This adhesive can be considered as an underfill to bond the chip and the substrate together and to encapsulate the interconnects. The whole process may be implemented on a modified wire bonder with a reel-to-reel operation, as shown in Fig. 12.14.

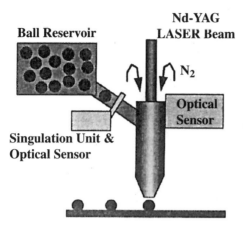

Figure 12.15 Schematic diagram for solder-ball bumping process.

Such a processing advantage is essential for low-cost and high-volume production.

The last step in fabrication is BGA ball attachment. The solder balls are deposited on the area-array bond pads at the bottom of the flex interposer by a high-speed solder-ball bumper [5]. Although this bumping process is suitable for solder balls of any desired alloy with a well-defined size, currently the conventional 63Sn/37Pb eutectic solder with a diameter of 0.3 mm is used. As illustrated in Fig. 12.15, a singulation mechanism regulates the flow of solder balls from a reservoir to a capillary. For each bumping, nitrogen gas (N_2) is employed to shoot a single solder ball onto the Cu bond pad. The solder ball is instantly reflowed by an Nd:YAG laser pulse. During the laser reflow, the nitrogen gas acts as a shielding atmosphere, resulting in a smooth, round solder bump. It should be noted that the SBB process requires neither masking nor fluxing for solder-ball attachment. The cross section of a completed *flex*PAC package is presented in Fig. 12.16. The geometry of the interconnects can be observed clearly. It should be noted that the package encapsulation is optional for the *flex*PAC.

12.5   Qualifications and Reliability

A series of qualification tests have been conducted to verify the reliability of the *flex*PAC package. One of the major concerns is the reliability of the chip-level interconnects. A thermal cycling test following MIL-STD-883C was performed. The specimens were subjected to a temperature change between −55 and 125°C. The inner leads of the interposer have a metallization of Cu/Ni/Au. *Flex*PAC packages with and

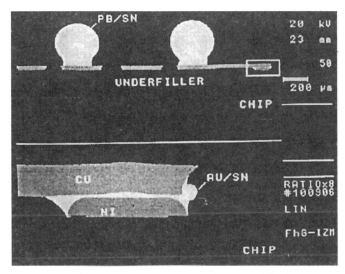

Figure 12.16 Cross section of jointed meniscus solder bump and BGA solder balls.

TABLE 12.2 Thermal Cycling Test Results for *flex*PAC Package

Lead	Underfill	No. of cycles	Failure
Cu/Ni/Au	Adhesive	1000	No
Cu/Ni/Au	No	1000	No

without die-attach underfill were tested. After 1000 temperature cycles, no failure could be observed in either type of specimen, as shown in Table 12.2. The cross section presented in Fig. 12.17 indicates that the solder joint between the chip and the interposer is still intact after 1000 cycles. Although an intermetallic compound such as Ni_3Sn_2 was detected by BSE at the Ni-solder interface, it was reported that this intermetallic layer was too thin to affect the interconnect reliability.

In addition to thermal cycling, high-temperature storage and popcorn tests were also performed to investigate the package reliability. The testing conditions and results are summarized in Table 12.3. No failures were observed in all cases, indicating good package reliability. Also, an ultrasonic image by scanning acoustic microscopy (SAM) after the popcorn test is presented in Fig. 12.18. No delamination or popcorning was detected. Therefore, it may be concluded that the *flex*PAC is qualified for JEDEC level 1 (moisture insensitive).

The *flex*PAC package was also investigated for board-level solder joint reliability. Specimens with daisy chain design were mounted on

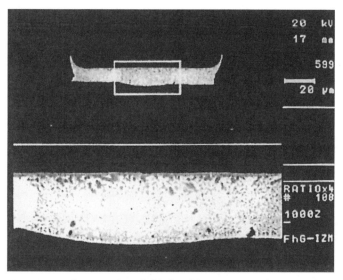

Figure 12.17   Cross section of chip-level interconnect after 1000 thermal cycles (with underfill adhesive).

TABLE 12.3   Qualification Data for *flex*PAC Package

Sample	Test	Duration	Evaluation	Result
CSP	Thermal cycling	1000 cycles	Shear test and cross section	No failure
CSP	Storage at 150°C	1000 h	Shear test and cross section	No failure
CSP	JEDEC level 1	24 h@125°C 168 h@85/85 3 reflows	Shear test and cross section	No failure
CSP on board	Thermal cycling	1000 cycles	Electrical shear test, and cross section	No failure

a test board and then subjected to temperature cycling. No significant change in electrical resistance could be detected after 1000 cycles of testing. Furthermore, SEM inspection of the cross section found no failure in the solder joint, as shown in Fig. 12.19. Therefore, the board-level reliability for the *flex*PAC assembly was assured to a certain extent.

12.6   Applications and Advantages

The package lineup and targeted application of IZM's *flex*PAC have not been reported yet. Based on the BGA configuration, this package should be used for IC devices with a low to medium number of I/Os.

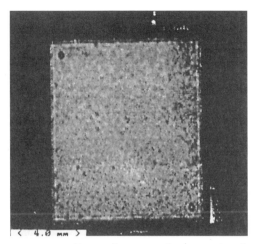

Figure 12.18 Image from scanning acoustic microscopy showing no delamination after JEDEC level 1 popcorn test.

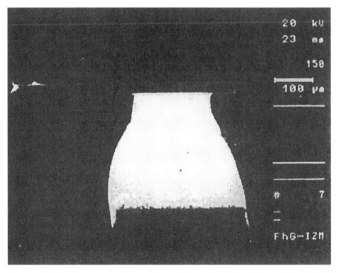

Figure 12.19 Cross section of board-level solder joint after 1000 thermal cycles.

Because of the low package profile, compact package size, and peripheral bond pads, the *flex*PAC may find application in SRAMs, flash memory, or ASICs in portable equipment.

According to the manufacturer, the major advantages of *flex*PAC are low production cost and high package reliability. It is claimed that the electroless Ni plating and the meniscus soldering are very cost-effective. In addition, the fabrication of *flex*PAC allows a reel-to-reel operation, which is an important advantage for low-cost mass production. The package reliability of *flex*PAC has been verified by several qualification tests. The JEDEC level 1 moisture resistance can be achieved. Such good performance makes this package stand out from other CSPs in this category.

12.7   Summary and Concluding Remarks

The *flex*PAC is a flex-based CSP featuring several unconventional packaging technologies, namely, meniscus solder bumping at the wafer level, laser fiber-push connection at the chip level, and high-speed solder-ball bumping at the package level. The wafer is treated with electroless Ni/Au UBM followed by eutectic Au/Sn soldering to form the meniscus solder bumps. The chip-to-substrate interconnection is implemented by FPC technology. The interposer is a flex circuit with a single layer of Cu traces. The package terminals are BGA solder balls which are attached to the bottom of the interposer by SBB. The ball diameter and ball pitch are 0.3 mm and 0.8 mm, respectively. Including the solder balls, the total thickness of this package is less than 1.0 mm. The package size is 1.0 mm larger than the chip size. Note that the encapsulation molding is optional for *flex*PAC packages.

It has been reported that the bumping metallurgy of *flex*PAC is relatively cost-effective and the manufacturing process can be operated in a reel-to-reel fashion. Such advantages are essential for low-cost mass production. However, because of the adoption of unconventional packaging technologies, initial investment in modifying existing infrastructure may be a concern. Several qualification tests have been conducted to investigate the reliability of *flex*PAC. This package has been qualified for JEDEC level 1 moisture resistance, and good package reliability under thermal cycling and high-temperature storage has been reported. Based on the low profile and compact size, the *flex*PAC may find suitable applications in memory and ASIC devices in portable equipment.

12.8   References

1. C. Kallmayer, E. Jung, P. Kasulke, R. Azadeh, G. Azdasht, E. Zakel, and H. Reichl, "A New Approach to Chip Scale Package Using Meniscus Soldering and FPC Bonding," *Proceedings of the 47th ECTC,* San Jose, Calif., May 1997, pp. 114–119.
2. A. Ostmann, J. Kloeser, E. Zakel, and H. Reichl, "Implementation of a Chemical Wafer Bumping Process," *Proceedings of the IEPS,* San Diego, 1995, pp. 354–366.
3. A. Ostmann, G. Motulla, J. Kloeser, E. Zakel, and H. Reichl, "Low Cost Techniques for Flip Chip Soldering," *Proceedings of the SMI,* San Jose, Calif., 1996, pp. 319–328.

4. G. Azdasht, E. Zakel, and H. Reichl, "A New Chip Packaging Method Using Windowless Flip TAB Laser Connection on Flex Circuits," *Proceedings of the ITAP,* San Jose, Calif., pp. 237–244.
5. P. Kasulke, G. Azdasht, E. Zakel, and H. Reichl, "A New Solution for Solder Application to FCA, BGA and CSP Challenges," *Proceedings of Micro Systems Technologies,* 1996.
6. A. Ostmann, J. Simon, and H. Reichl, "The Pretreatment of Aluminum Bondpads for Electroless Nickel Bumping," *Proceedings of the IEEE Conference on MCM,* Santa Cruz, Calif., March 1993, pp. 74–78.
7. C. Kallmayer, R. Azadeh, K-F. Becker, S. Anhock, E. Busse, H. Oppermann, G. Azdasht, R. Aschenbrenner, and H. Reichl, "A Low Cost Approach to CSP Based on Meniscus Bumping, Laser Bonding through Flex and Laser Solder Ball Placement," *Proceedings of the EPTC,* Singapore, October 1997, pp. 34–40.

Chapter

13

NEC's
Fine-Pitch Ball Grid Array
(FPBGA)

13.1 Introduction and Overview

NEC Corp. developed the fine-pitch ball grid array (FPBGA) package in 1995 [1]. This package uses a two-layer (polyimide/copper) carrier tape as the interposer and belongs to the category of flex-based CSP. The FPBGA features NEC's through-hole bonding process and micropunching technology. The former is used for establishing chip-level interconnection, while the latter is used to form solder bumps for board-level assembly. The FPBGA has a very simple package structure with three variations in encapsulation. The interconnects between the die and the interposer are preformed metal bumps on the flex substrate. The package I/Os to the PCB are BGA solder bumps. The bump height is 0.1 mm with a standard pitch of 0.5 mm. The qualification testing results show that the FPBGA has the same package reliability as conventional plastic packages. This package is developed for devices with medium pin counts, such as ASICs. Currently the FPBGA with 208 pins is in production for digital camcorders. A 204-pin version for cellular phones is under way.

13.2 Design Concepts and
Package Structure

The FPBGA is a new chip scale package developed by NEC based on its through-hole bonding process and micropunching technology. This package has a very simple structure, as shown in Fig. 13.1. A two-layer polyimide/Cu carrier tape is used as the package interposer. The

Chip   Adhesive

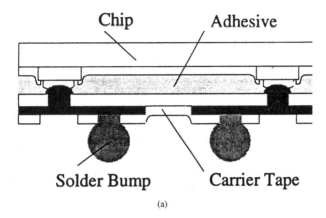

Solder Bump   Carrier Tape

(a)

Mold Resin

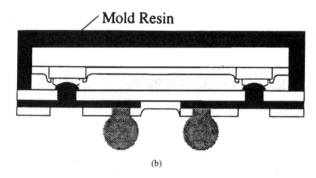

(b)

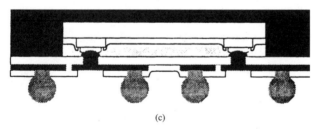

(c)

Figure 13.1   Cross section of package structure: (*a*) bare
FPBGA, (*b*) molded FPBGA, (*c*) molded and pad-extended
FPBGA.

die-attach adhesive is a thermoplastic polyimide film installed on one
side of the base tape opposite to the Cu traces. The die is mounted on
the flex substrate with its active side facing toward the interposer.
The chip-level interconnection is established by inner-bump bonding.
The inner bumps are prefabricated by electroplating in the through-

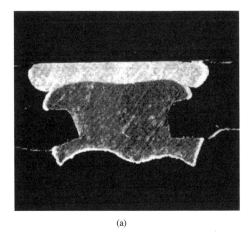

(a)

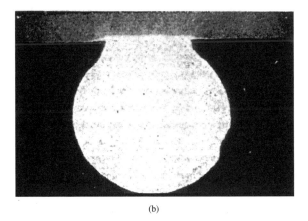

(b)

Figure 13.2 Cross-sectional view of interconnects: (*a*) double-bump bonding (chip side at the top), (*b*) BGA solder ball.

hole vias on the polyimide tape. The bump has a Cu core with Au plating on the surface. There are two variations for the inner-bump bonding. One is to joint the inner bumps with the Al bond pads on the die directly (single-bump bonding). The other is to form an Au ball bump on each die pad first. Then the inner bumps are pressed to the Au ball bumps on the die to form the joint (double-bump bonding). The cross-sectional picture of an interconnect with double-bump bonding is shown in Fig. 13.2*a*. According to NEC, the double-bump bonding can provide higher joint strength and better bump height control. However, the tradeoff is the higher processing cost.

Figure 13.3   Bottom view of NEC's FPBGA.

The package-to-board interconnects of FPBGA are small solder bumps at the bottom of the package, as shown in Fig. 13.3. These bumps are not made from conventional solder spheres. Instead, disclike solder pieces are punched from a thin tape to the preopened bond pads at the back of the interposer and then reflowed to form the solder bumps. The cross-sectional view of a solder bump is shown in Fig. 13.2b. Because the volume of the punched solder disc can be precisely controlled, a bump height of 0.1 mm ± 5 μm can be achieved. The standard ball pitch is 0.5 mm.

The FPBGA can be encapsulated by a conventional transfer molding process. There are three types of encapsulation configuration for this package. Bare FPBGA is a configuration without any encapsulation, as shown in Fig. 13.1a. Molded FPBGA has a thin layer of encapsulation around the chip, as shown in Fig. 13.1b. It should be noted that both of these types of FPBGA have only fan-in circuitry at the bottom of the interposer. If the encapsulation is expanded laterally to accommodate additional fan-out I/Os, as illustrated in Fig. 13.1c, NEC calls this configuration molded and pad-extended FPBGA.

From the cross sections shown in Fig. 13.1, it can be seen that the package size of FPBGA is either equal to or slightly larger than the die size. The package outline of FPBGA is not yet standardized. However, some general design rules for packages with fan-in configuration can be induced, as presented in Fig. 13.4 and Table 13.1. From those guidelines, it is clear that FPBGA complies with the dimension requirements of CSP.

In general, FPBGA has been developed for IC devices with a medium range of pin counts. However, the number of I/Os depends on various factors, such as the chip size and the ball pitch. Therefore, it is rather difficult to derive a general formula for the pin counts. Figure 13.5

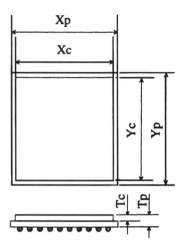

Figure 13.4 Definition of package outline for FPBGA.

TABLE 13.1   Package Outline of FPBGA

	Chip	Package
Size	X_c	$X_p < X_c + 0.1$
Size	Y_c	$Y_p < Y_c + 0.1$
Thickness	T_c	$T_p < T_c + 0.1$

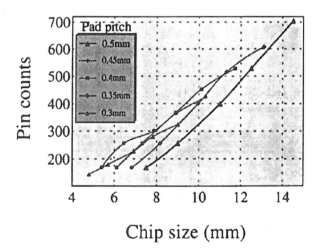

Chip size (mm)

Figure 13.5 Maximum allowable pin counts versus the chip size.

shows the relationship between the maximum allowable pin count and the chip size for FPBGA with various ball pitches based on a specific design rule. The Cu trace on the interposer is single-layer and has a fan-in pattern with a line/space width of 30 μm. It is obvious that FPBGA can easily accommodate hundreds of I/Os. Such capability makes this package suitable for a wide spectrum of applications.

13.3 Material Issues

FPBGA is a flex-based BGA package without conventional wire or TAB bonding. In addition to the silicon chip, the package materials include the metallized flex substrate, the die-attach adhesive, the molding encapsulant, and the BGA solder balls. Of these, the flex interposer is the most significant element for FPBGA. The detailed structure of this interposer is illustrated in Fig. 13.6. The base substrate is a polyimide carrier tape. Cu traces are patterned on one side of the tape for circuit redistribution. A thin polyimide cover coat is deposited over the top of the metal circuitry for protection and insulation. This cover coat also serves as a solder resist layer and is patterned to form area-array openings for solder-bump formation. A number of through-hole vias are made on the carrier tape to accommodate inner bumps for chip-level interconnection. Cu cores are formed in those vias by electroplating. The surface of the Cu core is subsequently plated with Au in order to improve the bondability to the bond pads on the die.

The die-attach adhesive is a thermoplastic polyimide film which has good adhesion strength and high heat resistance. This adhesive film can be mounted on the chip during the fabrication of FPBGA or precoated on the flex substrate. In order not to interfere with the inner-bump bonding, the film should be made slightly smaller than

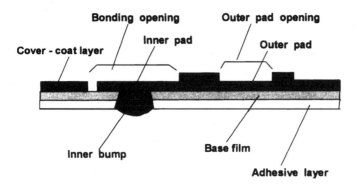

Figure 13.6 Detailed structure of the flex interposer for FPBGA.

the peripheral die-pad array. Encapsulation is optional for FPBGA. For molded FPBGA and its pad-extension version, a conventional transfer molding process is followed. However, warpage due to the thermal mismatch between the encapsulant and the die is a serious concern because the coplanarity of the BGA solder bumps would be affected, as shown in Fig. 13.7. A study was performed to investigate the effect of the coefficient of thermal expansion (CTE) of the encapsulant. The experimental results are presented in Fig. 13.8. The warpage shown in this figure is induced by the cooling of the package from the molding temperature to the room temperature. It can be seen that the CTE of the molding compound plays a significant role in the warpage. Therefore, a low-CTE encapsulant should be used for transfer molding.

Although FPBGA uses BGA as the package-to-board interconnects, the solder balls are not formed by the conventional solder sphere attachment approach. Instead, the micropunching technology developed by NEC is used to form solder bumps for FPBGA [2]. Small disclike

Figure 13.7   Warpage of molded FPBGA.

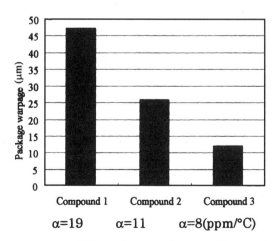

α=19   α=11   α=8(ppm/°C)

Figure 13.8   Effect of encapsulant coefficient of thermal expansion (α) on the package warpage.

solder pieces are punched out of a thin solder ribbon by an automated punching mechanism. The solder pieces can be mounted directly on the bond pads at the bottom of the package or via a temporary transferring plate. Once the solder pieces are deposited in place, the solder bumps can be formed by regular reflow heating. According to NEC, the micropunching technology can be applied to various solder materials. However, for FPBGA, conventional eutectic solder has been chosen as the material for solder bumps.

13.4 Manufacturing Process

Two manufacturing flowcharts are presented in Fig. 13.9a and b. The former is for FPBGA with double-bump bonding, and the latter is for single-bump bonding. The double-bump bonding requires an Au ball bump on the die pad. This ball bump is formed by the first bond of

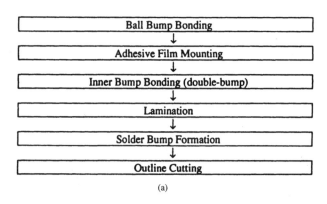

(a)

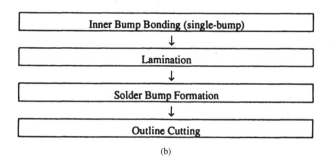

(b)

Figure 13.9 Flowcharts for the fabrication of FPBGA (a) with double-bump bonding and (b) with single-bump bonding.

conventional Au wire bonding. Although double-bump bonding can provide higher joint strength and better bump height control, it is obvious that this process increases the manufacturing cost. Therefore, NEC decided to adopt single-bump bonding for FPBGA. The second step in Fig. 13.9*a* is optional and is not restricted to double-bump bonding only. If the polyimide adhesive is precoated on the flex substrate, this step can be omitted.

Figure 13.10 provides a graphical illustration of the fabrication of FPBGA with single-bump bonding. The inner bumps are bonded to the Al bond pads on the die directly with a probe from the back side. This is a thermosonic single-point bonding process which is called, in NEC's term, the "through-hole bonding process." Afterwards, the die-attach adhesive is melted and cured by a hot-press process (called

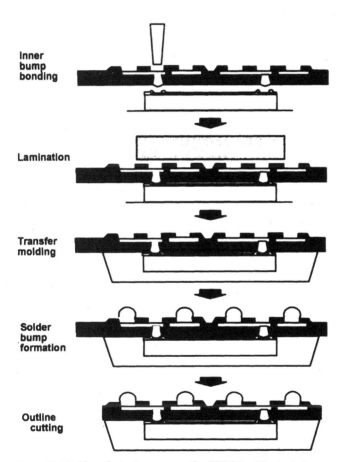

Inner bump bonding

Lamination

Transfer molding

Solder bump formation

Outline cutting

Figure 13.10 Manufacturing process for FPBGA with single-bump bonding and precoated die-attach adhesive.

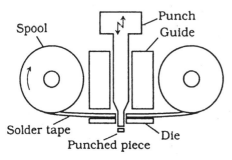

Figure 13.11   Mechanism for micropunching technology.

lamination) to provide adhesion between the die and the interposer as well as to encapsulate the interconnects formed from the through-hole bonding. The next step is chip encapsulation, which is an option for FPBGA. For molded FPBGA and its pad-extended version, the conventional transfer molding process is employed.

Following the encapsulation is solder-bump formation. This process is performed with NEC's proprietary micropunching technology [2]. This technology is implemented by the mechanism shown in Fig. 13.11. A thin solder ribbon runs between two spools. Small disclike solder pieces are punched out of the solder ribbon by a cemented carbide head and a ruby die. The punch is driven by a piezoelectric actuator and runs at a frequency of 10 punches per second. In order to maintain a consistent volume and a good shape for the punched-out solder pieces, a set of optimized processing parameters needs to be determined. For the 63Sn/37Pb solder bumps used for FPBGA, the thickness of the sol-der ribbon is 100 μm. The diameter of the solder disc is 150 μm with a clearance of 5 μm. In addition, the axis alignment error for the punch head must be less than 2 μm.

Once the small solder pieces have been made, they are placed on the bond pads at the bottom of the interposer and then are reflowed to form the solder bumps. There are two ways to deposit the solder pieces. One is by direct punching, as shown in Fig. 13.12*a*. This method is straightforward and low-cost. However, the punching force (200 to 300 mN) may damage the flex substrate. A more sophisticated method is plate transferring, as shown in Fig. 13.12*b*. The solder pieces are punched to one side of a temporary transferring plate, which is precoated with a thin adhesive film. The plate is flipped over and the solder pieces are aligned with the corresponding bond pads at the bottom of the interposer, which is precoated with flux. Both heat and pressure are applied in this process. Because the adhesive film on

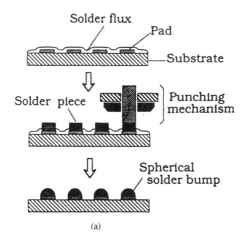

(a)

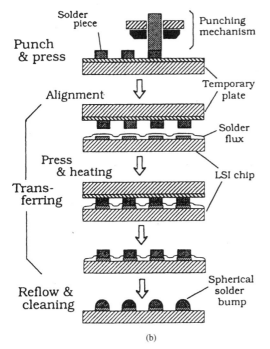

(b)

Figure 13.12   Solder-bump formation by micropunching technology: (*a*) direct punching method; (*b*) plate transferring method.

the transferring plate loses its adhesion at elevated temperature and the flux is sticky, the solder pieces are transferred from the temporary plate to the bond pads. A set of optimized parameters for this process is given in Table 13.2. A picture of solder bumps formed by the plate transferring method is shown in Fig. 13.13.

The last step in the manufacturing process is package outline cutting. A fourth-harmonic-generation YAG (FHG-YAG) laser beam is used for singulation. The wavelength of this laser (0.266 μm) is close to that of a KrF excimer laser (0.248 μm), but its cost is much lower.

TABLE 13.2   Optimized Conditions for Solder-Bump Formation by Plate Transferring

Adhesive sheet	Initial adhesion	350 mN/mm
	Adhesion losing	120°C
Solder flux	Viscosity	20–100 Pa · s
	Coating thickness	10 μm
Punching	Pressure	200–300 mN
Transferring	Pressure	10–20 mN/bump
	Temperature	120°C (peak) at temporary plate

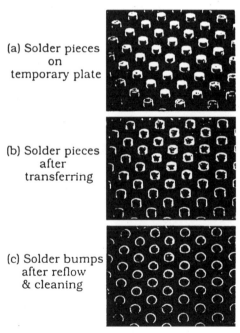

(a) Solder pieces on temporary plate

(b) Solder pieces after transferring

(c) Solder bumps after reflow & cleaning

Figure 13.13 Solder bumps formed by micropunching with plate transferring method.

It should be noted that the fabrication of FPBGA allows a reel-to-reel operation. This is an essential advantage because the manufacturing process can be fully automated and the production cost can be substantially reduced.

13.5 Performance and Reliability

The electrical performance of FPBGA has been characterized for a 208-pin configuration. The package size is 10×10 mm, and the BGA ball population is two-row fan-in/two-row fan-out [3]. The simulation data are given in Table 13.3. Such good electrical performance can be expected because the electrical passage in FPBGA is rather short. From these results, it may be concluded that FPBGA is suitable for high-speed ASIC applications. The thermal performance of FPBGA has not yet been reported. However, NEC claims that a heat sink may be installed at the back of a bare FPBGA if necessary. In this case, the heat dissipation capability should be good.

A series of qualification tests have been conducted to verify the package reliability of FPBGA. These qualifications include temperature cycling (T/C), high temperature and humidity (HH), high-temperature storage (HT), and high-temperature, humidity, bias (HHBT) tests. The results are given in Table 13.4. From these data, it may be concluded that the FPBGA has the same level of package reliability as conventional plastic packages such as PQFP.

The chip-level interconnects of FPBGA are formed by the through-hole bonding process, which is a relatively new technology. The relia-

TABLE 13.3 Simulated Electrical Performance for FPBGA

	100 MHz	300 MHz
L (nH)	0.14–0.64	0.13–0.60
R (mΩ)	4.08–21.46	7.04–35.65
C (pF)	0.13–0.33	0.13–0.33

TABLE 13.4 Package Qualification Data for FPBGA

Test conditions	Failures/samples
T/C (−65 to 150°C), 600 cycles	0/17
HH (85°C, 85% RH), 1000 h	0/19
HT (150°C), 1000 h	0/20
HHBT (85°C, 85% RH, 3.6 V), 1000 h	0/24

bility of these joints should be a major concern with FPBGA. Since the inner bumps are covered by the flex substrate, it is rather difficult to test (or even to inspect) these interconnects. NEC has developed certain techniques such as LED reflection (Fig. 13.14) and mechanical peel (Fig. 13.15) for geometry inspection and strength characterization, respectively. However, these results have not yet been reported. On the other hand, a surface analysis was performed on the cross section of a single-bump bonding joint, as illustrated in Fig. 13.16. Auger electron spectroscopy (AES) was used to investigate the diffusion phe-

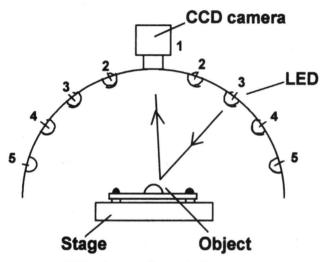

Figure 13.14   LED reflection technique for bump geometry inspection.

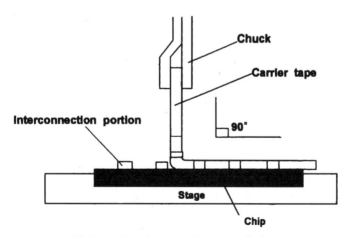

Figure 13.15   Mechanical peel test for joint strength characterization.

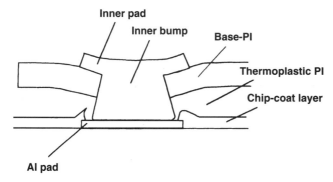

Figure 13.16   Cross-sectional view of a single-bump bonding joint.

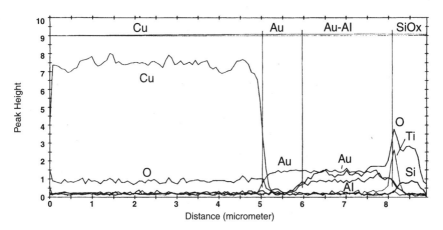

Figure 13.17   Surface analysis results by AES.

TABLE 13.5   Solder Joint Qualification Data for FPBGA

	Condition	Failures/samples
Thermal cycling	−40 to 125°C, 200 cycles	0/5
HAST	110°C, 85% RH, 288 h	0/5

nomenon in the inner bump. From the results presented in Fig. 13.17, it is clear that the Au plating on the surface of the Cu core and the Al metallization on the die pad have diffused into each other [3]. Therefore, it may be concluded that the joint has been formed properly by the through-hole bonding process.

The solder bump of FPBGA is relatively small. The bump height is only 0.1 mm. Therefore, the board-level solder joint reliability could be rather critical. Two qualification tests have been performed for the FPBGA assembly. The testing results are given in Table 13.5 and

TABLE 13.6   Comparison of Thermal Stresses for FPBGA and
Direct Chip Attach (DCA)

Portion	DCA maximum stress (kgf/mm²)	FPBGA maximum stress (kgf/mm²)
Solder bump	15.6	7.1
Outer pad	62.1	13.4
Substrate pad	8.6	5.4

Package Size	Pin Count	Assembled LSI
3.35 × 3.89 mm	20	Video signal processing LSI
7.3×7.3 mm	160	ASIC
9.7×9.7 mm	232	TEG

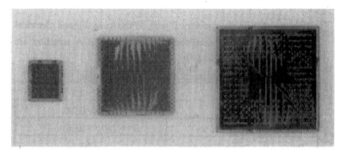

Figure 13.18   Specifications for various FPBGA samples.

seem to be satisfactory. Further investigation was conducted using finite-element analysis to compare the maximum stresses with FPBGA
and DCA. From the data given in Table 13.6, it is clear that the solder
joints of FPBGA are much more reliable than those of DCA. This con-
clusion should be anticipated because the flex interposer can substantially reduce the thermal mismatch between the die and the PCB
and, hence, relax the stresses in solder joints.

13.6   Applications and Advantages

FPBGA is a flex-based chip scale package with fine-pitch (0.5-mm)
BGA solder bumps. The circuitry on the substrate may have both fan-
in and fan-out patterns. Therefore, this package can easily accommodate hundreds of I/Os and any depopulated configurations. Such capability makes this package suitable for a wide spectrum of
applications. Figure 13.18 presents three typical samples packaged by
FPBGA. Besides, since FPBGA has very compact package size and

rather small solder bumps, it can be used for MCM applications, as shown in Fig. 13.19.

It is a general trend for CSPs to use fan-in circuits to minimize the package footprint. It is also popular to use full-grid BGA to maximize the package I/Os. Recently NEC proposed a two-row fan-in/two-row fan-out configuration for its molded and pad-extended FPBGA [3]. With such a configuration, the total pin count is actually larger than the full-grid fan-in only pattern, as illustrated in Fig. 13.20. However, the tradeoff is a certain increase in the package size. NEC has used this configuration to package a 208-pin ASIC for digital camcorders. A similar version with 204 pins is being developed for application to cellular phones.

Figure 13.19 FPBGA for MCM application (1-MPU and 10-SRAM MCM).

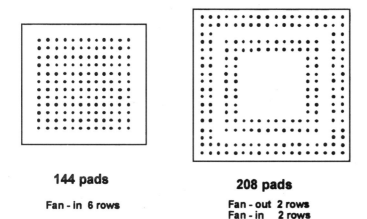

144 pads

Fan - in 6 rows

208 pads

Fan - out 2 rows
Fan - in   2 rows

Figure 13.20 Comparison of BGA pattern for full-grid and perimeter FPBGA.

The major advantages of NEC's FPBGA include simple package structure, compact package size, and flexible configurations. Above all, the manufacturing process allows a reel-to-reel operation, which can substantially reduce the production cost and increase the productivity. This package should find plenty of applications in portable devices.

13.7 Summary and Concluding Remarks

FPBGA is a flex-based chip scale package with fine-pitch BGA solder bumps. This package features NEC's through-hole bonding process and micropunching technology. The former is used for establishing chip-level interconnection, while the latter is used to form solder bumps for board-level assembly. FPBGA has a very simple package structure, and there are three encapsulation variations. This package may have both fan-in and fan-out circuitry on the interposer, making the configuration very flexible. The fabrication of FPBGA allows a reel-to-reel operation. Therefore, the manufacturing process may be fully automated and the production cost can be considerably reduced.

The qualification testing data indicated that FPBGA has the same package reliability as conventional plastic packages. Good electrical performance was reported as well. The board-level solder joint reliability has been verified to some extent. However, comprehensive testing data are not yet available. In general, the FPBGA package was developed for devices with medium pin counts such as ASICs. Currently the FPBGA with 208 pins is in production for digital camcorders. A 204-pin version for cellular phones is under development. It should be noted that this package can be applied to MCMs as well. With its compact package size and flexible configurations, the FPBGA should find extensive applications in portable equipment.

13.8 References

1. S. Matsuda, K. Kata, and E. Hagimoto, "Simple-Structure, Generally Applicable Chip-Scale Package," *Proceedings of the 45th ECTC,* Las Vegas, Nev., May 1995, pp. 218–223.
2. Y. Kato, Y. Ueoka, E. Kono, and E. Hagimoto, "Solder Bump Forming Using Micro Punching Technology," 1996.
3. S. Matsuda, K. Kata, H. Nakajima, and E. Hagimoto, "Development of Molded Fine-Pitch Ball Grid Array (FPBGA) Using Through-Hole Bonding Process," *Proceedings of the 46th ECTC,* Orlando, Fla., May 1996, pp. 727–732.

14

Nitto Denko's Molded Chip Size Package (MCSP)

14.1 Introduction and Overview

The resin-molded chip size package (MCSP) was developed by Nitto Denko Corp. in 1995 [1]. This package belongs to the category of flex-based CSP because a metallized polyimide (PI) tape is used as the interposer. This package features Nitto Denko's ASMAT as the interposer and TAPI as the bonding layer between the die and the interposer. The former is a PI/Cu tape carrier, while the latter is a thermoadhesive polyimide. The package structure of the MCSP is very simple. The chip-level interconnects are metal bumps formed on the vias of the flex substrate. The joints are made by thermocompression. The board-level interconnects area BGA solder balls. The ball diameter and pitch are 0.2 mm and 1.0 mm, respectively. The MCSP can be encapsulated by either TAPI or conventional molding compound. Qualification testing results indicated that the MCSP has package reliability similar to that of the PQFP. Although this package has very small solder balls, a computational analysis revealed that the MCSP has better solder joint reliability than the flip chip because of the existence of the flexible substrate.

14.2 Design Concepts and Package Structure

The MCSP is a flex-based chip scale package implemented with Nitto Denko's proprietary materials, namely, ASMAT and TAPI. The former is a polyimide tape carrier with patterned Cu circuits and is employed

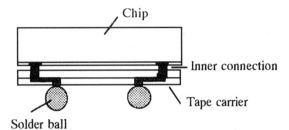

Figure 14.1 Cross-sectional view of Nitto Denko's MCSP structure.

as the package interposer. The latter is a thermoadhesive polyimide which serves as the die-attach material. The total thickness of ASMAT plus the die-attach TAPI is 53 μm. This package has a very simple structure, as shown in Fig. 14.1. The silicon chip is placed face down toward the flexible substrate. The first-level interconnection (or inner connection, to use Nitto Denko's term) is made by the metal bumps on the vias of the flex substrate. These interconnects are either Au bumps or Cu bumps with Au plating [2]. The bump height is at least 10 μm. The bond pads on the die are equipped with conventional Al metallization. No additional surface modification is needed. For the test element group (TEG), the chip has peripheral bond pads which are located 1 mm inside the chip edge. The joints are formed by thermocompression. The board-level interconnects of the MCSP are BGA solder balls. The ball diameter is 0.2 mm, and the pitch is 1.0 mm. This package can be encapsulated by either TAPI or conventional molding compound around the die. Since the encapsulation is very thin (100 μm if by TAPI), the package size is very close to the die size. Therefore, Nitto Denko's MCSP can be considered as a true chip size package which conforms to the dimension requirements of CSP.

14.3 Material Issues

The MCSP is a plastic-encapsulated package with flexible substrate. The major package materials include the polyimide tape carrier, the die-attach adhesive, the chip encapsulant, and the eutectic BGA solder balls. The flex interposer of the MCSP is Nitto Denko's proprietary material called ASMAT, as shown in Fig. 14.2. The base substrate is a polyimide tape; its material properties are given in Table 14.1. The thickness of this tape is 25 μm, including the die-attach adhesive film (10-μm-thick TAPI, to be discussed later) on one side. On the other side of the tape carrier is a layer of 18-μm-thick Cu traces which is patterned by photolithography. A 10-μm-thick overcoat layer made of the same polyimide as the substrate is deposited on top of the circuitry

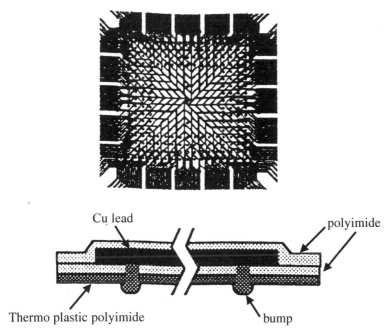

Figure 14.2   A 160-pin ASMAT with inner bumps (chip side at the bottom).

TABLE 14.1   Material Properties of Polyimide and Epoxy Encapsulant for Nitto Denko's MCSP

	Carrier tape PI	TAPI	Primer PI	Epoxy composite
Elastic modulus (kg/mm^2)	700	120	200	1450
Thermal expansivity (10^{-6}/°C)	20	98	50	19
Glass transition temperature (°C)	275	150	245	160

for protection and insulation. It should be noted that the polyimide used in ASMAT has a coefficient of thermal expansion (CTE) similar to that of Cu traces. Therefore, this flex interposer has very good dimensional stability under high temperature.

A number of through-hole vias are made by chemical etching on the tape carrier. These vias have a diameter of 60 μm and are filled with Au or Cu cores which are formed by electroplating. The bottom of these metal cores is connected with the Cu traces. The other end of the metal cores should be made at least 10 μm higher than the surface of the flex substrate to form inner bumps for chip-tape carrier interconnection. In addition, if Cu cores are used, the top of the cores must be plated with Au in order to ensure bondability to the die pads. A similar process needs to be applied to the overcoat layer at the other side of the Cu traces. The vias are filled with Au to form flat lands which are bond pads for BGA solder-ball attachment.

According to Nitto Denko, the MCSP has two types of encapsulation, as illustrated in Fig. 14.3. The first type is called interface encapsulation, and the second is chip encapsulation. The former is in fact a die-attach process, and the latter may be either lamination or molding. Since the inner bumps on the tape carrier are rather small, the gap between the chip and the substrate can be as narrow as 10 μm. As a result, it is very difficult to fill this gap after the chip-tape carrier joints are formed. Therefore, a precoated encapsulant on the substrate must be used. The main issue of concern with such an encapsulant is viscosity. On the one hand, a low-viscosity encapsulant will more easily form voids, which in turn affect the package reliability. On the other hand, excessive viscosity may result in poor wettability and, consequently, reduce the adhesion strength. The performance of a number of polymeric materials has been investigated, as shown in Fig. 14.4. It was determined that, in general, a polymer with viscosity higher than

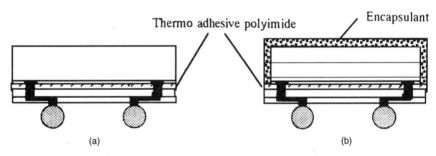

Figure 14.3 Two types of encapsulation for MCSP: (a) with interface encapsulation only; (b) with both interface and chip encapsulation.

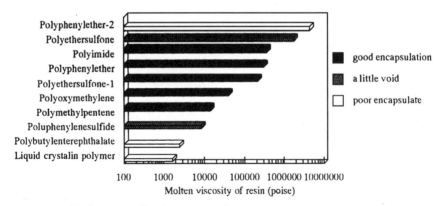

Figure 14.4 Performance of various encapsulants.

5000 poise should be used. After all, Nitto Denko developed a new thermoadhesive polyimide called TAPI for the interface encapsulation of MCSP. This material has a viscosity of 10,000 to 10,000,000 poise at 350°C, which is the temperature at which chip-level interconnects are formed by thermocompression. The other properties of TAPI are given in Table 14.1. It was reported that the viscosity of TAPI can be controlled while maintaining its adhesion strength. This is implemented by tuning the copolymerization of TAPI with organic silicon [1]. Currently diaminosilicone is used. The chemical structure of TAPI copolymerized with silicone is presented in Fig. 14.5.

In order to ensure the performance of TAPI, a mechanical peel test as shown in Fig. 14.6a was performed. It was found that delamina-

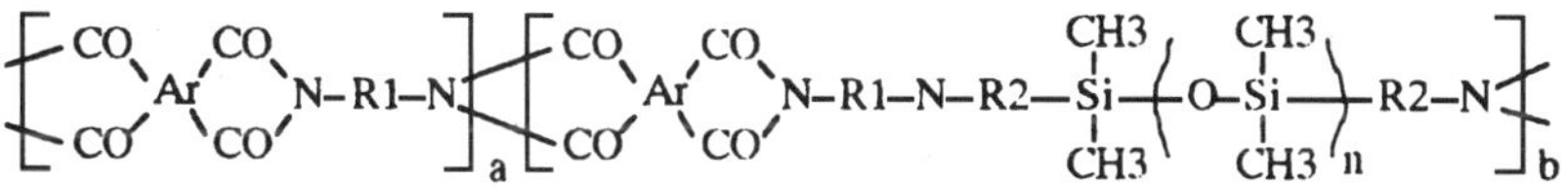

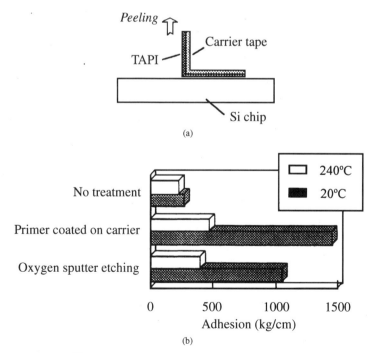

Ar: Aromatic tetra-carbonylacid   R1: Aromatic diamine containing oxygen   R2: Alkyl

Figure 14.5 Chemical structure of TAPI copolymerized with silicone.

Figure 14.6 Characterization of adhesion strength for TAPI: (a) peel test configuration; (b) comparison of performance.

tion may occur between TAPI and the tape carrier. Consequently, the adhesion strength was affected. Two attempts were made to improve this situation. The first approach was to perform oxygen sputter etching. The second method was to apply primer to the surface of the carrier tape. From the results shown in Fig. 14.6*b*, both treatments could improve the adhesion strength. However, it seems that the primer was more effective than the oxygen sputtering. The applied primer was a polyimide without silicone. After using the primer, the interfacial delamination was suppressed and turned into cohesive failure in the TAPI layer. Further investigation indicated that the composition of the primer could affect the bonding strength, as shown in Fig. 14.7. When the oxygen content in the primer is increased, the chemical conformation of the primer becomes more flexible and, hence, the performance improves. Currently the primer C is adopted for the MCSP. The material properties of primer C are given in Table 14.1.

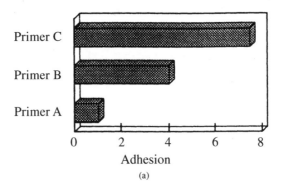

Figure 14.7 Comparison of various primers: (*a*) adhesion strength; (*b*) chemical structures.

For the chip encapsulation of the MCSP, both TAPI and epoxy composite can be used. The former is a 100-μm-thick polyimide sheet which is laminated onto the chip by thermocompression at 350°C. The latter is a regular molding compound implemented by conventional transfer molding. There are a few advantages associated with TAPI for chip encapsulation. Among them are mold-free process, reel-to-reel operation, less material waste, and high performance (heat stable). However, with this approach, voids form more easily inside the package. On the other hand, the epoxy composite has a CTE matching that of the tape carrier, as shown in Table 14.1. Therefore, the MCSP molded by epoxy resin has very little warpage (<20 μm), and this may compensate for the requirement for a mold.

14.4 Manufacturing Process

Because of the simple package structure, the fabrication of the MCSP is very straightforward. The manufacturing process begins with a polyimide tape carrier with Cu metallization (ASMAT). The Cu traces are patterned by photolithography and then covered by a thin polyimide overcoat for protection and insulation. Another thin layer of thermoadhesive polyimide (TAPI) is precoated on the side opposite to the circuitry for die attach. Primer should be used in this coating process in order to enhance the adhesion between the TAPI and the polyimide base tape. Through-hole vias with a diameter of 60 μm are made on the flex substrate by chemical etching. These vias are filled with Au or Cu cores by electroplating to form inner bumps for chip-tape carrier interconnection. If Cu cores are used, the top of the cores must be plated with Au to ensure bondability to the die pads. It should be noted that the aforementioned inner bumps must be at least 10 μm higher than the surface of the interposer. Detailed discussion of the bondability of inner bumps is given in Sec. 14.5.

Once the tape carrier has been prepared, the die is mounted with the active face of the silicon chip toward the flex substrate. This is a gang-bonding process performed with a flip-chip bonder at 350°C. The pulse-heating method is employed. The joints between the inner bumps on the substrate and the bond pads on the die are formed simultaneously with the bonding between the TAPI layer and the silicon chip (interface encapsulation, according to Nitto Denko's term).

Following the die mounting is the chip encapsulation. The MCSP may be encapsulated by either TAPI or conventional resin epoxy. The former is a lamination process at 350°C. A 100-μm-thick TAPI sheet is hot-pressed onto the back of the die, as illustrated in Fig. 14.8. In order to suppress the formation of voids, a soft backing material such as silicone rubber should be used during this thermocompression

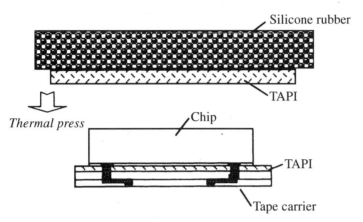

Figure 14.8 Fabrication of MCSP.

process. If an epoxy composite is used to encapsulate the MCSP, a regular transfer molding process is followed. Because of the CTE matching between the epoxy and the polyimide carrier tape, this approach will minimize the warpage of MCSP. The last step of the manufacturing process is solder-ball attachment. Via holes are prefabricated on the overcoat layer at the bottom of the polyimide tape carrier. These vias are filled with Au to form bond pads for BGA ball attachment.

14.5   Qualifications and Reliability

A series of qualification tests were performed to investigate the package reliability of the MCSP. For the study of moisture sensitivity, various configurations of the MCSP were subjected to 85°C/85 percent RH and a pressure cooker test (PCT). The specimens included the MCSP without any encapsulation, the MCSP with interface encapsulation (die mounting) only, and the MCSP with both interface and chip encapsulation. In addition, a 64-pin PQFP was tested for benchmarking. The ratio of connected electrodes (I/Os) was used as an index to evaluate the moisture sensitivity. A 100 percent ratio indicated that no interconnects were damaged due to the absorption of moisture. The experimental results are presented in Fig. 14.9. It is obvious that the encapsulation plays an important role for the package reliability. Without any encapsulation, the MCSP is very vulnerable. Although the MCSP with interface encapsulation only can survive an 85°C/85 percent RH environment, it is very fragile under PCT. As long as the package has both interface and chip encapsulation (either by TAPI lamination or by epoxy molding), the MCSP has the same reliability as the PQFP.

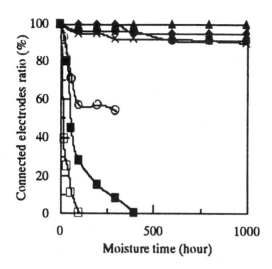

■ MCSP without any encapsulation (85/85)

● MCSP with interface encapsulation only (85/85)

▲ MCSP with interface encapsulation and transfer molding (85/85)

◆ MCSP with interface encapsulation and sheet molding (85/85)

✕ QFP-64 pins (85/85)

□ MCSP without any encapsulation only (PCT)

○ MCSP with interface encapsulation only (PCT)

Figure 14.9   Moisture sensitivity of MCSP.

An IR reflow qualification test was also performed for fully encapsulated MCSP specimens with various preconditioning. The test results are given in Table 14.2. Under 30°C/85 percent RH and 85°C/85 percent RH preconditioning, the MCSP has reasonable resistance to delamination, indicating that TAPI is an effective adhesive for die attach. However, with the PCT condition, cohesive failure is found in TAPI after the IR reflow test. This issue will need to be addressed in the future material development for MCSP.

The joint reliability of the inner bumps in the MCSP was investigated by a comprehensive experimental testing program [2]. An ASMAT substrate with 32 inner bumps (in two rows) was mounted on a silicon chip with unpatterned Al metallization by thermocompression and then subjected to mechanical peel loading. For each loading

TABLE 14.2   Package Qualification Data for Nitto Denko's MCSP

Preconditioning	Time (h)	After IR reflow
30°C/85% RH	200	Pass
30°C/85% RH	500	Pass
85°C/85% RH	200	Pass
85°C/85% RH	500	Delamination
PCT 121°C	200	Delamination

TABLE 14.3   Bonding Conditions for Inner-Bump Peel Strength Tests

Conditions	Values
Temperature (°C)	250, 300, 350, 400
Pressure (gf/bump)	50, 100, 200, 400, 800
Bonding time (s)	10 (fixed)

stroke, only 16 joints were broken. The ultimate loads were taken as the peel strength for comparison. Note that the TAPI die-attach adhesive was not used in this test. Various bonding conditions were investigated, as shown in Table 14.3. The tests were performed for specimens with different inner-bump heights. The results for inner bumps with all Au cores are presented in Fig. 14.10. It was found that for a bonding temperature of 350°C or above and a pressure of 200 g/bump or above, more consistent and acceptable peel strength could be achieved for bump heights of 30 μm and 50 μm. For a bump height of 10 μm, the peel strength was always unsatisfactory. Figure 14.11 shows the relative contribution of various bonding parameters to the bondability of inner bumps. It can be seen that the temperature is the most significant factor for inner-bump bonding. The bump height is an influential parameter as well. The bonding pressure is not as critical as the other two factors.

The peel strength for inner bumps with Cu cores and Au plating was evaluated as well. The bonding temperature was 400°C. Three bump heights with various Au plating thicknesses, as shown in Table 14.4, were investigated. The results are presented in Fig. 14.12. The strength values and the data trends are very similar to those observed for all Au inner bumps. In addition, the peel strength seemed to be insensitive to the thickness of the Au plating. Therefore, it may be concluded that the inner bumps with all Au and Cu/Au metallization have the same bondability as long as the appropriate bonding conditions

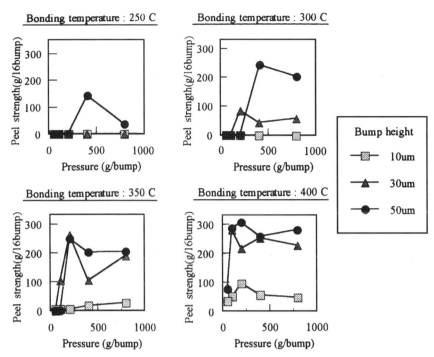

Figure 14.10   Peel strength test results at various bonding temperatures for inner bumps with all Au cores.

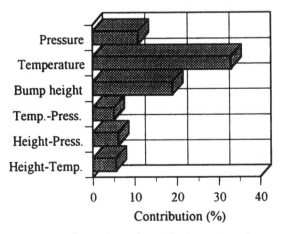

Figure 14.11   Comparison of contribution to inner-bump bondability.

TABLE 14.4   Au Plating Thickness for Various Inner-Bump Heights

Bump height (μm)	Gold plating thickness (μm)					
10	1	5	10	—	—	—
30	1	5	10	20	30	—
50	1	5	10	20	30	50

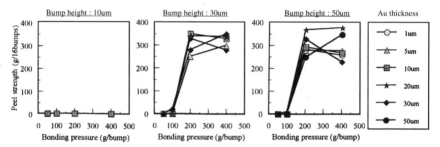

Figure 14.12   Peel strength test results at various bump heights for inner bumps with Cu cores and Au plating.

are applied and a sufficient bump height is maintained. Further study by Nitto Denko indicates that even with a bump height of 10 μm, satisfactory bondability may be achieved if certain measures such as bond pad scrubbing are taken. Detailed discussion on this subject is given in Ref. 2.

The MCSP has very small BGA solder balls, the diameter of which is close to that of flip-chip solder joints. Therefore, the board-level reliability of the MCSP is a main concern. A computational analysis was performed to investigate the solder joint reliability using the finite-element method. The package has a 15×15 mm chip. The solder-ball diameter and pitch are 0.2 mm and 1.0 mm, respectively. A uniform thermal loading of 220°C is applied. It is found that the most critical solder joint is at the corner of the package, as shown in Fig. 14.13. The maximum stress in this solder ball due to thermal mismatch between the package and the PCB is 104 MPa. A flip-chip assembly with all dimensions and conditions equal to those in the aforementioned analysis was investigated as well for comparison; the maximum thermal stress was 189 MPa. This result is reasonable because the flex substrate of the MCSP can relax stresses in the solder balls. Therefore, it may be concluded that the MCSP has better solder joint reliability than direct chip attach (DCA).

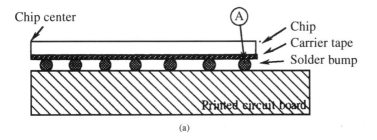

(a)

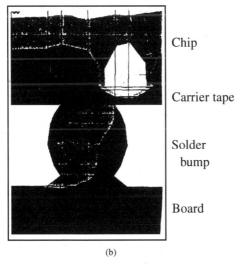

(A) enlargement scheme
[Thermal stress distribution]

Chip

Carrier tape

Solder
bump

Board

(b)

Figure 14.13 Schematic diagram of computational analysis for solder joint reliability: (a) position of most critical solder joint; (b) typical finite-element result.

14.6   Applications and Advantages

The MCSP is a flex-based chip scale package implemented with Nitto Denko's proprietary materials, namely, ASMAT and TAPI. So far there is no report on the targeted devices and applications. Judging from the package design, the MCSP has the advantages of simple structure, low profile, and compact size. This package has relatively small solder balls (0.2 mm) and a coarse ball pitch (1.0 mm). It should be designed for devices with low to medium pin counts.

The MCSP has certain unique features in fabrication. The inner bump joints and the die attach (interface encapsulation) are accomplished simultaneously by thermocompression. Also, the chip encapsulation may be performed by lamination using TAPI sheet. The whole manufacturing process allows reel-to-reel operation so that the production cost can be reduced. However, Nitto Denko did not address the material cost issue in its literature.

14.7 Summary and Concluding Remarks

The MCSP is a flex-based plastic-encapsulated chip scale package. This package features Nitto Denko's ASMAT as the interposer and TAPI as the bonding layer between the die and the interposer. The former is a PI/Cu tape carrier. The latter is a thermoadhesive polyimide for interface encapsulation (die attach) as well as chip encapsulation. The MCSP has a very simple package structure. The chip-level interconnection is made by Au or Cu/Au bumps preformed on the ASMAT substrate. The inner-bump joints and the die attach are accomplished simultaneously by gang bonding with a flip-chip bonder. The board-level interconnects are BGA solder balls. The MCSP can be encapsulated by either TAPI or conventional molding compound.

The MCSP has a very low profile and compact size. It can be considered as a true chip size package. Because of the relatively coarse ball pitch, this package should be designed for low- to medium-pin-count applications. From the results of qualification tests, if the MCSP is fully encapsulated, this package should have the same moisture resistance capability as the PQFP. However, the MCSP is somewhat susceptible to the IR reflow test with PCT preconditioning. The reliability of the inner bumps has been investigated comprehensively. It is concluded that the bondability is satisfactory if the appropriate bonding conditions are applied and a sufficient bump height is maintained. Although the MCSP has very small solder balls, a computational analysis revealed that this package has better solder joint reliability than the flip chip because of the existence of the flexible substrate. Nevertheless, experimental testing should be performed for verification, and the results need to be benchmarked with other CSPs instead of DCA.

14.8 References

1. S. Tanagawa, K. Igarashi, M. Nagasawa, and N. Yoshio, "The Resin Molded Chip Size Package (MCSP)," *Proceedings of the International Electronics Manufacturing Technology Symposium,* October 1995, Austin, Tex., pp. 410–415.
2. M. Nagasawa, S. Tanagawa, N. Yoshio, and K. Igarashi, "Inner Lead Bonding for a Resin Molded Chip Size Package," *Proceedings of the International Electronics Manufacturing Technology Symposium,* October 1996, pp. 386–392.

15

Sharp's Chip Scale Package

15.1 Introduction and Overview

Sharp Corporation developed a new flex-based chip scale package (Sharp's CSP) using the company's existing facilities for IC packaging [1]. This package uses a single-sided polyimide flex circuit as the interposer. The chip-level interconnects are Au wire bonds. The die is overmolded with epoxy encapsulant. The thickness of the molding compound over the die is only 0.6 mm. The package size is 1.0 mm larger than the chip size. The board-level interconnects are BGA solder balls. The ball diameter is 0.45 mm if the ball pitch is 1.0 mm or 0.35 mm if the ball pitch is 0.8 mm. Including the solder balls, the total package thickness is less than 1.2 mm. The major features of this package are ultra-low loop height (160 μm), very short (0.5 mm) wire bonds, and small vent/via holes on the flex substrate. Sharp's CSP was proven to be a very reliable package and has been in production since August 1996. The production volume in 1997 was about one to two million packages per month. The main applications are ASICs and memory devices for portable equipment. Currently there are more than 10 companies worldwide incorporating Sharp's CSPs into their products.

15.2 Design Concepts and Package Structure

Sharp's CSP is a compact BGA package with flexible interposer. The package structure is illustrated in Fig. 15.1. The substrate is a polyimide tape with Cu traces on one side. The metal circuit is bonded to the carrier tape by polyimide adhesive, and the die-attach area is coated with a heat-resisting insulation layer. These designs are im-

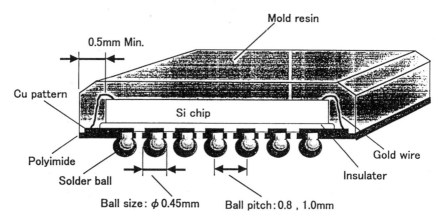

Figure 15.1   Package structure of Sharp's CSP.

plemented to improve the thermal performance of the package. The line/space width of the Cu trace pattern is 40 μm. With this design rule, at most three lines can pass between two solder pads for a 0.8-mm area-array pitch. In order to enable wire bonding for interconnection, the surface of the Cu traces is treated by a proprietary electroless plating process. To improve the popcorning resistance, small vent holes are made between the Cu lines on the flex substrate so that steam pressure cannot build up in the package. In addition, through-hole vias are prefabricated along the periphery of the carrier tape to enhance the bonding between the encapsulant and the flex substrate.

The first-level interconnects of Sharp's CSP are Au wire bonds. No modification on the die pads is required. With special capillary material and tooling, Sharp is able to perform wire bonds with loop height of 160 μm and short wire length of 0.5 mm (minimum 0.4 mm), as shown in Fig. 15.2 [2]. Note that the loop shape is trapezoidal to prevent edge contact. This configuration allows the flex substrate to be only 0.5 mm longer than the chip along each of the four edges. The die is overmolded by transfer molding. The encapsulant over the die is very thin (0.6 mm) so that the package warpage can be minimized. The second-level interconnects of Sharp's CSP are BGA solder balls. These balls are formed by placing solder spheres on the lands at the bottom of the flex substrate followed by reflow heating. The cross section of a solder ball is shown in Fig. 15.3. After solder-ball attachment, the total package thickness is less than 1.2 mm. There are two ball pitches, namely, 0.8 mm and 1.0 mm, for Sharp's CSP. The solder-ball diameter for the former is 0.35 mm, while that for the latter is 0.45 mm. A ball pitch of 0.5 mm is under development. In general, the area-array pattern is either full-grid or perimeter. However, for some

Figure 15.2 Short, low-loop wire bonds of Sharp's CSP.

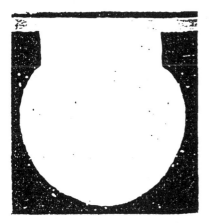

Figure 15.3 Cross section of a solder ball after attachment to the flexible substrate.

packages, such as memory devices with overhang, extra dummy balls may be arranged as illustrated in Fig. 15.4 in order to enhance the solder joint reliability in board assembly as well as to match the common land pattern on the PCB [3].

Sharp's CSP was developed for ICs with low to medium pin counts. The current lineup is 28 to 280 pins with a package size less than or equal to 16×16 mm, as presented in Table 15.1 and Fig. 15.5. The shape of the die can be either square or rectangular, but the wire-bond pads must be distributed along the periphery. From the data given in Table 15.1, the IC devices packaged by Sharp's CSP include

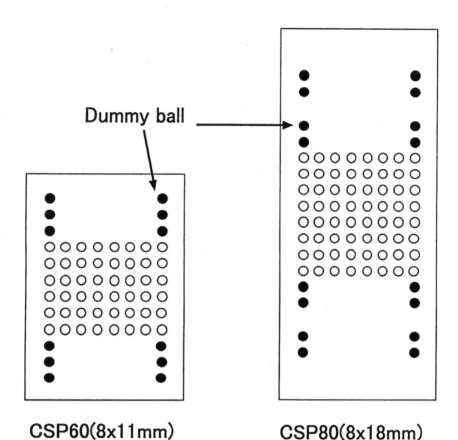

Dummy ball

CSP60(8x11mm)   CSP80(8x18mm)

Figure 15.4 BGA pattern with extra dummy balls for Sharp's CSP.

TABLE 15.1   Sharp's CSP Family

	Square type				
CSP size	6 × 6	8 × 8	10 × 10	12 × 12	16 × 16
Ball pitch (mm)	0.8	1.0 0.8	0.8	0.8	0.8
Pin counts	28	42 48	108	160 180	280
Device	SRAM	Flash memory	ASIC	ASIC	ASIC
Status ('97/6)	MP.	MP.	MP.	MP.	Dev.

	Rectangular type				
CSP size	6 × 6	6 × 10	8 × 10	8 × 11	9 × 15
Ball pitch (mm)	0.8	0.8	0.8	0.8	1.0
Pin counts	48	32	48	48 64	64
Device	Flash memory	SRAM	Flash memory	Flash memory	VRAM
Status ('97/6)	MP.	MP.	MP.	MP.	Dev.

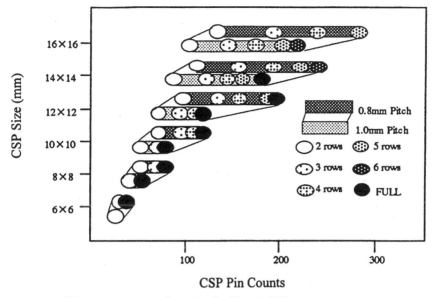

CSP Pin Counts

Figure 15.5 Pin count versus package size for Sharp's CSP.

flash memory, SRAMs, VRAMs, and ASICs. Most of them are in mass production already. Since 16M and 64M DRAMs have center bond pads, they do not appear in the product list of Sharp's CSP.

15.3 Material Issues

The CSP developed by Sharp is a plastic-encapsulated package with flexible interposer. The major package constituents in addition to the silicon chip include the gold bonding wires, the die-attach adhesive, the flexible substrate, the encapsulation molding compound (EMC), and the BGA solder balls. The interposer of this CSP is a single-sided flex circuit. The base film is a polyimide carrier tape. A Cu foil is bonded to one side of the carrier by polyimide adhesive and patterned to form the desired circuitry. Since wire bonds are the chip-level interconnects, the top of the Cu traces is treated with a proprietary electroless plating to provide an Au-wire-bondable surface. In addition, the die-mounting area is coated with a heat-resisting insulation layer to increase the thermal resistance of the flex substrate. Since popcorning is a major concern, small vent holes are built in on the carrier tape so that the steam pressure can be relieved during reflow heating. Furthermore, in order to enhance the bonding between the encapsulant and the flex substrate, through-hole vias are made along the periphery of the carrier tape.

TABLE 15.2   Comparison of Various Molding Compounds

Mold resin/ use	Resin type	Filler content (vol %)	Viscosity (175°C) (poise)	Spiral flow (175°C) (in)	Flexural strength (25°C) (kgf/mm²)	Flexural modulus (25°C) (kgf/mm²)
A/ CSP	Biphenyl + EOCN	α	β	γ	η	λ
B/ CSP	Biphenyl + EOCN	$\alpha - 4$	β	γ	$\eta - 1$	$\lambda - 180$
C/ CSP	Biphenyl	$\alpha + 13$	β	γ	η	$\lambda + 250$
D/ TSOP	Biphenyl	$\alpha + 4$	2β	0.8γ	η	$\lambda + 400$

Sharp's CSP is overmolded on only one side of the substrate. Because of the asymmetric package structure, warpage could be a critical issue. One way to minimize the warpage is to make the encapsulation as thin as possible. However, the thin encapsulant needs to be formed by a narrow mold passage, causing another concern about the viscosity of the molding compound. Therefore, the material properties of the encapsulant should be optimized. In order to identify the most suitable encapsulant, several molding materials were evaluated, as shown in Table 15.2. It was reported that a thin encapsulation of 0.6 mm could be achieved using mold resin A. Currently Sharp uses this encapsulant for molding its CSP [1]. However, considering the possible chip size variation, which could affect the package warpage, mold resin C may be used in the future [2].

Unlike most other BGA-type packages, Sharp's CSP does not use conventional 63Sn/37Pb for the solder balls. Instead, a precipitation-hardening solder which contains a small amount of Ag was adopted [1]. Because of the addition of silver, the elastic modulus and the tensile strength are increased. It is believed that this modification may improve the thermal fatigue resistance of the solder. Note that this material has a melting point similar to that of conventional eutectic solder. During the ball attachment for Sharp's CSP, a flux with proper viscosity and thixotropy needs to be transferred to the solder balls. A nitrogen gas environment is required for the reflow of these solder balls in order to reduce the oxidation.

15.4   Manufacturing Process

The package structure of Sharp's CSP is relatively simple, and the manufacturing process is quite straightforward. The flowchart is given in Fig. 15.6. Once the flex substrate has been prepared, the silicon chip is mounted on the Cu trace side of the interposer and wire bonding is performed. The next step is encapsulation by transfer

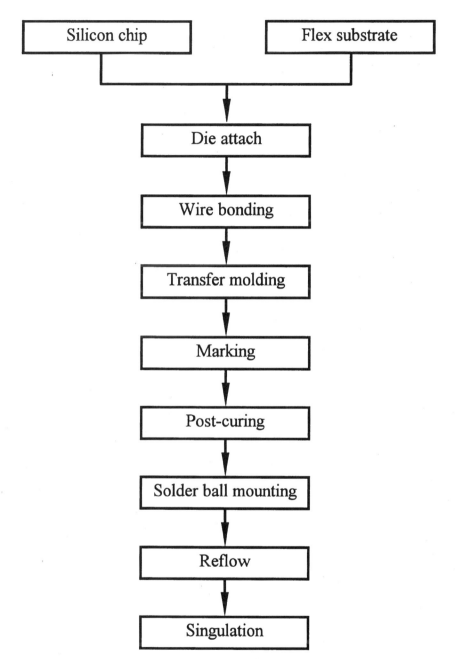

Figure 15.6 Manufacturing process for Sharp's CSP.

TABLE 15.3   Tolerance for Various Dimensions of the
CSP Outline

Item	Tolerance
Package outline	± 0.1 mm
Ball location A*	± 0.15 mm
Ball location B†	± 0.075 mm
Ball coplanality	≤ 0.1 mm
Package total height	< 1.2 mm

*Determine the central coordinates of the package.
†Determine the central coordinates of the entire ball's position.

molding. After marking and postcuring, solder spheres are attached to the lands on the bottom surface of the interposer and then reflowed to form BGA interconnects. It should be noted that in this ball-attachment process, a flux with proper viscosity and thixotropy needs to be transferred to the solder spheres before mounting. Also, in order to reduce the oxidation, a nitrogen gas environment is required during the reflow. According to Sharp, with the aforementioned procedures, a high yield can be achieved for a ball pitch of 0.8 mm.

The last step in fabrication is package singulation. A YAG laser is employed for this task in order to minimize the mechanical stress exerted on the CSP. It was reported that a processing precision of ± 50 μm could be achieved. The tolerance for various dimensions is given in Table 15.3 [2].

15.5   Performance and Package Qualifications

The electrical and thermal performance of Sharp's CSP have been evaluated by simulation. The inductance and thermal resistance θ_{ja} of a 42-pin flash memory device (CSP42) are given in Table 15.4 and benchmarked with the performance of another package (TSOP40). It was found that the inductance of Sharp's CSP is much lower than

TABLE 15.4   Electrical and Thermal Performance of Sharp's CSP

	Thermal resistance		Inductance (nH)
Package	Condition	θ_{ja} (°C/W)	(10 MHz)
CSP42-P-0808	Mounted on PCB	54.5	4.015
	Not mounted	148.4	
TSOP40-P-1010	Mounted on PCB	123.0	5.695
	Not mounted	126.0	

that of TSOP. This superior electrical performance is attributed to the rather short electrical passage in the CSP. With regard to thermal performance, it seems that the package thermal resistance of TSOP is slightly lower than that of the CSP. However, when the package is mounted on the PCB, the heat-dissipation capability of Sharp's CSP is much better [1].

The package reliability of Sharp's CSP has been verified by a series of qualification tests, as shown in Table 15.5. From the results summarized in Table 15.6, it may be concluded that this CSP has the same package reliability as conventional plastic packages. The detailed test data are given in Tables 15.7 to 15.12. In general, Sharp's CSP can be qualified as a JEDEC level 2 package, as shown in Table 15.7. The main reason this CSP can resist popcorning is the existence of vent holes on the flex substrate. A study was conducted to investigate the

TABLE 15.5   Package Qualification Tests for Sharp's CSP

Test item	Condition
1. Popcorn test	a. 30°C/70% RH × 120 h, 2 Reflows
	b. JEDEC level 2
2. PCT	121°C/100% RH × 300 h
3. HT/HH storage	60°C/90% RH × 1000 h
4. HT/HH operation	85°C/85% RH × 1000 h
5. HT operation	125°C × 1000 h
6. HT storage	150°C × 1000 h
7. LT storage	−65°C × 1000 h
8. Temperature cycling	−65°C ⇔ 150°C, 300 cycles

(JEDEC level 2: 85°C/60% RH × 168 h)

TABLE 15.6   Summary of Results for Package Qualification Tests

	Package	CSP28	CSP32	CSP48	CSP42	CSP80	CSP160	CSP176
	Size (mm)	6 × 6	6 × 10	6 × 8	8 × 8	8 × 8	12 × 12	12 × 12
T	1a.	0/11	0/10	0/10	0/22	0/22	0/32	0/10
e	1b.	—	0/10	0/10	—	—	—	0/10
s	2.	0/22	0/22	0/15	0/22	0/22	0/22	0/22
t	3.	0/22	—	—	—	0/22	0/20	In test
	4.	0/22	0/45	0/50	—	0/22	—	In test
i	5.	0/45	0/76	0/100	—	0/45	—	In test
t	6.	0/11	—	—	—	0/11	—	In test
e	7.	0/11	—	—	—	0/11	—	In test
m	8.	0/22	0/15	0/20	0/22	0/22	0/22	0/20

TABLE 15.7   Qualification Data for Popcorn Tests

Condition	Package	Size (mm × mm)	Devices	First reflow	Second reflow	Third reflow
	CSP42	8 × 8	8M flash	0/22	0/22	—
A	CSP42	8 × 8	4M flash	0/10	0/10	—
	CSP80	8 × 8	Microprocessor	0/22	0/22	—
	CSP108	10 × 10	Gate array	0/10	0/10	—
B	CSP160	12 × 12	ASIC	0/32	0/32	—
	CSP160	12 × 12	ASIC	0/32	0/32	—
C	CSP42	8 × 8	4M flash	0/18	0/18	—
D	CSP42	8 × 8	4M flash	0/30	0/30	0/30

Conditions:
A: (30°C/70% RH × 120 h   →   240°C maximum reflow) × 2 times
B: (30°C/70% RH × 96 h   →   230°C maximum reflow) × 2 times
C: (35°C/85% RH × 192 h   →   240°C maximum reflow) × 2 times
D: (85°C/65% RH × 192 h)   →   (220°C maximum for 10 s minimum and reflow 3 times)

TABLE 15.8   Effect of Number of Vent Holes on Popcorn Resistance

Package	Number of vent holes	Popcorn or delamination		
		Assembly complete	After first reflow	After second reflow
	15	0/11	0/11	0/11
CSP28-P-0606	12–14	0/20	0/20	0/20
	9–11	0/20	0/20	0/20
	6–8	0/14	0/14	1/14
	3–5	1/11	3/11	4/11

effect of the number of vent holes, and the results are presented in Table 15.8 [2]. It is clear that more vent holes definitely can improve the capability to resist popcorning and delamination.

Table 15.9 shows the qualification data from the pressure cooker test (PCT). The testing conditions were 121°C, 100 percent RH, and 2 ATM. All Sharp's CSPs could sustain the imposed environment for at least 300 h. In particular, the CSP100 package could go through 500 h of PCT without any failure. The testing data for storage under high temperature (60°C) and high humidity (90 percent RH) are given in Table 15.10, and the qualification results of temperature (85°C), humidity (85 percent RH), and bias (V_{cc} max.) tests (THB) are presented in Table 15.11. All specimens could survive for at least 1000 h. Such package reliability performance is considered outstanding. The last test of package reliability is the temperature cycling (TC) test. The

TABLE 15.9   Qualification Data for Pressure Cooker Test

Package	Size (mm × mm)	Devices	100 h	200 h	300 h	400 h	500 h
CSP28	6 × 6	256K SRAM	0/22	0/22	0/22	—	—
CSP32	7 × 13	1M SRAM	0/22	0/22	0/22	—	—
CSP42	8 × 8	8M flash	0/22	0/22	0/22	—	—
CSP80	8 × 8	Microprocessor	0/22	0/22	0/22	—	—
CSP100	12 × 12	TEG	0/10	0/10	0/10	0/10	0/10
CSP160	12 × 12	TEG	0/20	0/20	0/20	—	—
CSP160	12 × 12	ASIC	0/22	0/22	0/22	—	—

Condition: 121°C/100% RH.

TABLE 15.10   Qualification Data for High-Temperature and High-Humidity Storage

Package	Size (mm × mm)	Devices	240 h	500 h	1000 h
CSP28	6 × 6	256K SRAM	0/22	0/22	0/22
CSP32	7 × 13	1M SRAM	0/14	0/14	0/14
CSP80	8 × 8	Microprocessor	0/22	0/22	0/22
CSP160	12 × 12	ASIC	0/20	0/20	0/20

Condition: 60°C/90% RH.

TABLE 15.11   Qualification Data for Temperature, Humidity, and Bias Test

Package	Size (mm × mm)	Devices	Voltage	240 h	500 h	1000 h
CSP32	7 × 13	1M SRAM	3.3 V	0/22	0/22	0/22
CSP80	8 × 8	Microprocessor	5.5 V	0/22	0/22	0/22

Condition: 85°C/85% RH.

TABLE 15.12   Qualification Data for Temperature Cycling Test

Package	Size (mm × mm)	Devices	100 cycles	200 cycles	300 cycles	400 cycles	500 cycles
CSP32	7 × 13	1M SRAM	0/22	0/22	0/22	0/22	0/22
CSP42	8 × 8	4M flash	0/22	0/22	0/22	—	—
CSP42	8 × 8	8M flash	0/22	0/22	0/22	—	—
CSP80	8 × 8	Microprocessor	0/22	0/22	0/22	—	—
CSP100	12 × 12	TEG	0/10	0/10	0/10	0/10	0/10
CSP160	12 × 12	ASIC	0/22	0/22	0/22	—	—
CSP160	12 × 12	TEG	0/20	0/20	0/20	—	—
CSP160	12 × 12	TEG	0/30	0/30	0/30	—	—

Condition: −65 to 150°C/cycle/h.

testing condition is cycling between −65 and 150°C with 20 min of dwell time at each temperature. The qualification data are given in Table 15.12. No open circuits or bulging was observed after 300 cycles (two CSPs even lasted to 500 cycles). According to Sharp, such performance is acceptable for these products [1].

15.6 Solder Joint Reliability

The solder joint reliability of Sharp's CSP has been investigated comprehensively. Table 15.13 shows the results of a ball shear test before the CSP is assembled to the PCB [1]. The data distribution for a typical case is presented in Fig. 15.7 [2]. All failures occurred in a cohesive mode. Therefore, the strength of the joint between the solder ball and the bond pad on the interposer has been verified.

The surface-mounting quality of Sharp's CSP was evaluated through two items, namely, solder joint defects and self-alignment capability [2]. In both cases, a CSP160 package with ball diameter of 0.35 mm and ball pitch of 0.8 mm was investigated. Various sizes of

TABLE 15.13 Results of Ball Shear Tests

		Ball shear strength (N/ball)		
Package	Ball size	Average	Maximum	Minimum
CSP42-P-0808	0.45 mm	6.69	7.99	5.75
CSP108-P-1010	0.35 mm	4.70	5.39	4.17
CSP160-P-1212	0.35 mm	4.81	5.49	4.02

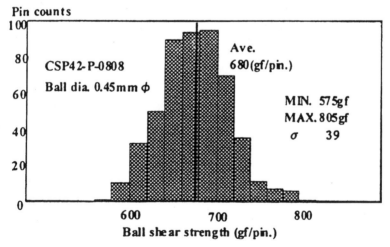

Figure 15.7 Data distribution for solder-ball shear test.

TABLE 15.14   Evaluation for Solder Joint Defects

Diameter (mm)			Evaluation items	
Land	Solder mask	Stencil	Open	Bridge
0.23	0.33	0.28	0/64	0/64
0.30	0.40	0.28	0/72	0/72
0.38	0.48	0.28	0/72	0/72
0.45	0.55	0.35	0/64	0/64

TABLE 15.15   Evaluation for Self-Alignment Capability

(200-μm shift to X direction from mount center)

Diameter (mm)			Mounting method (recognition)	
Land	Solder mask	Stencil	Balls	Body
0.23	0.33	0.28	0/8	0/8
0.30	0.40	0.28	0/8	0/8
0.38	0.48	0.28	0/8	0/8
0.45	0.55	0.35	0/8	0/8

TABLE 15.16   Characterization for Solder Joint Standoff Height

		Standoff (mm)	
Pitch	Ball diameter	Before mounting	After mounting
1.0 mm	0.45 mm	0.33–0.35	0.28–0.30
0.8 mm	0.35 mm	0.24–0.26	0.21–0.23

solder land, solder mask, and stencil opening were evaluated. The thickness of the stencil was 150 μm (6 mil). The results are given in Tables 15.14 and 15.15. In all cases, no opening or bridging of solder joints was observed. In addition, it was determined that a 200-μm off-set in mounting position could be recovered by the self-alignment effect. The standoff height of the solder joint was evaluated as well. The results of measurement are given in Table 15.16. The cross section of a typical solder joint is presented in Fig. 15.8.

The board-level reliability of Sharp's CSPs was evaluated by thermal cycling between −40 and 125°C [2]. An open circuit was identified by an increase of 1 kΩ in electrical resistance. The qualification data are given in Table 15.17. It can be seen that all but one of the specimens could sustain more than 500 temperature cycles. In the failed case, a crack could be observed, as shown in the cross-sectional picture in Fig. 15.9.

In order to identify the factors influencing solder joint reliability, a

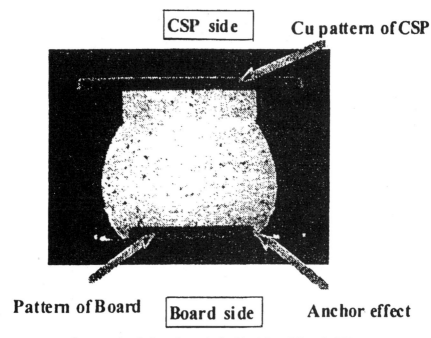

| CSP side | Cu pattern of CSP

Pattern of Board | Board side | Anchor effect

Figure 15.8 Cross-sectional view of a typical solder joint of Sharp's CSP.

TABLE 15.17 Thermal Cycling Test Data for Board-Level Reliability

Package	Size (mm × mm)	Ball (mm)	200 cycles	300 cycles	400 cycles	500 cycles	600 cycles	700 cycles	800 cycles	900 cycles
CSP42	8 × 8	0.45	← ————————— 0/10 ————————— → —
CSP48	6 × 8	0.45	← ————————— 0/20 ————————— →
CSP108	10 × 10	0.35	← ————————— 0/5 ————————— →
CSP160	12 × 12	0.35	← —— 0/5 ——→ 1/5	—	—	—	—
CSP180	12 × 12	0.45	← ——————— 0/22 ——→ ← ——————— In test ——————→

series of thermal cycling tests was performed to investigate various design parameters [3]. Figure 15.10 shows the effect of chip size. It is clear that the thermal fatigue life of solder joints decreases when the chip size becomes larger. This result is reasonable and can be explained by the DNP (distance from the neutral point) effect [4]. The effect of solder-ball size is presented in Fig. 15.11. It can be seen that large solder balls can substantially enhance the solder joint reliability. Therefore, Sharp plans to use 0.45-mm solder balls for all its CSPs

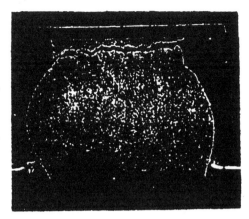

Figure 15.9 Failure mode of solder joint after thermal cycling test.

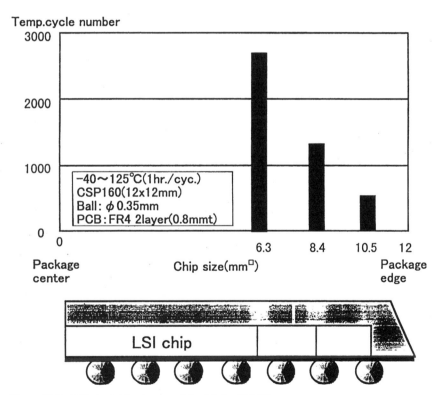

Temp.cycle number

-40~125°C(1hr./cyc.)
CSP160(12x12mm)
Ball: φ0.35mm
PCB:FR4 2layer(0.8mmt)

Package center   Chip size(mm□)   Package edge

LSI chip

Figure 15.10   Effect of chip size on solder joint reliability.

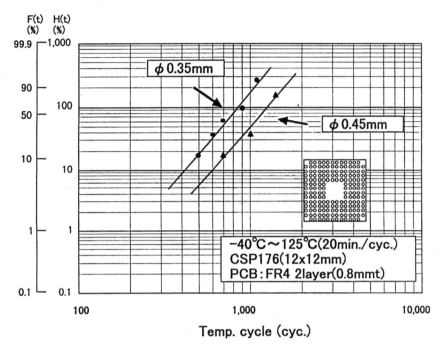

Figure 15.11   Effect of solder-ball size on solder joint reliability.

[3]. Figures 15.12 and 15.13 illustrate the effectiveness of dummy balls. With extra solder joints, the thermal fatigue life is certainly extended.

Figure 15.14 shows the influence of on-board land diameter. It seems that a smaller land diameter could improve the thermal fatigue life. This phenomenon may be attributed to the variation in the solder joint standoff height due to the change of land diameter. The effect of the mounting configuration on the PCB is presented in Fig. 15.15. The test results indicated that better solder joint reliability could be achieved by avoiding double-side mounting. This is because bending of the assembly board can relieve the thermal mismatch between the CSP and the PCB. It was also found that the configuration of the PCB may affect the solder joint reliability. Figure 15.16 shows that a two-layer PCB is better than a four-layer PCB with the same thickness. A possible reason for this phenomenon is that the CTE of the former is smaller than that of the latter. Consequently, the thermal mismatch between the CSP and the PCB is reduced. In addition to the design parameters, the testing temperature profile has some influence on the thermal fatigue life. Figure 15.17 presents the results of thermal cycling with two different testing periods per cycle. It is observed that

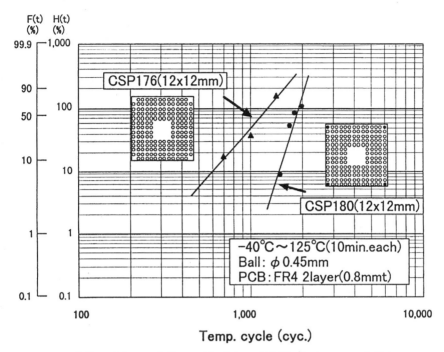

Figure 15.12 Effect of dummy ball on solder joint reliability (extra corner balls).

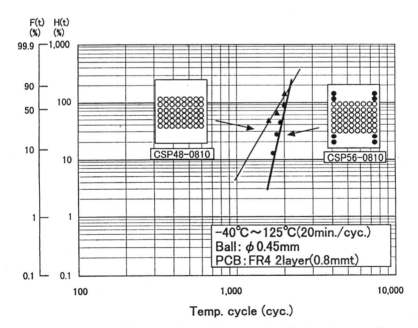

Figure 15.13 Effect of dummy ball on solder joint reliability (overhang configuration).

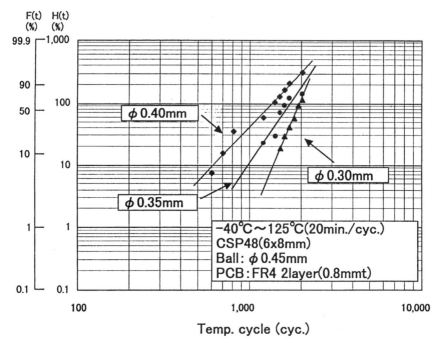

Figure 15.14   Effect of on-board land diameter on solder joint reliability.

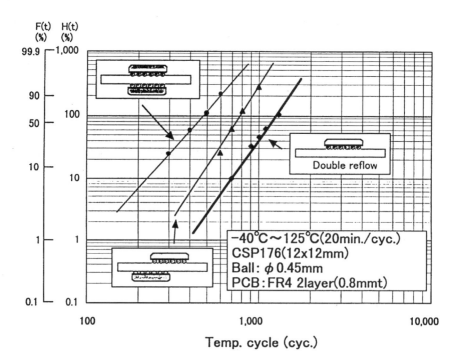

Figure 15.15   Effect of mounting configuration on solder joint reliability.

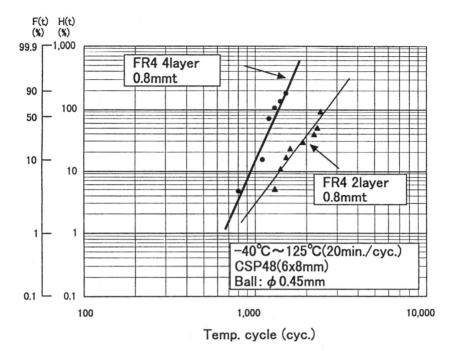

Figure 15.16   Effect of PCB configuration on solder joint reliability.

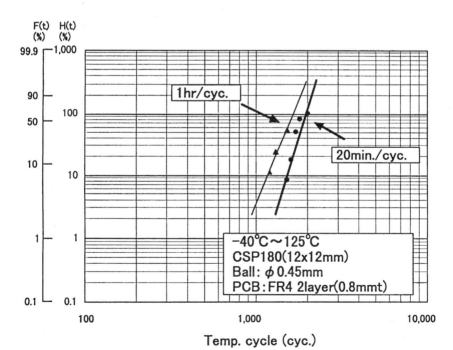

Figure 15.17   Effect of testing condition on solder joint fatigue cycles.

the longer period leads to fewer fatigue cycles. This is reasonable, since with a longer period, more inelastic deformation is accumulated per cycle. However, it should be noted that fewer fatigue cycles do not imply a shorter lifetime [5]. In fact, the total testing time required for the temperature with a 20-min period per cycle is shorter than that for the other profile.

Since the major applications of Sharp's CSP are for portable equipment, the solder joint reliability under mechanical loading such as bending and impact is as critical as that under thermal cycling. A series of mechanical tests, as shown in Fig. 15.18, was conducted to evaluate the board-level reliability of Sharp's CSP when it was subjected to various loading conditions. A daisy chain test element group (TEG) was used to detect the opening of solder joints during each test. The bending test evaluated performance under the mechanical loading that might be incurred during manufacturing and handling. The CSP-PCB assembly was installed on a three-point bending test rig. Two ends of the PCB were simply supported and subjected to repeated stroke in one direction. The results of the bending tests are given in Table 15.18. No failure was observed in any specimen.

In addition to these testing results, some parametric studies were conducted to investigate the effects of various factors on the board bending test under a fixed stroke. Figure 15.19 presents the influence of dummy balls on the bending fatigue life. It is obvious that the dummy balls could enhance the solder joint reliability under mechanical loading. The effect of PCB features is shown in Fig. 15.20. In general, the preflux land finishing, the buildup structure, and the via

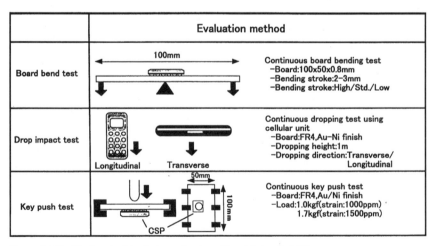

Figure 15.18   Mechanical tests to evaluate solder joint reliability.

TABLE 15.18   Experimental Results of Board Bending Tests

Package	Size (mm × mm)	Ball pitch/ diameter (mm)	Bending direction	Bending depth	Result 50 cycles	100 cycles
CSP42	8 × 8			2 mm	0/20	0/20
			+	3 mm	0/20	—
		1.0/0.45		4 mm	0/20	—
			−	4 mm	0/20	—
CSP64	9 × 15		+	2 mm	—	0/5
CSP160	12 × 12	0.8/0.35	+	2 mm	—	0/10

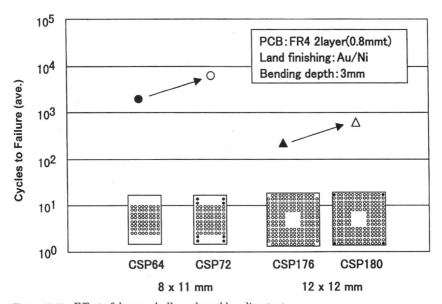

Figure 15.19   Effect of dummy balls on board bending test.

hole on land can improve the solder joint reliability. Figure 15.21 shows the bending fatigue life of solder joints subjected to various periods of high-temperature aging before mounting. However, no obvious effect could be identified.

The drop impact test simulates the loading from the accidental falling of portable equipment. The CSP-PCB assembly was fixed in the housing of a cellular phone. The whole specimen was repeatedly dropped from a height of 1.0 m with two orientations, namely, longitudinal and transverse. A strain gauge was installed on the PCB underneath the CSP mounting site to monitor the strain during loading. The results of the drop impact test for a CSP48 package are given in

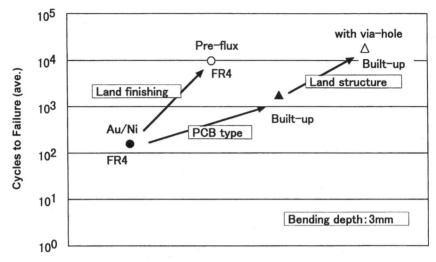

Figure 15.20   Effect of PCB features on board bending test.

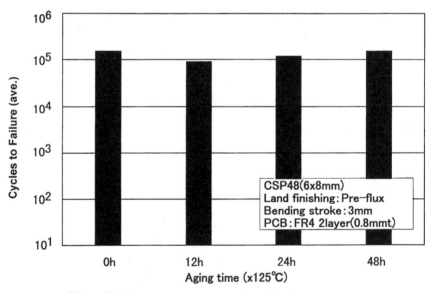

Figure 15.21   Effect of high-temperature aging on board bending test.

Fig. 15.22. The data from corresponding bending tests with various loading levels and strain rates are also presented for comparison. It was found that, compared to a longitudinal impact, a transverse impact could induce a larger strain on the PCB, resulting in shorter fatigue life. Further investigation by bending tests indicated that, along

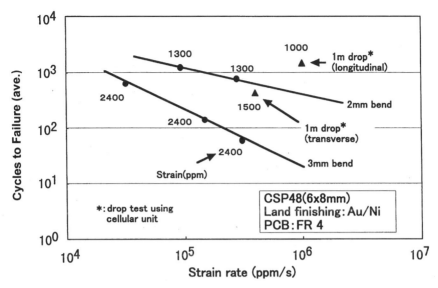

Figure 15.22   Experimental results of drop impact test.

with loading levels, the strain rate played a significant role in the fatigue life of solder joints. In general, for the same strain level, a higher loading rate leads to shorter fatigue life.

It is quite usual to mount electronic components on the back side of keypads in cellular phones. To investigate the effect of key-push loading on solder joint reliability, a key-push test was performed, as shown in Fig. 15.18. The results, given in Fig. 15.23, showed that a CSP64 assembly could sustain about 10,000 keystrokes. If extra dummy balls were added, the fatigue life could be extended to two times that or even one order of magnitude longer, depending upon the loading level.

15.7   Applications and Advantages

Sharp's CSP is a compact-size (chip size + 1.0 mm), low-profile (<1.2 mm), and fine-pitch (0.8 mm/1.0 mm) BGA package. Currently the CSPs in this family have a package size less than or equal to 16×16 mm. The package pin count ranges from 28 to 280. The packaged chip should have peripheral bond pads and may be square or rectangular in shape. Sharp's CSPs have been adopted for IC devices including flash memory, SRAMs, VRAMs, and ASICs, as shown in Table 15.19. Most of them are used in portable equipment such as camcorders and cellular phones. Since 16M and 64M DRAMs have center bond pads, they do not appear in the list of products using Sharp's CSP. This

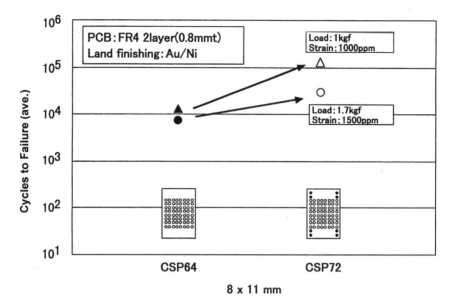

Figure 15.23 Experimental results of key-push test.

TABLE 15.19 Applications and Advantages of Sharp's CSP

Device	I/O	Size (mm × mm)	Pitch (mm)	Product	Conventional package	Ratio CSP/conventional package Area	Weight
256K SRAM	28	6 × 6	0.8	Cellular phone	SOP28	—	0.08
1M SRAM	32	6 × 10	—	Cellular phone	TSOP32	0.38	0.26
4M flash 8M flash	42	8 × 8	1.0	Cellular phone	TSOP40	0.32	0.24
4M flash 8M flash	48	6 × 8	0.8	—	TSOP40	0.24	0.24
Video RAM	64	9 × 15	1.0	Camcorder	SSOP64	0.36	0.14
ASIC	108	10 × 10	0.8	Cellular phone	QFP100	0.51	0.27
ASIC	176	12 × 12	0.8	Camcorder	QFP176	0.21	0.09

family of CSPs has been in mass production since August 1996. The production volume in 1997 was about one to two million packages per month. Currently more than 10 companies have adopted Sharp's CSPs for their products.

The major advantages of Sharp's CSP are small form factors and high reliability. The ratios of this CSP's area and weight to those of the conventional package are given in Table 15.19. It is clear that a great reduction in mounting area and package weight can be achieved with Sharp's CSP. The package reliability and solder joint reliability of this CSP have been evaluated by comprehensive qualification tests, as presented in the previous section. Good performance in both areas has been verified. It may be concluded that Sharp's CSP is as reliable as the conventional plastic packages.

15.8   Summary and Concluding Remarks

Sharp's CSP is a compact BGA package with flexible substrate. This package uses a polyimide carrier tape with single-side Cu traces as the interposer. The chip-level interconnects are Au wire bonds. No modification is required for the bond pads on the die. The silicon chip is overmolded with a thin layer of epoxy encapsulant. The board-level interconnects are solder balls with 0.8- or 1.0-mm area-array pitch. The package size is very compact (chip size + 1.0 mm). Including the solder balls, the total package thickness is less than 1.2 mm. The current family of this CSP has pin counts ranging from 28 to 280. The package size is less than or equal to 16×16 mm. The IC devices packaged by this CSP include flash memory, SRAMs, VRAMs, and ASICs.

The major features of Sharp's CSP are low loop height, short wire bonds, and through-hole vias/vents on the flex substrate. The former lead to small form factors, while the latter enhance the package's resistance to damage. Sharp's CSP has been proven to be a very reliable package. This series of CSPs appeared in the commercial market in August 1996 and grew to a production volume of one to two million packages per month in 1997. The main applications of this package are for devices in portable equipment. Close to a dozen companies worldwide are using Sharp's CSPs in their products.

15.9   References

1. Y. Yamaji, H. Juso, Y. Ohara, Y. Matsune, K. Miyata, Y. Sota, A. Narai, T. Kimura, K. Fujita, and M. Kada, "Development of Highly Reliable CSP," *Proceedings of the 47th ECTC,* San Jose, Calif., May 1997, pp. 1022–1028.
2. T. Kimura, Y. Yamaji, H. Juso, Y. Ohara, Y. Matsune, K. Miyata, Y. Sota, A. Narai, T. Kimura, K. Fujita, and M. Kada, "CSP Packaging and Mounting Technologies for Mobile Apparatus," *Proceedings of the International Symposium on Microelectronics,* Philadelphia, October 1997, pp. 256–261.
3. H. Juso, Y. Yamaji, T. Kimura, K. Fuhita, and M. Kada, "Board Level Reliability of CSP," *Proceedings of the 48th ECTC,* Bellevue, Wash., May 1998, pp. 525–531.
4. J. H. Lau and Y. H. Pao, *Solder Joint Reliability of BGA, CSP, Flip Chip, and Fine Pitch SMT Assemblies,* McGraw-Hill, New York, 1997.
5. S-W. R. Lee and X. Zhang, "Optimization of Temperature Profile on the Thermal Cycling Test for BGA Solder Joint Reliability," *Journal of Surface Mount Technology,* vol. 10, no. 4, 1997, pp. 9–17.

Chapter

16

Tessera's Micro-Ball Grid Array (μBGA)

16.1 Introduction and Overview

Tessera's micro-ball grid array (μBGA) is probably the most publicized chip scale package on the market [1]. This package belongs to the category of CSPs with flexible interposer. The unique features of the μBGA are the ribbonlike flexible leads for chip-level interconnection and the compliant elastomer between the interposer and the chip to relieve the thermal mismatch. The leads are bonded to the Al bond pads on the die one at a time by a thermosonic bonder. The bonding location is not restricted to the perimeter and may be anywhere within the die. Encapsulation is required to protect the bonded leads. The package terminals of the μBGA may be plated bumps, solder balls, or solid-core metal spheres. The array pitch may be 0.5, 0.75, or 1.0 mm, and the number of I/Os ranges from 46 to 300. The major application of μBGA packages is for memory devices, especially DRAMs and flash memory. Currently Tessera has high-volume production lines in California and Singapore to fabricate μBGA packages. Several major semiconductor manufacturers have adopted or licensed this CSP technology from Tessera, and many other companies are in collaboration with Tessera to develop or improve materials and processes for the μBGA.

16.2 Design Concepts and Package Structure

The μBGA package developed by Tessera is a CSP with flexible substrate. The basic package structure is illustrated in Fig. 16.1. The package interposer is a 25-μm-thick polyimide film with double-sided

Flex Circuit with Plated Contacts

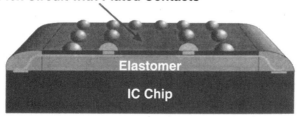

Figure 16.1 Sectional view of the μBGA package.

Cu. One side of the Cu serves as a ground plane, while the other side has signal traces for I/O redistribution. Between the chip and the substrate, there is a layer of silicone elastomer. This compliant layer has a thickness of 150 μm and serves three major purposes: It acts as a buffering zone to relieve the stresses due to thermal mismatch between the chip and the substrate, it protects the IC from the attack of α particles which may be emitted from the solder, and it provides the compressibility for testing the package in a socket [2].

The first-level interconnects of the μBGA are flexible ribbons which are bonded on the Al die pads by a single-shot thermosonic process. The bonding is performed one lead at a time and may be implemented by a modified wire bonder such as the K&S 1484 or 1488. The ribbons are 25-μm-wide soft Au leads with a thickness of 20 to 25 μm. The practical limit to the bond pitch may be as low as 45 μm. The bonded leads are in a lazy "S" shape (Fig. 16.2) so that they may accommodate any deformation due to thermal expansion. It should be noted that the bonding of these flexible ribbon leads is not restricted to the peripheral die pads; they may be located anywhere within the chip. In order to protect the bonded leads, encapsulation is required for μBGA packages. The encapsulant is dispensed from the back side (between the chip and the interposer) after the lead bonding is completed.

The package terminals of the μBGA may be plated Ni/Au bumps, solder balls, or solid-core metal spheres clad with solder. The array pitch may be 0.5, 0.75, or 1.0 mm, and the number of I/Os ranges from 46 to 300. Figure 16.3 shows the bottom view of a 0.5-mm-pitch μBGA with 188 I/Os. This is a daisy chain test module with depopulated package terminals. Since this picture was taken prior to encapsulation, the bonded ribbon leads can be clearly observed along the perimeter of the package. Figure 16.4 presents a 0.75-mm-pitch μBGA with 46 I/Os. This is a standard module which has (almost) fully populated package terminals [3].

The inherent structure of the μBGA makes it a very slim and compact package. The nominal package thickness is 0.7 mm, which is much thinner than most other CSPs. With the fan-in configuration,

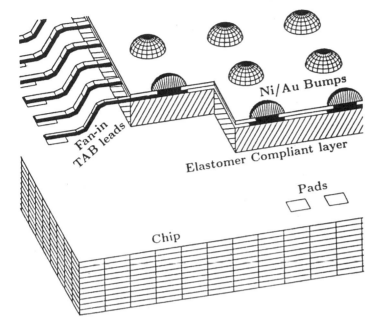

Figure 16.2 Anatomy of a package structure for the μBGA.

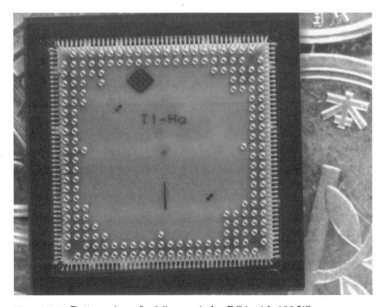

Figure 16.3 Bottom view of a 0.5-mm-pitch μBGA with 188 I/Os.

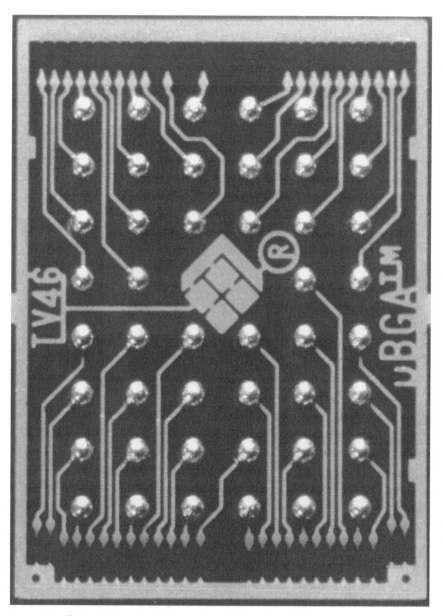

Figure 16.4   Bottom view of a 0.75-mm-pitch μBGA with 46 I/Os.

the footprint of μBGA is actually smaller than the chip size. As a result, the μBGA can be considered a true chip size package. It should be noted that this package has other variations in configuration, as shown in Fig. 16.5. Therefore, the package-to-chip size ratio of some μBGA variations may not comply with the strict definition of CSPs.

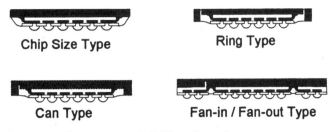

Chip Size Type   **Ring Type**

Can Type   **Fan-in / Fan-out Type**

Figure 16.5   Various types of μBGA configurations.

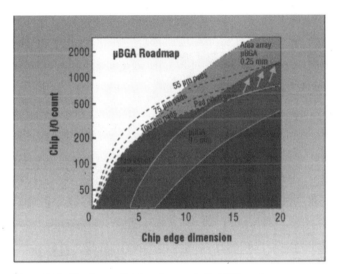

Figure 16.6   Road map for development of μBGA packages.

However, the fine-pitch I/Os can always qualify μBGAs as chip scale packages. Figure 16.6 illustrates the road map for the development of μBGA packages. From this figure, it can be seen that the family of μBGA packages is mainly for IC devices with low to medium pin counts [4].

16.3   Material Issues

Because of the existence of the polyimide film, the μBGA package is classified as a CSP with flexible interposer. The packaging materials in addition to the silicon chip include the ribbon leads, the compliant layer, the flexible substrate, the encapsulant, and the package terminals. To ensure the highest reliability for chip-level interconnection, the ribbon leads of μBGA are made of high-purity (99.99 percent) Au. The soft Au leads can be easily bonded to the Al die pads by a modified wire bonder. However, since the material cost is rather high, new efforts are under way to develop Au-plated Cu leads to replace them. The compliant layer of the μBGA is made of silicone elastomer. This is a high-temperature elastomeric material filled with 50 percent of pure silica. As illustrated in Fig. 16.7, the major function of the S-shaped ribbon leads and the compliant layer is to relieve the stresses due to the thermal mismatch between the silicon chip and the substrate. As well as the intrinsic material properties, as indicated in Fig. 16.8, the thickness of the compliant layer may influence the effectiveness of this function as well. In general, a thicker elastomer yields a higher effective CTE but lower package stiffness.

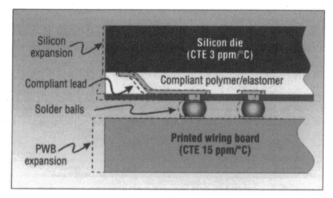

Figure 16.7 Compliant elastomer and flexible lead to relieve the stress caused by thermal expansion mismatch.

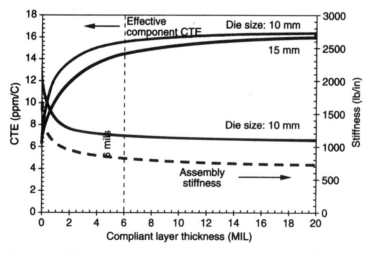

Figure 16.8 Effect of compliant-layer thickness on the effective CTE of the μBGA.

Consequently, more stresses due to thermal mismatch may be relieved [5].

The interposer of the μBGA is a polyimide film with two layers of Cu. One side of the Cu serves as a ground plane, while the other side forms signal traces for I/O redistribution. The 5-μm-thick ground plane establishes a controlled impedance of 65 Ω for a trace width of 25 μm over the 25-μm-thick polyimide dielectric layer. In addition to controlling the impedance, the ground plane provides a degree of shielding for the circuitry on the chip from both the signal traces and the substrate. The encapsulant of the μBGA is a silicone material. Its main functions are to protect the bonded ribbon leads and to improve the package reliability. The package terminals of the μBGA may be plated bumps, solder balls, or solid-core metal spheres. For plated bumps, the metallurgy is electroplated Ni coated with Au flash. The Ni bump height is 85 to 90 μm, and the thickness of the Au flash is 0.3 to 0.5 μm. For solder-ball terminals, eutectic solder is preferred for low-cost assembly. The purpose of using solid-core metal spheres is to maintain sufficient bump height for better board-level reliability. The core is usually made of Cu, and the spheres should be clad with solder for joint formation [6].

16.4 Manufacturing Process

Since the μBGA has a unique package structure, the fabrication process is quite different from that of other CSPs. The first step is to prepare the so-called Tessera-compliant mounting tape (TCMT). The flex circuit interposer is made by a TAB tape manufacturer and then converted into TCMT by another company, such as 3M or Flexera. This conversion process involves tailoring tapes from a reel and mounting the strips on metal frames. The metal frames have outside dimensions similar to those of conventional lead frames. The elastomer layer is applied to the tape, and an adhesive material is deposited for die attachment.

Once the TCMT is prepared, die attachment is performed with an automated pick-and-place machine. Subsequently, the ribbon leads are bonded on the Al die pads by a modified wire bonder, as shown in Fig. 16.9. Such bonding is a thermosonic process and is conducted one lead at a time. Figure 16.10 presents close-up views of bonded ribbon leads. It should be noted that, for better interconnect reliability, an angled lead at the corner, as shown in Fig. 16.10a, should be avoided. The preferred corner configuration should be orthogonal leads, as shown in Fig. 16.10b. Once the lead bonding is finished, a dry film resist is laminated to the interposer using a vacuum system. Afterwards, encapsulant is dispensed from the back side (the chip

Figure 16.9 Modified thermosonic bonder to bond flexible leads on the die.

side), and then curing is performed to complete the encapsulation [7]. The subsequent procedures include dry film exposure and developing, solder-ball attachment and reflow, cleaning, marking, and package singulation. During the μBGA fabrication process, a panel form as shown in Figs. 16.11 and 16.12 is maintained for ease of handling and to permit high-volume production. It should be noted that, after the packaging procedures are completed, testing may be performed in the panel form or on individual packages after singulation [8].

16.5 Electrical and Thermal Performance

The electrical performance of the μBGA has been evaluated by numerical simulation and experimental measurement. The device under investigation was a high-frequency test module, as shown in Fig. 16.13. The results of simulated capacitance, inductance, and resistance are presented in Tables 16.1 and 16.2. For comparison, the measured values of resistance and inductance are given in Table 16.3. Furthermore, the S-parameter of the trace and beam lead is shown in Fig. 16.14. To evaluate the package impedance of the μBGA, the method in Fig. 16.15 was employed. The measurement results for the characteristic impedance are presented in Fig. 16.16 [9].

The thermal performance of μBGA packages was investigated by computational modeling. The module under consideration was a package with an Al fin heat spreader, as illustrated in Fig. 16.17. The values of thermal resistance θ_{ja} obtained from simulation are presented in Figs. 16.18 and 16.19. A comparison between the μBGA and the PQFP is given in Fig. 16.20. It is interesting to note that the PQFP outperformed the μBGA in heat dissipation when a two-layer board was used. However, if the number of layers is increased from two to

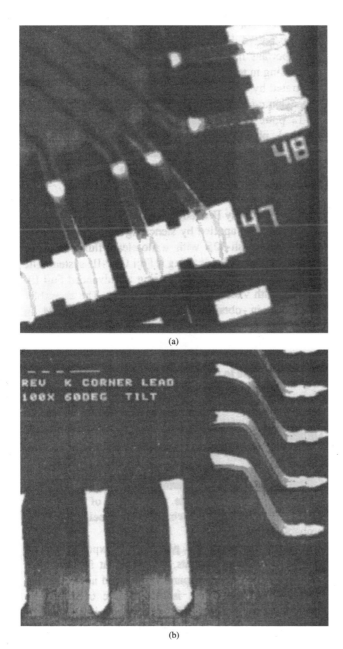

(a)

(b)

Figure 16.10 Close-up views of bonded flexible leads: (a) poor design with angled bonding at the corner; (b) improved design with orthogonal leads.

Figure 16.11 Panel-form fabrication process for high-volume production.

Figure 16.12 Close-up view of the μBGA in panel form.

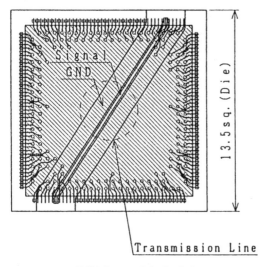

Figure 16.13 μBGA designed for high-frequency test.

TABLE 16.1   Simulation Results for Capacitance

Trace length		Loading C	Lead-to-lead C
Shortest	1.00 mm	160 fF	11.4 fF (19–18 pin)
Longest	2.89 mm	262 fF	23.4 fF (168–169 pin)

TABLE 16.2   Simulation Results for Resistance and Inductance

MHz		10	50	100	150	200
Shortest trace	R	0.021	0.023	0.026	0.031	0.037
	L	0.666	0.653	0.643	0.642	0.639
Longest trace	R	0.063	0.073	0.089	0.116	0.153
	L	2.51	2.50	2.47	2.47	2.46

Units: R, ohms; L, nH.

TABLE 16.3   Measurement Results for Resistance and Inductance

MHz		1	10
Shortest trace	R	0.0176	0.0270
	L	0.746	0.547
Longest trace	R	0.0497	0.1085
	L	1.91	1.52

Units: R, ohms; L, nH.

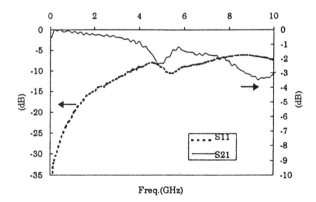

Figure 16.14   S-parameter of trace and beam lead.

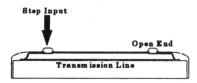

Figure 16.15 Measurement method
for characteristic impedance.

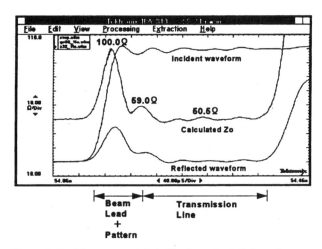

Figure 16.16 Measurement data for characteristic impedance.

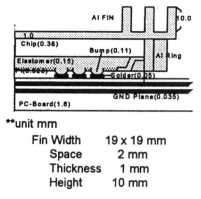

**unit mm

Fin Width	19 x 19 mm
Space	2 mm
Thickness	1 mm
Height	10 mm

Figure 16.17 Model for the analysis of
thermal performance.

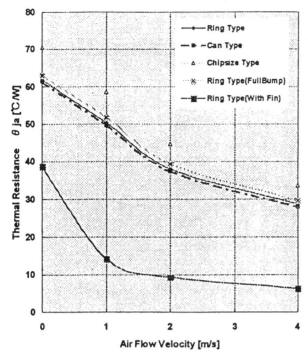

Figure 16.18 Simulation results of thermal resistance for the μBGA on a two-layer board.

four, the difference between the PQFP and the μBGA is not significant [9].

16.6 Qualifications and Reliability

A series of qualification tests was conducted to investigate the package reliability of the μBGA. The test conditions and requirements are listed in Table 16.4. From the qualification testing results given in Table 16.5, it may be concluded that the μBGA has package reliability similar to that of conventional plastic-encapsulated modules [10].

In addition to package qualifications, the board-level reliability is also a major concern for any new package. A series of experimental studies was performed to investigate this issue for the μBGA-PCB assembly. The SMT mounting conditions and reliability test requirements are given in Tables 16.6 and 16.7, respectively. The testing results are given in Tables 16.8 to 16.11. It is obvious that the μBGA has outstanding board level reliability. Such results are expected because of the existence of the compliant layer between the chip and the interposer [10].

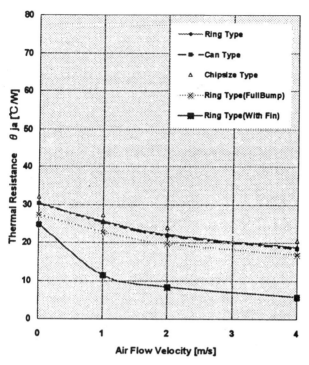

Figure 16.19   Simulation results of thermal resistance for the μBGA on a four-layer board.

TABLE 16.4   μBGA Package Qualification Test Requirements

Test	Test condition	Passing criteria
Preconditioning	− 55°C/ + 125°C, 5 cycles + 125°C, 24 h + 30°C/70% RH, 192 h + 3 times IR Reflow	Popcorn, die delamination and crack, structural integrity; SAT, cross section, open check (dog-bone pad die)
Thermal shock with preconditioning	− 55°C, 5 min/ + 125°C, 5 min, 300 cycles	Die delamination and crack, structural integrity; SAT, cross section, open check (dog-bone pad die)
Temperature cycling with preconditioning	− 65°C, 30 min/ + 150°C, 30 min, 1000 cycles	Die delamination and crack, structural integrity; SAT, cross section, open check (dog-bone pad die)
BIAS HAST with preconditioning	+ 130°C/85% RH + 10 V dc bias	Maximum leakage current 10^{-10} A at 10 V dc, inter inner lead
PCT with preconditioning	+ 121°C/100% RH (2 atm), 168 h	Open check (dog-bone pad die)

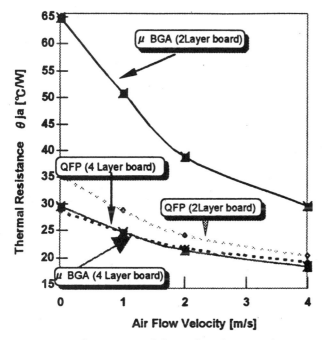

Figure 16.20 Comparison of thermal performance between μBGA and PQFP.

TABLE 16.5   Summary of Qualification Test Results for μBGA

Test	Conditions	Duration	Sample/fails
Thermal shock	-55 to $+125°C$, method A110	100 cycles	76/0
Thermal cycling	-55 to $+125°C$	1000 cycles	45/0
Pressure pot	121°C/100% RH, 2 atmospheres	168 h	45/0
HAST	130°C/85% RH, 5.5 V bias	96 h	45/0
High temperature storage	150°C	1000 h	45/0
Mechanical shock	600G, 2.5 ms, method 2002	6 axis	32/0
Solderability	JEDEC 22-B	—	3/0
Physical dimensions	JEDEC 22-B	—	
Mass permanency	Method 2015	—	

TABLE 16.6   Surface-Mount Conditions for Qualification Testing

Test board	FR-4, metal layers, 181 $\times$ 130 mm, 1.6-mm thickness
Solder paste	63Sn/37Pb, RMA 400 mesh
Printing mask	0.1-mm thickness, additive
Reflow	IR

TABLE 16.7   μBGA Board-Level Qualification Test Requirements

Test	Test condition	Passing criteria
Thermal shock	−55°C, 5 min/+125°C, 5 min 100, 200, 300 cycles	Open check [resistance (room temperature) maximum + 5%], cross section
Temperature cycling	−55°C, 30 min/+125°C, 30 min 100, 300, 500, 1000 cycles	Open check [resistance (−55, 125°C) maximum + 5%], cross section
BIAS HAST	+130°C/85% RH + 5 V dc bias 50, 100, 150, 200 h	Maximum leakage current 10^{-10} A at 5 V dc, cross section
High-temperature storage	+125°C/in the air 100, 200, 300, 500, 1000 h	Open check [resistance (room temperature) maximum + 5%], cross section

TABLE 16.8   Solder Joint Qualification Test Results for Various Bond-Pad Configurations under Thermal Shock

Land/resist opening	200 cycles	300 cycles	700 cycles
0.300 mm/0.300 mm	0/6	0/6	0/6
0.300 mm/0.250 mm	0/6	0/6	0/6
0.300 mm/0.200 mm	0/6	0/6	0/6
0.200 mm/0.300 mm	0/6	1/5	0/5

TABLE 16.9   Solder Joint Qualification Test Results for Various Bond-Pad Configurations under Temperature Cycling

Land/resist opening	100 cycles	200 cycles	300 cycles
0.300 mm/0.300 mm	0/6	0/6	0/6
0.300 mm/0.250 mm	0/6	0/6	0/6
0.300 mm/0.200 mm	0/6	0/6	0/6
0.200 mm/0.300 mm	0/6	0/6	0/6

TABLE 16.10   Solder Joint Qualification Test Results for Various Bond-Pad Configurations under High-Temperature Storage

Land/resist opening	100 h	200 h	300 h
0.300 mm/0.300 mm	0/6	0/6	0/6
0.300 mm/0.250 mm	0/6	0/6	0/6
0.300 mm/0.200 mm	0/6	0/6	0/6
0.200 mm/0.300 mm	0/6	0/6	0/6

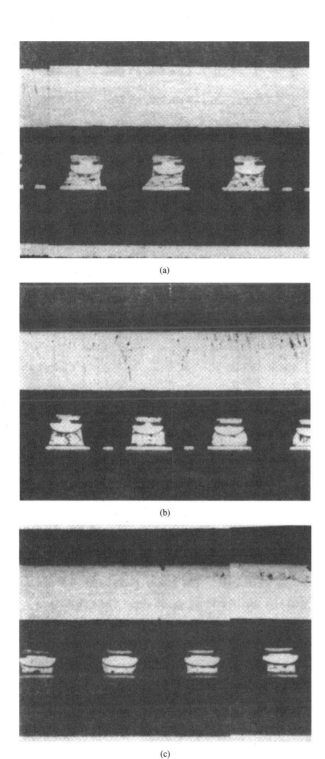

(a)

(b)

(c)

Figure 16.21 Cross-sectional views of solder joints in a μBGA assembly: (a) with solder-mask opening of 0.3 mm; (b) with solder-mask opening of 0.35 mm; (c) with solder-mask opening of 0.2 mm (land diameter in all cases is 0.3 mm).

TABLE 16.11   Summary of Board-Level Qualification Tests for μBGA

Test	Conditions	Duration	Samples/fails
Thermal shock	−196°C/+160°C (liquid to liquid)	130 cycles	3/0
Thermal cycling	0°C/+100°C (air to air)	1163 cycles	46/0
Thermal cycling	−55°C/+150°C (air to air)	100 cycles	48/0

TABLE 16.12   Material Properties for Computational Modeling

Component	Material	Young's modulus (GPa)	Poisson's ratio	TCE [μ→ppm/°C]
PC board	FR-4	18.6	0.19	15
Solder	Sn63/Pb37	30.8	0.4	24.5
Film	PI	8.82	0.3	12
Elastomer	—	0.0061	0.49	200
Lead	Au	76.5	0.44	15.3
Lead	Cu	117.7	0.34	17.2
Bump	Ni	206.9	0.3	15.2
Chip	Si	200.1	0.33	3.4
Ring	Al	255	0.34	24
Epoxy	—	7.8	0.3	15

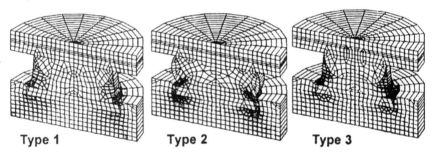

Type 1   Type 2   Type 3

Figure 16.22   Computational analysis for three types of μBGA solder joints.

Further study was conducted to compare various bond pad conditions on the PCB for better board-level reliability. Three types of configuration as presented in Fig. 16.21 were considered. It can be seen that the geometry of the solder joints is affected by the relative size of the land and the solder mask opening. A finite-element model with material properties given in Table 16.12 was established to perform stress analysis. The typical strain contours are shown in Fig. 16.22 for comparison. It was identified that the Type 1 configuration led to the smallest inelastic strain and, hence, the best reliability [9].

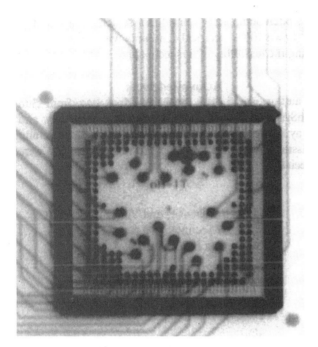

Figure 16.23 TI die in the μBGA package assembled on a test board (x-ray image).

16.7   Applications and Advantages

The μBGA was developed for packaging devices with low to medium pin counts. The typical I/O number is between 46 and 300. The main applications are for memory modules such as DRAMs and flash memory. The μBGA technology has been successfully applied to package commercial IC devices, as presented in Figs. 16.23 and 16.24. The packaged modules also have been used in commercial products such as the PC card and miniature memory cards shown in Fig. 16.25. Currently Tessera maintains high-volume production lines in California and Singapore to fabricate μBGA packages. Several major semiconductor manufacturers have adopted or licensed this CSP technology from Tessera, and many other companies are in collaboration with Tessera to develop or improve materials and processes for the μBGA [11].

Small form factors and board-level reliability are the two major advantages of μBGA packages. Figures 16.26 and 16.27 show comparisons of the package size of μBGA and other packages. The superiority in form factor of the former is obvious. On the other hand, with a similar footprint, the μBGA has better solder joint reliability than the flip chip on board (FCOB). This can be illustrated by the model case comparison presented in Fig. 16.28. A stress analysis by a finite-ele-

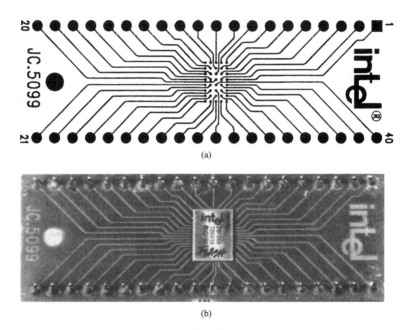

(a)

(b)

Figure 16.24 Intel's flash memory with μBGA package.

Figure 16.25 Applications of the μBGA to PC card and miniature memory cards.

Figure 16.26 Comparison between μBGA and TSOP with the same flash memory chip.

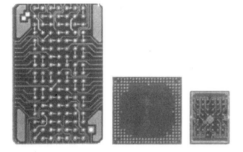

Figure 16.27 Comparison between PBGA and μBGA.

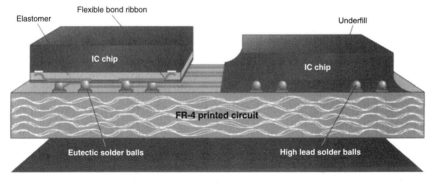

Figure 16.28 Comparison between μBGA assembly and flip chip on board.

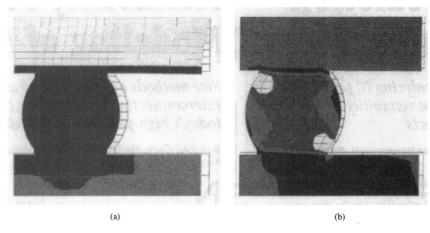

(a)   (b)

Figure 16.29   (a) Strain contour in the solder joint of the μBGA assembly and (b) flip chip on board (without underfill).

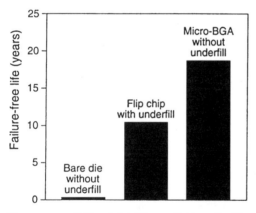

Figure 16.30   Solder joint life predictions for flip chip and μBGA.

ment method was performed to investigate the solder joint reliability. The resulting strain contours are shown in Fig. 16.29. It can be seen that the solder joint of the FCOB has much more inelastic strain, which would result in much poorer reliability. It should be noted that the FCOB in this comparison was not underfilled. However, from the results of a further study (Fig. 16.30), the μBGA still has better board-level reliability than the FCOB with underfill [12].

In summary, the merits of the μBGA may be seen in the diagram shown in Fig. 16.31. This CSP is fully SMT compatible and has form factors similar to those of the FCOB. However, the board-level relia-

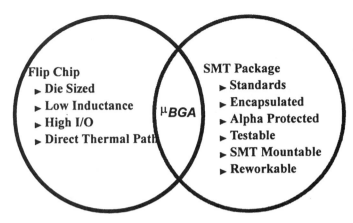

Figure 16.31   Merits of μBGA package.

bility of the μBGA is much better than that of the FCOB (not to mention the repair issue due to underfill). However, the μBGA technology is relatively expensive at present. Also, the requirement for ad hoc tooling in the fabrication process may prolong the lead time for new package production. Therefore, there is still room for improvement.

16.8   Summary and Concluding Remarks

Tessera's μBGA is probably the most publicized chip scale package on the market. This package belongs to the category of CSPs with flexible interposer. The unique features of the μBGA are the ribbonlike flexible leads for chip-level interconnection and the compliant elastomer between the interposer and the chip to relieve the thermal mismatch. The leads are bonded to the Al bond pads on the die one at a time by a thermosonic bonder. The bonding location is not restricted to the perimeter and may be anywhere within the die. Encapsulation is required to protect the bonded leads.

The package terminals of the μBGA may be plated bumps, solder balls, or solid-core metal spheres. The array pitch may be 0.5, 0.75, or 1.0 mm, and the number of I/Os ranges from 46 to 300. The major application of μBGA packages is for memory devices, especially DRAMs and flash memory. The packaged modules have been used in commercial products such as PC cards and miniature memory cards.

16.9   References

1. T. W. Goodman and E. J. Vardaman, *CSP Markets and Applications,* TechSearch International, Inc., Austin, Tex., 1998, pp. 37–43.
2. J. Fjelstad, T. DiStefano, B. Faraji, C. Mitchell, and Z. Kovac, "μBGA Packaging Technology for Integrated Circuits," *Proceedings of NEPCON-East,* Boston, Mass., 1995, pp. 221–227.

3. C. Mitchell, "Assembly and Reliability Study for the Micro-Ball Grid Array," *Proceedings of the IEEE/CPMT IEMTS,* International Electronics Manufacturing Technology Symposium, LaJolla, Calif., September 1994, pp. 344–346.

4. D. Numakura, K. Matsuo, M. Mizoguchi, K. Hsu, and M. Chen, "Micro BGA on Flex Circuits," *Proceedings of the IEPS Conference,* San Diego, Calif., September 1995, pp. 608–619.

5. T. DiStefano, "The μBGA as a Chip Size Package," *Proceedings of NEPCON-West '95,* Anaheim, Calif., February 1995, pp. 327–333.

6. V. Solberg, "Chip-Scale Ball Grid Array Packaging for Surface Mount Technology," *Proceedings of the SMI Conference,* San Jose, Calif., September 1996, pp. 221–228.

7. J. Fjelstad, T. DiStefano, and M. Perry, "Compliancy Modeling of an Area Array Chip Scale Package," *Proceedings of the SMI Conference,* San Jose, Calif., September 1996, pp. 236–243.

8. J. Fjelstad, T. DiStefano, and J. Link, "Reworkable, Socketable Chip Scale Area Array Package Utilizing a Solid Core Ball," *Proceedings of NEPCON-West '97,* Anaheim, Calif., February 1997, pp. 1069–1075.

9. T. Koyama, K. Abe, N. Sakaguchi, and S. Wakabayashi, "Reliability of Micro-BGA Mounted on a Printed Circuit Board," *Proceedings of the SMI Conference,* San Jose, Calif., August 1995, pp. 43–56.

10. V. Solberg, "Low Cost Chip-Scale BGA Packaging for Portable and Other Miniature Product Applications," *Proceedings of the International Symposium on Microelectronics,* 1997, pp. 116-121.

11. U. D. Perera, "Evaluation of Reliability of μBGA Solder Joints through Twisting and Bending," *Proceedings of the International Symposium on Microelectronics,* 1997, pp. 402–407.

12. J. Fjelstad, "Novel Approaches to Compliant Area Array Chip Interconnection," *Proceedings of SEMICON Europa,* April 1998.

17

TI Japan's
Micro-Star BGA (μStar BGA)

17.1 Introduction and Overview

Texas Instruments (TI) Japan's Micro-Star ball grid array (μStar BGA) package was released in late 1996 [1]. This package uses a TAB tape carrier as the interposer. The connection between the chip and the substrate is by wire bonding. The die and wire bonds are over-molded for protection. The board-level interconnects are BGA. The μStar BGA is designed for applications with medium pin counts. Both 0.5- and 0.8-mm ball pitch are used. This package is considered a CSP both because of the fine pitch and because the package size is only 2 mm larger than the die size. The package thickness is 1.2 mm. According to TI Japan, reasonable package reliability and solder joint reliability can be achieved [2]. The μStar BGA is already in mass production. A flip-chip version is under consideration. The main applications are for mobile homes and camcorders. This package is expected to be used in digital personal assistants.

17.2 Design Concepts and
Package Structure

The μStar BGA is a flex-based CSP. This package uses a three-layer tape carrier as the interposer. The backside of the die is mounted to the flexible substrate by nonconductive adhesive. Thermosonic wire bonding is employed for the first-level interconnection. The Au wires are directly bonded to the Ni/Au-plated Cu trace on the polyimide tape. The line width and space are 40 μm for the Cu trace pattern. The IC and wire bonds are overmolded with encapsulant for protection. The board-level I/Os are BGA. The eutectic solder balls are attached to

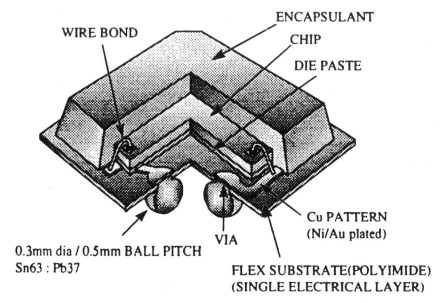

ENCAPSULANT

WIRE BOND

CHIP

DIE PASTE

0.3mm dia / 0.5mm BALL PITCH

VIA

Cu PATTERN
(Ni/Au plated)

Sn63 : Pb37

FLEX SUBSTRATE(POLYIMIDE)
(SINGLE ELECTRICAL LAYER)

Figure 17.1   Package structure of TI Japan's μStar BGA.

the bottom of the Cu trace through 200-μm vias on the tape carrier. The package structure of the μStar BGA is illustrated in Fig. 17.1. In addition to the current configuration, a flip-chip version is under plan.

The μStar BGA is considered a CSP because of its fine pitch terminals and package outline. The package size is only 2 mm larger than the die size. The package thickness is 1.2 mm. The current lineup for the μStar BGA is 62 to 144 pins. Both 0.5- and 0.8-mm ball pitch are used. The solder-ball diameter is 0.3 mm. The ball grid array can be full grid or depopulated to a perimeter pattern.

17.3   Material Issues and Manufacturing Process

The μStar BGA is a wire-bonded and overmolded tape carrier package (TCP). Besides the silicon chip, the package materials include the die-attach adhesive, the gold bonding wires, the flexible polyimide substrate with Cu traces, the encapsulation molding compound (EMC), and the eutectic solder balls. It has been reported that the tape carrier is the most critical material for the μStar BGA package. TI Japan uses the TAB tape from Shindo and Sumitomo Metal Mining (SMM), as shown in Fig. 17.2. This flexible substrate is a three-layer tape consisting of one metal layer, one layer of polyimide base film, and the bonding adhesive between the metal layer and the

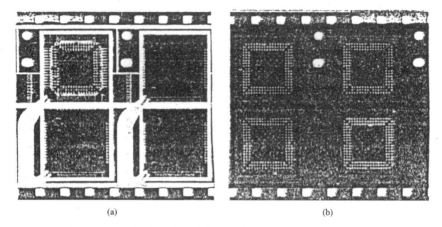

(a)   (b)

Figure 17.2   Tape carrier for the μStar BGA: (a) the top view; (b) the bottom view.

polyimide film. The metal layer is a patterned Cu trace. The line/space width is 40 μm. The top of the Cu trace is plated with Ni/Au to form a wire-bondable surface. On the polyimide film, an array of vias is drilled for solder-ball attachment. The diameter of the vias is 200 μm. It has been reported that the via drilling process is critical. According to TI Japan, the present tape carrier can provide reasonable package reliability (JEDEC level 3). If higher reliability is desired, a two-layer tape (without the adhesive layer) should be used. However, this will bring up other issues, such as material cost and compatibility.

The manufacturing process for the μStar BGA is very straightforward. The die is directly attached to the Cu trace side of the tape carrier by nonconductive adhesive. The Au wires are used to connect the bond pads on the chip and the Cu trace by thermosonic bonding. Then the IC, together with the wire bonds, is encapsulated with epoxy resin. After molding, the package is flipped over for ball deposition. The eutectic solder balls are reflowed by the IR oven and attached to the bottom of the Cu trace through the vias on the polyimide tape. Since a tape carrier is used, the manufacturing process can be operated in a reel-to-reel form. The packages are singulated from the carrier after ball placement.

17.4   Qualifications and Reliability

The μStar BGA has been evaluated for package reliability and solder joint reliability. The former was verified by a series of qualification tests, as shown in Table 17.1. The specimens were 144-pin μStar BGA

TABLE 17.1   Package Qualification Test Results for TI Japan's µStar BGA
Package

Test items	Conditions	Fail/sample
Temperature cycle (−65/150°C)	1000 cycles	0/78
Thermal shock (−65/150°C)	500 cycles	0/45
Autoclave (121°C/2 atm)	240 h	0/45
High-temperature operating life (125°C)	1000 h	0/96
THB (85°C/85%)	1000 h	0/80
High-temperature storage life (125°C)	1000 h	0/78
Salt atmosphere	MIL-STD-883D	0/15
Flammability	UL94V-0	0/5
Moisture sensitivity	JEDEC level 3	0/270

TABLE 17.2   Typical Requirements for Solder Joint Reliability under Thermal
Cycling

Customer	Temperature range (°C)	Criteria (cycles)	PCB material and thickness (mm)
A	0/100	1000	FR-4 (0.8)
B	−20/70	200	FR-4 (0.8)
C	−30/85	1000	FR-4 (0.8)
D	−40/80	600	Aramid (0.8)
E	−40/85	300	FR-4 (0.8)
F	−25/125	500	FR-4 (0.8)
G	−40/125	1000	FR-4 (0.8)
H	−40/125	1000	FR-4 (1.6)

with a ball pitch of 0.8 mm. The package size was 12×12 mm, while
the chip size was 9.5×9 mm. No cracks or delaminations were observed after those tests, indicating that the µStar BGA has the same
package reliability as PQFPs in most environments. The only exception is moisture sensitivity. Currently the µStar BGA is qualified for
JEDEC level 3 and can achieve level 2 in the laboratory. However,
there are difficulties in reaching level 1. The main problem is the separation between the adhesive and the polyimide of the three-layer
tape carrier. TI Japan claims that the situation can be improved by
using two-layer tape (without the adhesive layer). However, other issues, such as material cost and compatibility, will arise.

The solder joint reliability under thermal cycling is always a critical issue for microelectronics. However, there is no standard testing
specification in the industry so far. As a demonstration, Table 17.2

presents eight companies' requirements for solder joint reliability [2]. In general, the solder joints are considered to have acceptable reliability if they can sustain 1000 temperature cycles. Figure 17.3 shows the cross section of the solder joints of the μStar PCB assembly. It can be estimated that the standoff height of the solder joints is less than 0.3 mm, which is smaller than that of most other BGAs. Because of the short solder joints, the present assembly structure is more rigid. Consequently, the stresses induced by thermal mismatch become more severe [3].

In order to evaluate the solder joint reliability of the μStar BGA, a comprehensive study was conducted to investigate the effects of various design parameters. Both experimental testing and computational analysis were performed. Figure 17.4 shows the typical failure mode of a μStar BGA solder joint subjected to thermal cycling between −25 and 125°C. The crack propagates along the interface at the top of the solder joint. A computational simulation by finite-element analysis (FEA) confirms that the highest stress concentration occurs at this location. Furthermore, it was found that the crack could appear at either the top or the bottom of the solder joint, depending upon the relative size of the solder pads on the package substrate and the PCB. In principle, for the solder joints of the μStar BGA-PCB assembly, the failure occurs at the side with the smaller solder pad.

Another parameter under investigation is the configuration of PCB solder pads. In general, there are two types of solder pad, namely, non-solder-mask-defined (standard mask, in TI Japan's term) and sol-

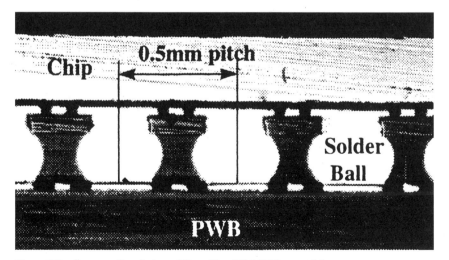

Figure 17.3   Cross-sectional view of the μStar BGA-PCB assembly.

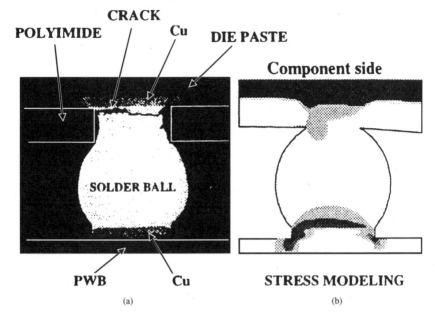

CRACK
POLYIMIDE   Cu   DIE PASTE

Component side

SOLDER BALL

PWB   Cu   STRESS MODELING

(a)   (b)

Figure 17.4   Typical failure mode of μStar BGA solder joint under thermal cycling: (a) cross section of testing specimen; (b) FEA results.

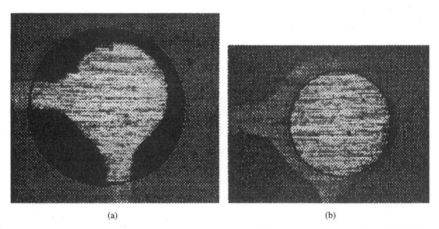

(a)   (b)

Figure 17.5   Various copper pad configurations on PCB: (a) non-solder-mask-defined; (b) solder-mask-defined.

der-mask-defined (overlap mask) [4]. Figure 17.5 shows the two solder-mask designs for the μStar BGA [1]. Currently the non-solder-mask-defined pad is more favored in the industry because it provides more space for Cu trace routing. However, this configuration may sometimes cause distortion of the solder joint, as shown in Fig. 17.6.

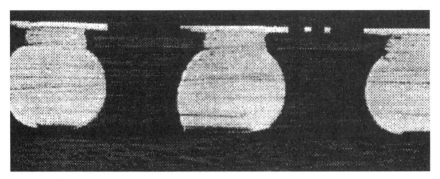

Figure 17.6 Distortion of solder joints due to non-solder-mask-defined pad.

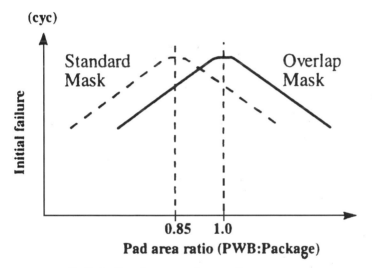

Figure 17.7 Optimization of copper pad area ratio.

This phenomenon is attributed to the exposed trace line next to the solder pad, as seen in Fig. 17.6. Once the solder joint is skewed, the thermal fatigue life will be substantially reduced. After numerous experiments, an empirical solder-pad design rule for the μStar BGA was determined; it is illustrated in Fig. 17.7. For the non-solder-mask-defined (standard mask) design, a solder-pad area ratio (PCB side to package substrate side) of 0.85 would result in the highest number of cycles before initial failure. On the other hand, if the solder-mask-defined (overlap mask) configuration is adopted, a unity solder pad area ratio should lead to the highest solder joint reliability.

In addition to solder-pad design, other parameters under investigation include the chip size, the ball pitch, and the number of solder balls. The results are summarized in Fig. 17.8. The general trends indicate that a smaller chip, a wider ball pitch, and more solder balls should lead to longer solder joint thermal fatigue life. Furthermore, the composition of the solder material was also taken into account for solder joint reliability. It was found that by adding various elements to the eutectic solder, the thermal fatigue life may be improved by 20 to 50 percent. However, 63Sn/37Pb is still the most favored solder. Figure 17.9 presents the Weibull plot for solder joint reliability. The specimens were 100-pin μStar BGA with a ball pitch of 0.5 mm. The package size was 10×10 mm. The temperature range was −25 to 125°C. It can be seen that the characteristic fatigue life is no longer than 1000 cycles, which is in general acceptable according to the criteria given in Table 17.2. Also, the improvement in the thermal fatigue life when more elements are added to the solder can be observed. However, according to TI Japan's report [2], if the criterion is 1000 cycles with a temperature range of −40 to 125°C, then underfill adhesive will be required for packages with ball pitch smaller than 0.8 mm.

Besides thermal cycling, the solder joint reliability was investigated under mechanical loading as well. The specimens were 79-pin μStar BGA with a ball pitch of 0.8 mm. The package size was 13×9 mm,

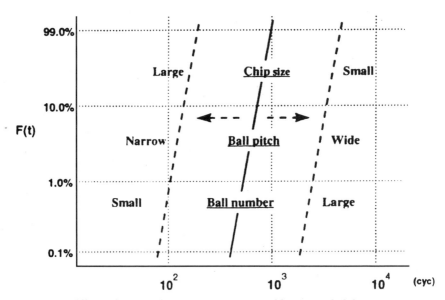

Figure 17.8   Effects of various design parameters on solder joint reliability.

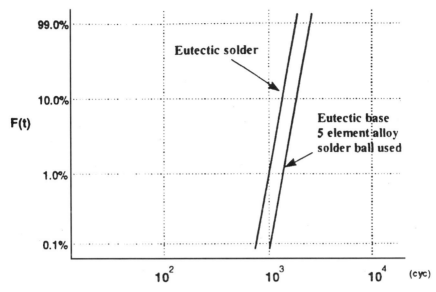

Figure 17.9   Thermal fatigue life distribution of μStar BGA solder joints.

and the PCB size was 100×50×0.8 mm. The assembly was tested under three-point bending with a deflection of 2 mm over a span of 80 mm. After 300 cycles of testing, no failure was observed. Therefore, the solder joint reliability of the μStar BGA under mechanical loading was verified.

17.5   Applications and Advantages

TI Japan's μStar BGA was developed for IC devices with a medium number of I/Os (62 to 144). The planned applications include memory modules, signal processors, and ASICs. Currently the μStar BGA package is in mass production for mobile phones and camcorders. This package is expected to be used in digital personal assistants.

The μStar BGA is an overmolded package with a flexible tape carrier. The major advantage of such a package is the feasibility of reel-to-reel operation for fabrication. The package size is only 2 mm larger than the die size. The ball pitch is less than 1.0 mm (0.5 mm and 0.8 mm). Therefore, the μStar BGA is recognized as a CSP. Because of the relatively small footprint, the assembly density can be increased. According to TI Japan, reasonable package reliability and solder joint reliability can be achieved for the μStar BGA.

17.6 Summary and Concluding Remarks

The μStar BGA is a flex substrate-based CSP. This package uses a three-layer tape carrier as the interposer. The chip-level interconnects are Au wire bonds. A flip-chip version is planned. The package is encapsulated by molding compound. The board-level I/Os are fine-pitch (0.5 mm and 0.8 mm) solder balls. The μStar BGA is intended for packaging IC devices with medium pin counts. The μStar BGA has a relatively compact package outline. The package size is only 2 mm larger than the die size. The package thickness is 1.2 mm.

It has been reported that the package reliability of the μStar BGA is similar to that of PQFPs. The moisture sensitivity is qualified for JEDEC level 3 and may reach level 2 in the laboratory. The thermal fatigue life of μStar BGA solder joints is in general acceptable. The solder joint reliability under mechanical loading has been verified by bending test. The μStar BGA is in mass production for mobile phones and camcorders. Another potential application is digital personal assistants.

17.7 References

1. TechSearch International, "1996 BGA Development—Developments in the Third Quarter," October 1996, p. 17.
2. K. Ano, T. Ohuchida, K. Murata, and M. Watanabe, "Reliability Study of the Chip Scale Package Using Flex Substrate," *Proceedings of the Chip Scale Packaging Symposium,* Surface Mount International Conference, San Jose, Calif., September 1997, pp. 44–47.
3. J. Lau and Y. H. Pao, *Solder Joint Reliability of BGA, CSP, Flip Chip, and Fine Pitch SMT Assemblies,* McGraw-Hill, New York, 1997.
4. J. Lau, *Ball Grid Array Technology,* McGraw-Hill, New York, 1995.

Chapter

18

TI Japan's Memory Chip Scale Package with Flexible Substrate (MCSP)

18.1 Introduction and Overview

The memory chip scale package (MCSP) was released by Texas Instruments (TI) Japan in 1997 [1]. There are two configurations of the MCSP. One is based on the tapeless lead-on-chip (LOC) technology, and the other uses a flexible substrate as the interposer. This chapter focuses on the MCSP with flexible substrate. The lead-frame-based MCSP is introduced in Chap. 8.

TI Japan's MCSP is a low-pin-count but large-lead-pitch package. The first- and second-level interconnects are wire bonds and BGA, respectively. The current lineup is 52 pins with a pitch of 1.27 mm. The MCSP uses a unique process for wafer preparation with a polyimide coating. This thin layer provides protection and adhesion. Good package and solder joint reliability have been reported. TI Japan has finished its internal qualifications for the MCSP. The targeted application of this package is DRAM devices.

18.2 Design Concepts and Package Structure

The tapeless LOC was developed by TI Japan in 1994 to replace the conventional LOC technology [2]. This method uses a unique wafer coating process with a thin layer of polyimide. TI Japan has implemented a lead-frame-based MCSP using tapeless LOC technology. Concurrently, another MCSP with flexible substrate was developed.

293

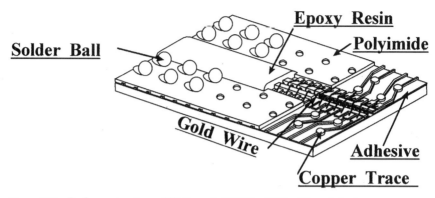

Figure 18.1   Package structure of TI Japan's MCSP with flexible substrate.

Although this package has no lead frame, the polyimide coating for the tapeless LOC is still employed for wafer preparation [1].

The MCSP with flexible substrate uses a polyimide tape carrier as the interposer. The first-level interconnects are conventional wire bonds. The Au wires are bonded directly to the Cu trace on the flexible substrate. Since the MCSP originated from the LOC packages, the current configuration is for devices with center bond pads only. The interconnects at the board level are BGA. The eutectic solder balls are attached to the Cu trace through the vias on the tape carrier. The package structure of the MCSP with flexible substrate is illustrated in Fig. 18.1.

The MCSP has limited molding area over the top of the wire bonds. The package thickness is 0.8 mm. It is a true chip scale package because the package size is exactly the same as the die size. The current lineup for the MCSP is 52 pins with a ball pitch of 1.27 mm. The solder-ball diameter is 1 mm. The ball population is arranged in the array pattern shown in Fig. 18.2. Compared to the lead-frame-based MCSP, this package is slightly smaller.

18.3   Material Issues

The MCSP with flexible substrate is a wire-bonded tape carrier package. In addition to the polyimide-coated silicon chip, the package materials include the gold bonding wires, the polyimide tape carrier with Cu traces, the thermoplastic polyimide (PI adhesive), the encapsulation molding compound (EMC), and the eutectic solder balls. It should be noted that there are three kinds of polyimide in this package: the polyimide coating on the surface of the die, the flexible polyimide tape as the package substrate, and the polyimide adhesive used to bond

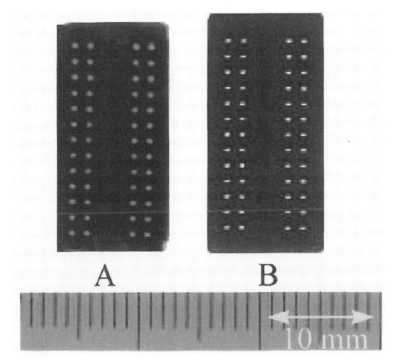

Figure 18.2   Bottom view of TI Japan's MCSP: (*a*) with flexible substrate; (*b*) with LOC design.

the former two materials together. The characteristics of the poly-imide coating were introduced in Chap. 8. The flexible substrate is a standard TAB tape carrier. The polyimide adhesive is a new material.

A new thermoplastic polyimide was developed by TI Japan as the adhesive material between the polyimide-coated silicon chip and the Cu-traced polyimide tape carrier. This polyimide adhesive is a silicone polyimide material; its chemical structure is given in Fig. 18.3. From an experimental investigation, this newly developed material has performance superior to that of the conventional polyimide at high temperature and in a humid environment. Figure 18.4 presents the comparison of shear strength. At room temperature (RT), the difference in shear strength between the conventional and the silicone polyimide is limited. However, when the specimens are aged at 85°C/85 percent RH for 72 h, the new PI adhesive outperforms the other one to a great extent. Figure 18.5 shows the comparison of peel strength at high temperature. Obviously the silicone polyimide is much better than its counterpart. The material properties of silicone polyimide are given in Table 18.1. With such an outstanding adhesive material, good package reliability under qualification tests can be anticipated.

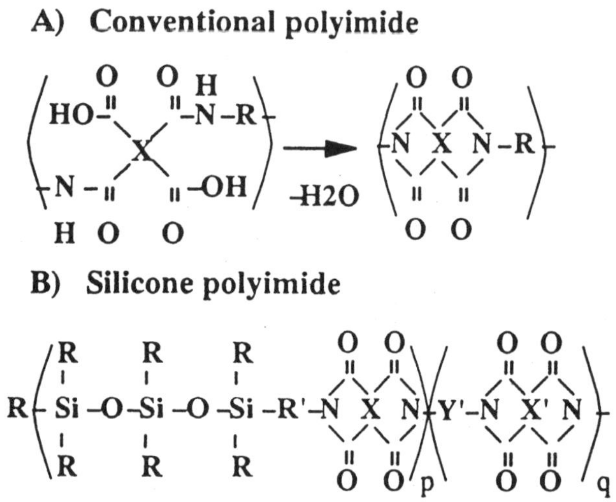

A) Conventional polyimide

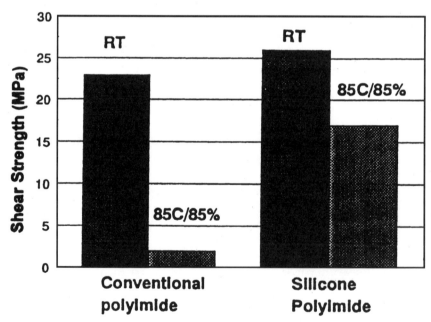

B) Silicone polyimide

Figure 18.3 Chemical structure of polyimide for TI Japan's MCSP.

Figure 18.4 Comparison of shear strength between conventional and silicone polyimide adhesives.

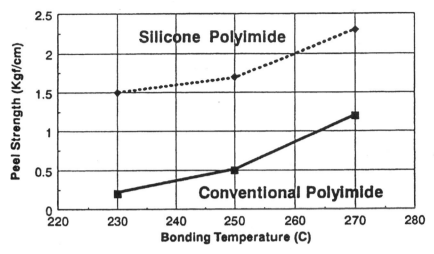

Figure 18.5 Comparison of peel strength between conventional and silicone polyimide adhesives.

TABLE 18.1 Material Properties of Silicone Polyimide

Properties	Value
Thickness (μm)	100
CTE (ppm/°C)	65
T_g (°C)	190
Elastic modulus (kg/mm^2)	300
Dielectric constant	2.9
Dielectric tangent	0.008
Volume resistance (Ω · cm)	2.0×10^{16}

18.4 Manufacturing Process

The manufacturing process for the MCSP with flexible substrate is similar to that for the lead-frame-based MCSP except that an additional polyimide adhesive is required for die mounting. Both packages feature a unique wafer-preparation process involving a polyimide coating. The procedures for applying the thin polyimide layer are illustrated in Fig. 18.6 [3]. Detailed descriptions for this process are given in Chap. 8.

Once the wafer preparation is completed, the process presented in Fig. 18.7 is performed to package the MCSP. First, the polyimide ad-

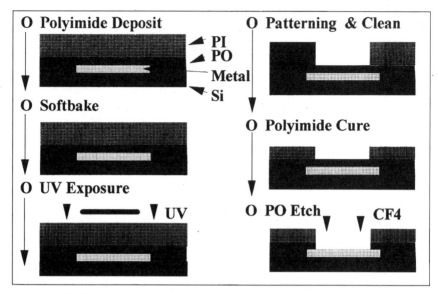

Figure 18.6 Wafer preparation process for TI Japan's MCSP.

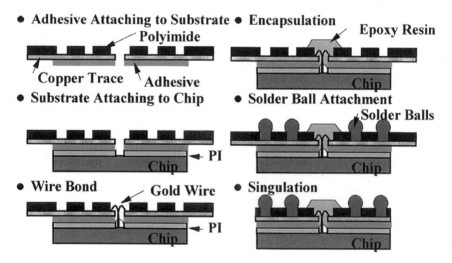

Figure 18.7 Package assembly process for TI Japan's MCSP with flexible substrate.

hesive is attached to the Cu trace side of the polyimide tape. The thickness of the adhesive is 100 μm. This relatively large thickness is for the relaxation of the stress induced by the thermal mismatch between the silicon chip and the PCB assembly. Then the precoated die is mounted to the tape carrier by thermal compression. The flexible

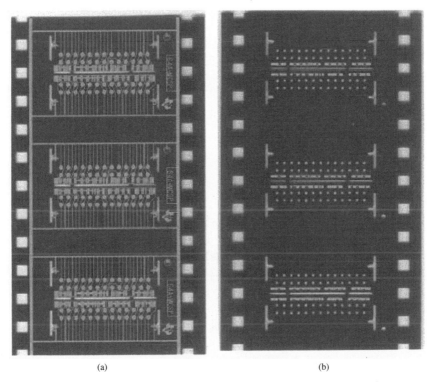

(a)   (b)

Figure 18.8 Tape carrier pattern for TI Japan's MCSP: (a) Cu trace side; (b) the bottom side.

substrate is in a reel form with the pattern shown in Fig. 18.8. The next steps are wire bonding and encapsulation. It should be noted that the molding area is limited to the top of the wire bonds. Also, an EMC with relatively low CTE is chosen in order to minimize the package warpage. Prior to the placement of solder balls, a Ni layer is plated on the designated solder lands as a diffusion barrier. Finally, the 63Sn/37Pb solder balls are attached to the plated Cu trace through the vias on the polyimide tape.

18.5 Qualifications and Reliability

The MCSP with flexible substrate has been evaluated for wire pull strength, package reliability, and solder joint reliability. The former two were investigated by experimental testing, while the last one was evaluated through computational analysis. The results of the wire pulling test are given in Table 18.2 and compared to the values for the lead-frame-based MCSP. The average strength is higher than the

TABLE 18.2   Wire Pull Strength of TI Japan's MCSP

	MCSP	
	Flexible substrate	Lead-on-chip
Minimum (g)	7.5	7.8
Maximum (g)	8.8	9.2
Average (g)	8.3	8.5
Ball off	0/20	0/20
Stitch off	0/20	0/20

Sample sizes: 20 wires.

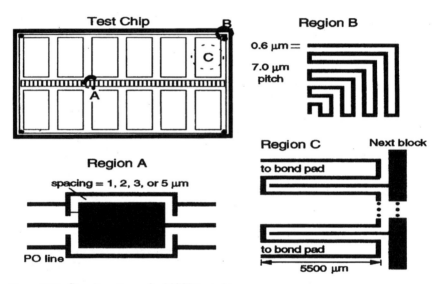

Figure 18.9   Circuit patterns for MCSP test chip.

common industrial requirement, and the deviation seems to be within a reasonable range. Furthermore, all failures occur in the wire instead of the bonding sites. This indicates that the wire bonding process for the MCSP is very reliable.

Various aspects of the package reliability of the MCSP were evaluated. A custom-designed test chip as shown in Fig. 18.9 was developed in order to determine the corrosion susceptibility of the bond pads (Region A), the stress concentration at corners (Region B), and the damage tolerance on the surface (Region C). A series of temperature- and humidity-related qualification tests was performed to evaluate the package reliability. The results are given in Table 18.3 and compared to those of the lead-frame MCSP [4]. Since no damage was

TABLE 18.3   Package Qualification Test Results for TI Japan's MCSP

	MCSP	
Test	LOC	Flexible
85/85 + IR	0/20	0/20
Temperature cycle	0/20	0/20
PCT	0/20	0/20
High-temperature storage	0/20	0/20

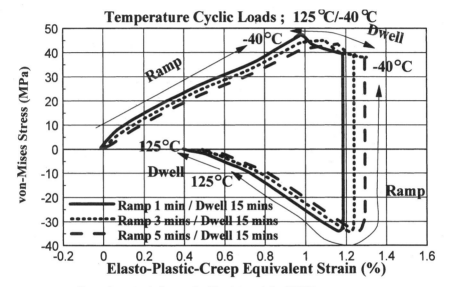

Figure 18.10   Creep hysteresis loops of solder joint of the MCSP.

observed in the specimens, the package reliability of the MCSP is verified.

A computational analysis was conducted to investigate the solder joint reliability of the MCSP using the finite-element method. An elastic-plastic-creep constitutive relation was adopted to model the solder material [5]. A series of analyses was performed for parametric study, as presented in Fig. 18.10. It was found that the highest stress concentration appears at the corner of the interface between the solder ball and the Cu trace. Furthermore, a Coffin-Manson type of formulation [6] was employed to predict the thermal fatigue life of solder joints. The results for a test temperature range of −40 to 125°C are presented in Fig. 18.11. It can be seen that the 50 percent failure rate for the MCSP with flexible substrate is close to 3000 cycles. This

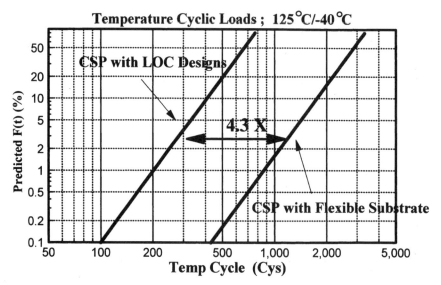

Figure 18.11   Thermal fatigue life prediction for the solder joints of the MCSP.

value is good enough for actual applications. On the other hand, the solder joint thermal fatigue life of the lead-frame MCSP is about four times lower. Therefore, the solder joint reliability could be a concern for the LOC MCSP under thermal cyclic loading.

18.6   Applications and Advantages

The MCSP was developed for low-pin-count ICs. The current lineup is 52 pins with a ball pitch of 1.27 mm. Since the MCSP originated from the LOC packages, the current configuration is for devices with center bond pads only. The targeted application is DRAM devices. The MCSP with flexible substrate is a true chip scale package because its outline is the same as the chip size. The objective of TI Japan is to use the MCSP to replace the SOJ and TSOP in the future.

The major advantage of the MCSP with flexible substrate is its reliability. From the results of qualification tests and computational analysis, both package reliability and solder joint reliability are outstanding. The BGA interconnects also provide self-alignment capability during surface-mount assembly. However, the BGA pattern of the MCSP seems to be different from the conventional ones. Therefore, custom solder pad design on the PCB may be required.

18.7 Summary and Concluding Remarks

TI Japan's MCSP with flexible substrate is a true chip scale package using a tape carrier as the interposer. It features a unique wafer-preparation process using a polyimide coating. The first- and second-level interconnects are wire bonds and BGA, respectively. The MCSP was developed to package low-pin-count ICs. The current lineup is 52 pins with a ball pitch of 1.27 mm. Good package reliability and solder joint reliability have been reported. Data on thermal performance have not yet been released. However, since the molding area is very limited, good heat dissipation capability should be expected. TI Japan has finished its internal qualifications for the MCSP. This package is aimed at replacing the SOJ and TSOP for DRAM applications.

18.8 References

1. M. Amagai, H. Sano, T. Maeda, T. Imura, and T. Saitoh, "Development of Chip Scale Packages (CSP) for Center Pad Devices," *Proceedings of the 47th ECTC,* San Jose, Calif., May 1997, pp. 343–352.
2. M. Amagai, R. Baumann, S. Kamei, M. Ohsumi, E. Kawasaki, and H. Kitagawa, "Development of a Tapeless Lead-on-Chip (LOC) Package," *Proceedings of the 44th ECTC,* Washington, D.C., May 1994, pp. 506–512.
3. M. Amagai, T. Saitoh, M. Ohsumi, E. Kawasaki, C. K. Yew, L. T. Chye, J. Toh, and S. Y. Khim, "Development of Tapeless Lead-on-Chip (LOC) Packaging Process with I-line Photosensitive Polyimide," *Proceedings of the IEEE/CPMT International Electronics Manufacturing Technology Symposium,* Austin, Tex., October 1997, pp. 237–244.
4. M. Amagai, "The Effect of Adhesive Surface Chemistry and Morphology on Package Cracking in Tapeless Lead-on-Chip (LOC) Packages," *Proceedings of the 45th ECTC,* Las Vegas, Nev., May 1995, pp. 719–727.
5. J. Lau and Y. H. Pao, *Solder Joint Reliability of BGA, CSP, Flip Chip, and Fine Pitch SMT Assemblies,* McGraw-Hill, New York, 1997.
6. J. Lau, *Solder Joint Reliability: Theory and Applications,* VNR, New York, 1991.

Chapter

19

Amkor/Anam's ChipArray Package

19.1 Introduction and Overview

The ChipArray package is a new chip scale technology developed by Amkor/Anam in 1996 [1]. The name ChipArray indicates that the packages are manufactured in an array format. This series of CSPs uses a rigid substrate as the package interposer. The substrate may be either organic laminate or thick-film ceramic. The first-level interconnects are wire bonds. The second-level interconnects may be BGA solder balls (for organic substrate) or LCC (leadless chip carrier) flat lands (for ceramic substrate). The chip is encapsulated with a liquid resin. The package size is 1.5 mm larger than the chip, and the package thickness is less than 1.5 mm. Amkor's ChipArray was developed for packaging devices with a low to medium number of I/Os. The current ChipArray family has pin counts from 32 to 128. The terminal pitch is between 0.5 and 1.0 mm. Flexibility is the most significant attribute of this technology. The applications of this package may include memory modules, analog devices, ASICs, and simple PLDs. The ceramic version with LCC format is ideal for disk drive applications.

19.2 Design Concepts and Package Structure

The ChipArray technology was developed by Amkor/Anam for its chip scale packages. The name of this technology reflects the fact that the packages are processed in an array format. The design concept of ChipArray is to implement near-die-size packages (NDSP) with an attribute of flexibility. With these guidelines, a series of CSPs using ChipArray technology have materialized.

Amkor/Anam's ChipArray package is a plastic-encapsulated CSP with rigid substrate. The interposer may be either organic laminate or thick-film ceramic. For the organic substrate, the nominal thickness is 0.34 mm. The typical chip thickness is 0.3 mm (12 mils). The minimum bond-pad pitch is 90 µm (3.5 mils). The silicon chip is mounted on the substrate by conductive adhesive. The die and the interposer are connected by wire bonding. The package is encapsulated by liquid resin. The total thickness of the encapsulation is 0.7 mm. The package size is 1.5 mm wider than the chip size. If the die is larger than 7.5×7.5 mm, the ChipArray package satisfies the dimension requirement for CSP.

The second-level interconnects of ChipArray may be either BGA solder balls (for organic substrate) or LCC flat lands (for ceramic substrate). The total package thickness is less than 1.5 mm (maximum, depending on the ball size) for the former and is 1.0 mm for the latter. The cross section of the package structure for both configurations is presented in Fig. 19.1. The detailed specifications for the LCC version have not been reported yet. For the ChipArray BGA, the package outline is illustrated in Fig. 19.2. The solder-ball pitch ranges from 1.0

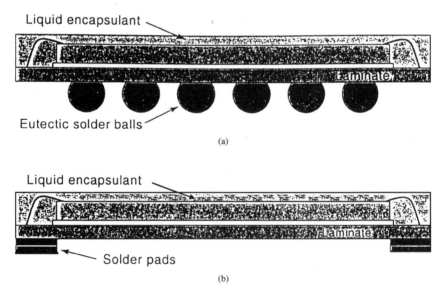

Liquid encapsulant

Laminate

Eutectic solder balls

(a)

Liquid encapsulant

Laminate

Solder pads

(b)

Figure 19.1   Package structure for Amkor/Anam's ChipArray: (*a*) BGA version; (*b*) LCC version.

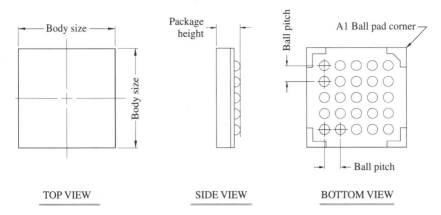

Figure 19.2   Outline of ChipArray BGA package.

mm to 0.5 mm, as presented in Table 19.1 [2]. The package may be square or rectangular in shape. The area array can be full-grid or perimeter. Amkor's ChipArray package was developed for devices with a low to medium number of I/Os. The current ChipArray lineup has pin counts from 32 to 128. Pictures of some members of the ChipArray BGA family are shown in Fig. 19.3.

19.3   Material Issues

The ChipArray package is a plastic-encapsulated CSP using a rigid substrate as the interposer. In addition to the silicon chip, the packaging materials include the rigid substrate, the die-attach adhesive, the bonding wires, the encapsulant, and the bottom solder terminals. The interposer of ChipArray may be organic laminate (for the BGA version) or thick-film ceramic (for the LCC version). The specifications for the latter have not yet been reported. The ChipArray BGA is basically a mini-PBGA package. The standard material for the organic substrate is Mitsubishi's BT (bismaleimide triazine). Other equivalent materials may be considered. The nominal thickness of the BT substrate is 0.34 mm. The Cu traces are finished with Ni/Au electroplating in order to provide an Au wire-bondable surface.

   The ChipArray uses a low-stress conductive adhesive such as QMI 596 for die attachment. The wire-bonding material is Au. This package is encapsulated with a liquid resin such as Hysol 4450. The solder for the BGA balls or LCC lands is 63Sn/37Pb.

TABLE 19.1 Amkor/Anam's ChipArray BGA Package Family

Body size (mm × mm)	Square options									Rectangular options		
	1.0-mm pitch			0.8-mm pitch			0.5-mm pitch			0.75-mm pitch		
	Solder-ball matrix	Maximum ball count	Ball matrix type	Solder-ball matrix	Maximum ball count	Ball matrix type	Solder-ball matrix	Maximum ball count	Ball matrix type	Body size (mm × mm)	Maximum array	Maximum ball count
5 × 5	N/A	N/A	N/A	N/A	N/A	N/A	8 × 8	48	2-row	5 × 8	4 × 8	32
6 × 6	N/A	N/A	N/A	6 × 6	36	Full	10 × 10	64	2-row	6 × 8	5 × 8	40
7 × 7	6 × 6	36	Full	7 × 7	49	Full	12 × 12	80	2-row	6 × 9	5 × 8	40
8 × 8	7 × 7	49	Full	8 × 8	64	Full	14 × 14	96	2-row	7 × 9	6 × 8	48
9 × 9	8 × 8	64	Full	9 × 9	81	Full	16 × 16	112	2-row	8 × 11	6 × 8	48
10 × 10	9 × 9	81	Full	10 × 10	100	Full*	18 × 18	128	2-row	8 × 13	6 × 8	48
11 × 11	10 × 10	100	Full	12 × 12	128	4-row	N/A	N/A	N/A	N/A	N/A	N/A

*Design-dependent.

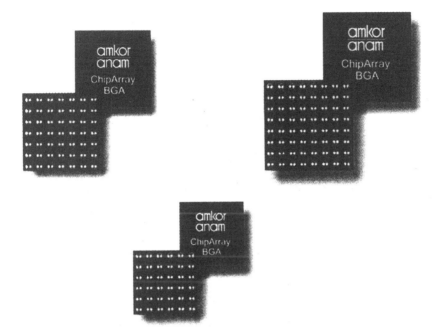

Figure 19.3   Top and bottom views of various ChipArray BGA packages.

19.4   Manufacturing Process

Although Amkor claims that the ChipArray packages have two configurations, it seems that the LCC version is still under development.
No detailed information about the ChipArray LCC has been released
so far. The fabrication of the ChipArray BGA is similar to that of
PBGA. The major steps in the manufacturing process are presented
in Fig. 19.4. Once the silicon chip and the rigid substrate have been
prepared, the die is attached to the interposer with a face-up configuration. Afterward, wire bonding is performed to establish the interconnection between the chip and the substrate.

The next step is encapsulation. It should be noted that the
ChipArray is so named because the fabrication of packages is done in
an array format. A number of chips are mounted on the organic substrate; this is in a strip panel form, and each row in the strip contains
several units. The whole strip is encapsulated with a liquid resin followed by laser marking. Afterward, solder balls are attached to the
bottom of the substrate by conventional IR reflow.

The last step in the manufacturing process is package singulation.
The whole strip is flipped over and mounted on the dicing tape (blue
mounting tape) with the solder balls facing up. The packages are sep-

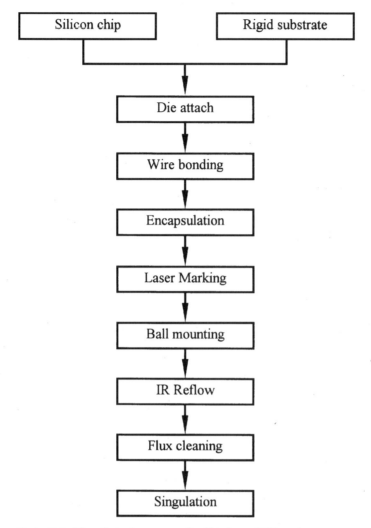

Figure 19.4 Manufacturing process for ChipArray BGA package.

arated by mechanical sawing. The conventional wafer saw can be used for this process. After singulation, the packages may be examined by test probes before packing.

19.5 Performance and Reliability

The electrical and thermal performance of the ChipArray BGA have been evaluated by simulation. Depending on the trace length, the self-inductance ranges from 1.4 to 7.8 nH. The package impedance is

170 Ω. The thermal resistance θ_{ja} under natural convection is 50 to 70°C/W, depending upon the package size. The simulated conditions include no thermal enhancement, two-layer PCB, and 1-W power test level.

The package reliability of the ChipArray BGA has been verified by a series of qualification tests. The moisture sensitivity has been classified as JEDEC level 3. Among 22 tested specimens, no delamination was found under 30°C/60 percent RH for 192 h. An autoclave test (PCT) was performed as well. Fifty specimens were tested under 121°C/100 percent RH/2 atm for 504 h without any failure. In addition, the qualification results for high-temperature (HT) storage (150°C, 1000 h) and temperature/humidity/bias (THB) test (85°C/85 percent RH, 1000 h) were satisfactory. The final package reliability test for the ChipArray BGA was temperature cycling (TC) between −55 and 150°C. The test results showed that this package can survive 1000 cycles of temperature cycling.

19.6 Applications and Advantages

Amkor/Anam's ChipArray is a near-die-size package with rigid substrate. This package was developed for IC devices with a low to medium number of I/Os. The current ChipArray family has pin counts from 32 to 128. The terminal pitch is between 0.5 and 1.0 mm. The possible applications of the ChipArray BGA package include memory modules, analog devices, ASICs, and simple PLDs which can be used in wireless telecom systems, notebook PCs, and PDAs. The ChipArray LCC may be mounted on the flex circuit tape and find suitable applications in disk drives.

Flexibility is the most significant attribute of the ChipArray packages. This series of CSPs can be offered with a standard package size/pin count or a custom-designed outline. For the latter, the body size is 1.5 mm larger than the chip size and the maximum package height is 1.5 mm. The BGA may be in any pattern with a minimum ball pitch of 0.5 mm. Currently Amkor/Anam is finishing its internal qualifications for the ChipArray BGA and is building risk production. The initial production volume is planned at 1 million units per month.

19.7 Summary and Concluding Remarks

The ChipArray technology was developed by Amkor/Anam to implement its chip scale packages. The ChipArray is so named to highlight that the packages are manufactured in an array format. This series of CSPs uses a rigid substrate such as organic laminate or thick-film ceramic as the package interposer. The first-level interconnects are wire bonds, while the second-level interconnects may be either BGA solder balls or LCC flat lands. The package encapsulation is performed with a liquid resin. The package size is 1.5 mm larger than the chip, and the maximum package thickness is 1.5 mm.

Amkor's ChipArray was developed for packaging devices with a low to medium number of I/Os. The current ChipArray family has pin counts from 32 to 128. The terminal pitch is between 0.5 and 1.0 mm. Flexibility is the most significant attribute of this technology. The applications of this package may include memory modules, analog devices, ASICs, and simple PLDs. The ceramic version with the LCC format is ideal for disk drive applications.

19.8 References

1. T. Glenn, "Chip Scale Packaging," *Proceedings of NEPCON–San Antonio,* San Antonio, Tex., October 1996, pp. 103–107.
2. Amkor ChipArray Data Sheet, January 1997.

20

EPS's
Low-Cost Solder-Bumped
Flip Chip NuCSP

20.1 Introduction and Overview

EPS's low-cost chip scale package, NuCSP, is for memory chips and not-so-high-pin-count ASIC applications. It is very similar to Motorola's slightly larger than IC carrier (SLICC) [1], Chapter 24, except that NuCSP uses 63wt%Sn/37wt%Pb eutectic solder–bumped (instead of the high-temperature 95wt%Pb/5wt%Sn or 97wt%Pb/3wt%Sn solder–bumped) chips and organic-coated copper (OCC) FR-4 or bismaleimide triazine (BT) substrates without 63wt%Sn/37wt%Pb solder coating. Also, NuCSP does not have solder balls and is an LGA package. Thus, NuCSP is lower cost than SLICC.

Another CSP which is very similar to NuCSP is Matsushita's LGA [2], Chapter 23. Matsushita uses gold stud bump bonding (SBB) technology with isotropic conductive adhesive to assemble the chip on the ceramic substrate. Thus, this CSP is more expensive than NuCSP and the CSP's solder joints on PCB are not as reliable as NuCSP's because of the large thermal expansion mismatch (TEM) between the ceramic substrate and the FR-4 PCB.

A wire-bonding version of NuCSP has been reported [3, 4]; it will not be mentioned in this book. It should be pointed out, however, that the electrical performance of the wire-bonding NuCSP is not as good as that of the solder-bumped flip-chip NuCSP.

20.2 Design Concepts and Package Structure

Figure 20.1 schematically shows a cross section of the NuCSP with solder-bumped flip-chip technology. It is for chips with low-power and low-pin-count applications. The design concept utilizes the FR-4 or

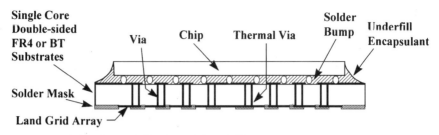

Figure 20.1   Solder-bumped flip-chip LGA—NuCSP.

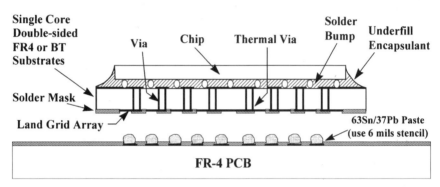

Figure 20.2   Solder-bumped flip-chip NuCSP on PCB.

BT substrate to redistribute the very fine-pitch peripheral pads on the chip to much larger-pitch area-array pads on the PCB. This is done by fanning the traces inward from the peripheral pads on the top layer of the substrate, then dropping vias to connect to the pads on the bottom side of the package substrate as illustrated in Fig. 20.1. NuCSP has the following specific design features:

- It is an LGA package that can be soldered onto the PCB through a 0.15-mm-thick (6-mil) solder paste, Fig. 20.2, and formed solder joints that are 3 mils (0.08 mm) high.

- It is a single-core and two-layer package.

- Traces from the peripheral pads are redistributed (fanning) inward under the die.

- Redistribution traces are connected to the copper pads on the bottom side of the package substrate through vias.

- The package copper pads are area-array at 0.5-, 0.75-, 0.8-, and 1-mm pitch.

- The package body size is about equal to die size + 1 mm.

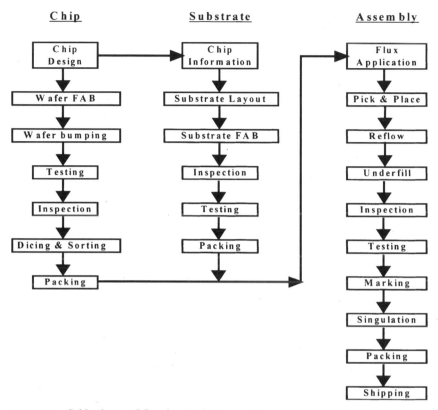

Figure 20.3 Solder-bumped flip-chip NuCSP assembly process.

- It is SMT compatible.
- It has self-alignment characteristics.
- With underfill, the flip-chip solder bumps are reliable.
- The solder joints on the PCB are reliable because of the small TEM.

NuCSP is a very simple and low-cost package. The assembly process is shown in Fig. 20.3. In the following sections, it will be demonstrated, as an example, by housing a 32-pin static random access memory (SRAM).

20.3 Material Issues

The functional 32-pin SRAM, Fig. 20.4a and b, is designed and manufactured at very high yield and low cost by United Microelectronics Corporation (UMC) on an 8-in wafer. This SRAM chip is primarily

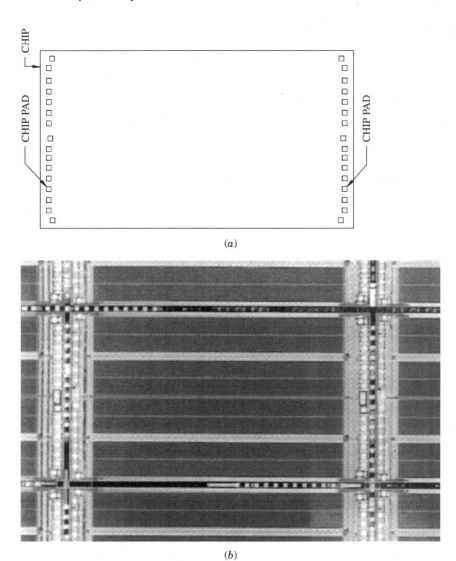

(a)

(b)

Figure 20.4 *(a)* 32-pin SRAM chip-pad geometry; *(b)* 32-pin SRAM chips.

used for very high-speed and low-power applications. The major characteristics of the chip for designing the NuCSP are as follows:

- Chip sizes are 3.556×6.3246 mm.
- Pad sizes are 0.09×0.09 mm.
- Pad pitch is 0.192 mm (minimum).

- Chip thickness is 0.675 mm.
- Chip pads are distributed on the two shorter sides.
- There are two pads for ground and two pads for power.

20.3.1 Wafer bumping and solder-bump characterizations

The 8-in wafers are solder-bumped by the electroplating method, Fig. 20.5. The under-bump metallurgies (UBM) of the wafers are titanium (Ti) and copper (Cu), Fig. 20.6. They are sputtered on the entire surface of the wafer, 0.1 to 0.2 μm of Ti first, followed by 0.5 to 0.8 μm of Cu. A 35-μm layer of resist is then overlaid on the Ti/Cu and a solder bump mask is used to define the bump pattern. The openings in

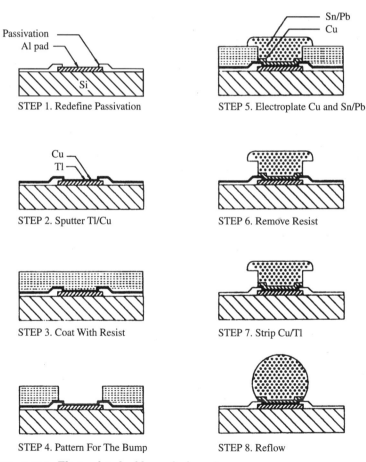

Figure 20.5 Electroplated solder wafer-bumping process.

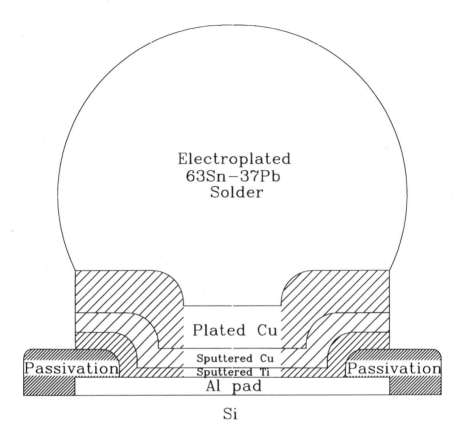

Figure 20.6 Schematic of a solder bump (not to scale).

the resist are 7 to 10 µm wider than the pad openings in the passivation layer. An 8- to 10-µm layer of Cu is then plated over the Ti/Cu, followed by electroplated 63wt%Sn/37wt%Pb solder. In order to plate enough solder to achieve a final solder-ball height of 90 µm, the solder is plated over the resist coating by about 12 µm to form a "mushroom" bump. The resist is then removed and the Ti/Cu is stripped off with a hydrogen peroxide etch. The wafers are then reflowed at 215°C, which creates smooth spherical solder bumps as a result of the surface tension, Fig. 20.6.

20.3.2 Solder-bump height measurements

The solder-bump height distribution on the wafers was investigated, since large variations can cause problems during the assembly process. Nikon microscope MM-40 was used for bump-height mea-

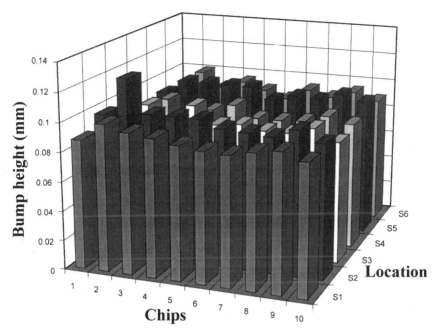

Figure 20.7   Solder-bump height distribution.

surement. The solder-bumped chip was made to stand up by a double-side adhesive tape on the base of the microscope. Two-corner and one-center bumps on both sides of the chip were measured. The locations S2 and S5 were the center bumps and had the largest values, since they had the largest pads. A three-dimensional plot of 10 different chips is shown in Fig. 20.7. It can be seen that the average height is about 90 μm.

20.3.3   Solder-bump strength measurements

The shear (push) strength of the solder bumps was measured with the Royce Instruments Model 550. The shear wedge was placed against the edge of the solder bump. The wedge position was about 45 μm from the top surface of the chip. The speed of the shear wedge was about 100 μm/s. A total of 46 solder bumps from several chips were measured; the results are shown in Fig. 20.8. The average bump strength was 32.7 g, and the standard deviation was about 4.3 g. The failure location is in the solder and not at the UMB.

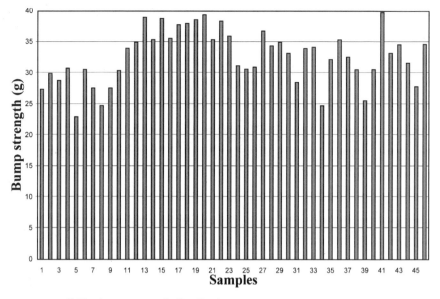

Samples

Figure 20.8 Solder-bump strength distribution.

20.4 NuCSP Substrate Design and Fabrication

Based on the above design concept and chip characteristics, a NuCSP has been designed for the 32-pin SRAM. Figures 20.9 and 20.10 show, respectively, the top and bottom sides of the substrate, which is a 22-mil-thick (0.55-mm) high–glass transition temperature organic material and has the following characteristics:

- The top-side pad diameter is 0.088 mm.
- The top-side pad pitch is 0.192 mm (minimum).
- The top-side solder-mask opening diameter is 0.1398 mm.
- The top-side solder-mask opening pitch is 0.192 mm (minimum).
- The trace width is 3.5 mils (0.0875 mm).
- The trace spacing is 3.5 mils (0.0875 mm) (minimum).
- The via diameter is 12.0 mils (0.3 mm).
- The via land is 20.0 mils (0.5 mm).
- The bottom-side land pad diameter is 0.5 mm.
- The bottom-side land pad pitch is 1.0 mm.
- The bottom-side solder-mask opening diameter is 0.4 mm.

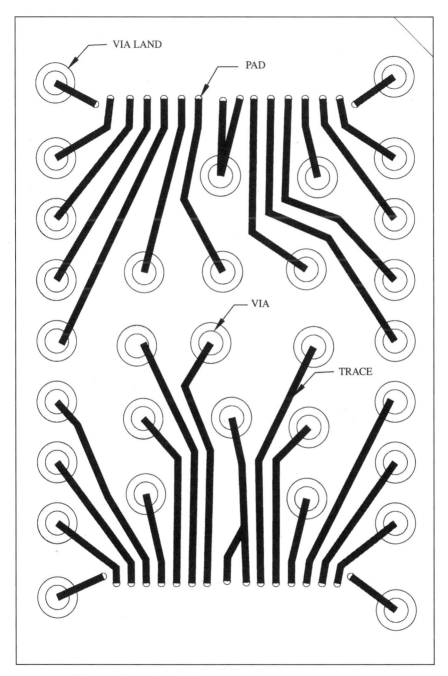

Figure 20.9 Substrate layout (top side).

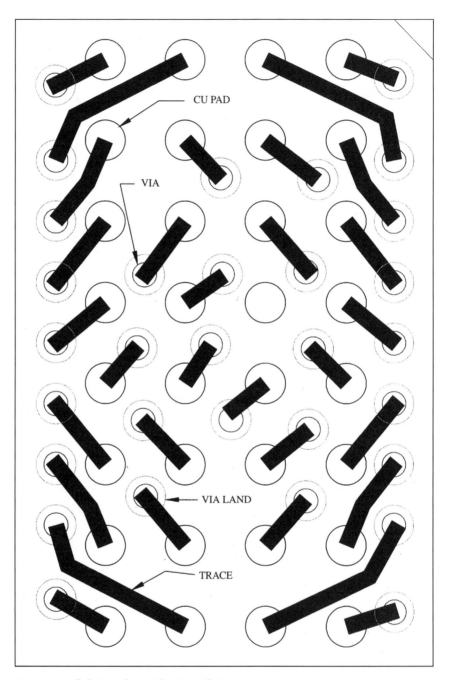

Figure 20.10   Substrate layout (bottom side).

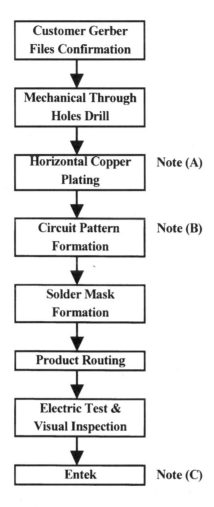

Note: (A) WWEI's unique plating process yields very smooth copper wall thickness in the drilled holes.
  (B) WWEI's unique pattern formation process yields 3/3 (mils) line width/space with more than 80% yield.
  (C) WWEI's Entek process uses Enthone 106 plus.

Figure 20.11 NuCSP substrate fabrication process.

The NuCSP substrate is manufactured by the process shown in Fig. 20.11. It can be seen that the process steps are very simple and conventional, except for the unique horizontal copper plating process, which yields very smooth copper wall thickness in the drilled holes, and the unique circuit pattern formation process, which yields fine line/space widths (3/3 mils or 0.08 mm) at more than 80 percent

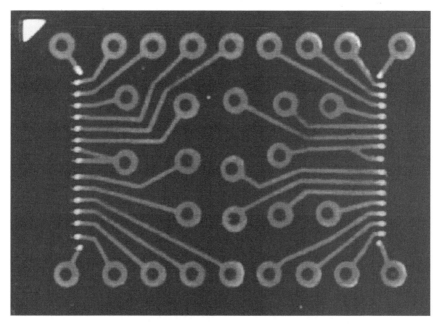

Figure 20.12   NuCSP substrate (top side) for the 32-pin SRAM.

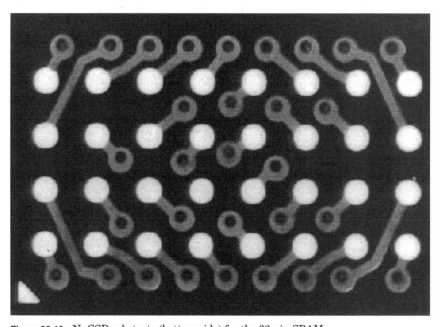

Figure 20.13   NuCSP substrate (bottom side) for the 32-pin SRAM.

yields. Figures 20.12 and 20.13 show, respectively, the top and bottom sides of the NuCSP substrate for the 32-pin solder-bumped flip-chip SRAM. It can be seen that they are very simple and low cost.

20.5 NuCSP Assembly Process

The assembly process for the NuCSP for the 32-pin SRAM is shown in Fig. 20.3. In the following sections, some of the major steps will be discussed.

20.5.1 Fluxing and pick and place

Before fluxing and pick and place, both the solder-bumped chip and the corresponding substrate are cleaned with isopropanol to reduce particle contamination. In this study, a no-clean flux is applied to the substrate only. The pick and place of the chip is performed on a Research Devices Flip Chip Aligner Bonder with a look-up and look-down camera for alignment. The alignment accuracy is within ± 12 μm. After alignment, the solder-bumped chip is flipped down to the substrate.

20.5.2 Solder reflow

After fluxing and pick and place, the flip chip on substrate is placed on the conveyor belt of a DIMA SMT system with nitrogen gas (50 ppm) environment for reflowing the solder bumps. The reflow temperature profile is shown in Fig. 20.14.

20.5.3 Inspection

Because of the self-alignment characteristic of solder bumps during reflow, assembly of the flip chip on the substrate is a very robust process [5–9]. Usually, for a mature manufacturing process it is not necessary to have an x-ray machine on the manufacturing floor to inspect every single flip-chip assembly. Occasional random inspection is more than adequate. For process development, however, the x-ray machine is a must for improving the process by checking the alignment, bridging, and voiding.

In this study, the x-ray machine was used to check the alignment between the bumped chip and the substrate and the solder-short between the bumps. Figure 20.15 shows the x-ray image of the solder bumps, copper pads, and traces by tilting the NuCSP assembly at an angle. It can be seen that the alignment between the solder bumps and copper pads is very accurate, and there is no solder bridging. Alignment can also be checked by cross-sectioning the encapsulated flip-chip NuCSP; see, for example, Fig. 20.16.

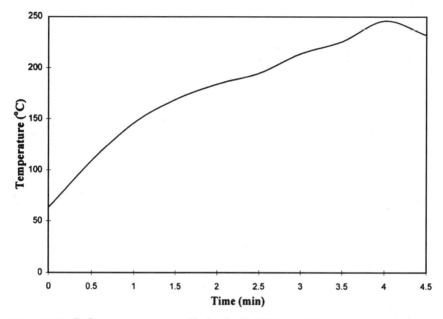

Figure 20.14 Reflow temperature profile for the NuCSP assembly.

Figure 20.15 X-ray image of solder joints (NuCSP is tilted with an angle).

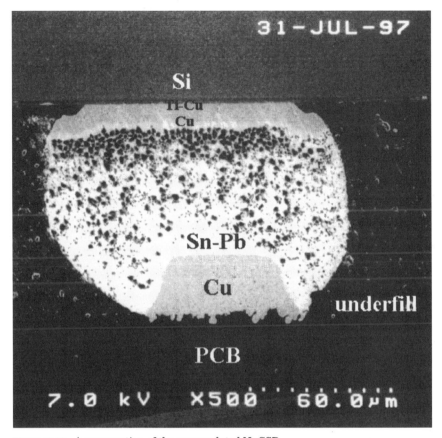

Figure 20.16   A cross section of the encapsulated NuCSP.

20.5.4   Underfill application

Because of the very large TEM between the silicon chip and the laminate organic substrate, the solder bumps are subjected to a very large deformation and may crack. One way to ensure solder-bump reliability is to use underfill encapsulant to cement the chip on the substrate. In this case, the encapsulant is not only preventing the solder bumps from cracking but also protecting the chip from moisture, ionic contaminants, radiation, and hostile operating environments such as mechanical shock and vibrations. Selection of underfill materials was discussed in Chap. 1 of this book. Table 20.1 shows the physical and mechanical properties of the underfill used for assembling the NuCSP.

TABLE 20.1   Curing Conditions, Flow Rate, Material Properties, and Mechanical
and Electrical Performance of the Underfill Material used for the NuCSP

Filler content (%)		60
Filler size (μm)		<20
Average filler size (μm)		4
Underfill resin		Bisphenol-type epoxy
Curing conditions: temperature (°C)/time (min.)		150/35
		160/15
		165/10
		170/7
TCE (ppm)		26.5
T_g (°C)		140
Viscosity (poise)		108
Storage modulus (GPa)		
25°C		6.73
55°C		6.55
110°C		4.44
Moisture uptake (%)		
Dry condition		0.053
After 20 h of steam aging		0.283
Flow rate (mm/s)		0.3
Shear strength (kgf)		
Dry condition		65.5
After 20 h of steam aging		62.2
Voltage readout (V)		
Without underfill	High	4.06
	Low	3.91
With underfill at dry condition	High	4.05
	Low	3.88
With underfill after 20 h of steam aging	High	4.04
	Low	3.85

20.6   NuCSP Mechanical and Electrical Performance

In this section, the effects of the underfill material on the mechanical
and electrical performance of the NuCSP will be discussed.

20.6.1   Mechanical performance of NuCSP

The Royce Instruments Model 550 was used to perform mechanical
shear tests of the NuCSP. The shear wedge was placed against one
edge of the solder-bumped flip chip with underfill on the BT substrate,

which is clamped on the stage. A push of the wedge was applied to shear the chip/bumps/underfill away. This is a destructive test! The average rest results for the solder-bumped flip-chip assembly with underfill are shown in Table 20.1. It can be seen that the shear force of the solder-bumped flip chip with the underfill drops only from 65.5 kgf to 62.2 kgf after 20 h of steam aging. For both dry and steam-aging conditions, the silicon chip is broken into many small pieces, with some remaining on the BT substrate; this shows good adhesion of the underfill material.

20.6.2 Electrical performance of NuCSP

The effect of the underfill material on the electrical performance of the 32-pin NuCSP was tested under 5 V dc using a Kepco power supply. The average voltage readout of the solder-bumped flip chip with and without the underfill is summarized in Table 20.1. It can be seen that under dry conditions, there is almost no difference in voltage readout between the chips with underfill and those without underfill. Also, for solder-bumped flip chips with underfill, the difference in voltage readout before and after a 20-h steam aging is insignificant.

20.7 NuCSP Solder Joint Reliability on PCB

What good is a new first-level package if the solder joints on the PCB are not reliable? The elastoplastic-creep strain in the second-level solder joints of the NuCSP on the PCB was determined by the finite-element method, and the thermal fatigue life of the solder joints was predicted from the average viscoplastic strain energy density and the linear fatigue crack growth rate equation.

Figures 20.17 and 20.18 show the finite-element model of the NuCSP. It consists of the chip, underfill, substrate, solder joints, and PCB. (It is noted that the focus of this analysis was the second-level solder joint reliability, not the solder-bump reliability; thus the solder bumps are not modeled.) Because of double symmetry, only one-quarter of the NuCSP is modeled. All materials in the model (Table 20.2) are assumed to be linear elastic except the 63wt%Sn/37wt%Pb eutectic solder joints, which are elastoplastic and obey Norton's steady-state creep relation.

$$d\gamma_{crp}/dt = B^* \exp\left(-\Delta H/k\mathrm{T}\right)\tau^n$$

In this equation, $d\gamma_{crp}/dt$ is the creep shear strain rate, τ is the shear stress, ΔH is the activation energy (= 0.49 eV), T is the absolute temperature, k is Boltzmann's constant (= 8.617×10^{-5} eV/K), and n is the stress exponent. For 63wt%Sn/37wt%Pb eutectic solder, the mate-

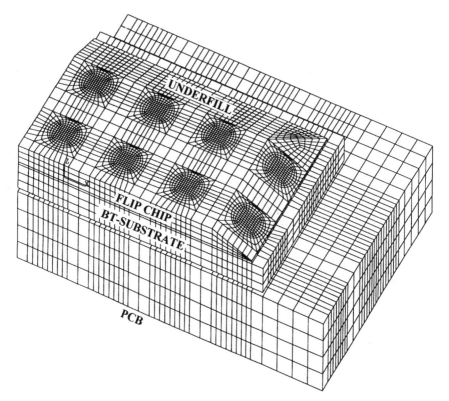

Figure 20.17   Finite-element model of ¼ of the 32-pin NuCSP PCB assembly.

rial constants of the Norton equation are given in Table 4.2 of [8] as n = 5.25 and $B^* = 0.205$ MPa$^{-5.25} \cdot$ s^{-1}. The temperature-dependent stress-strain curves and Young's modulus for the solder are given in Figs. 20.19 and 20.20.

The temperature loading imposed on the NuCSP assembly is shown in Fig. 20.21. It can be seen that for each cycle (60 min), the temperature condition was between -20 and $110°$C with 15 min ramp, 20 min hold at hot, and 10 min hold at cold. Two complete cycles were executed.

The whole-field deformation of the second-level solder joints is shown in Fig. 20.22. It can be seen that the maximum deflection is at the corner solder joint. This is due to the TEM in the X and Y directions between the chip, the BT substrate, and the FR-4 epoxy glass PCB. Also, the location of the maximum stress and strain hysteresis responses is at the corner solder joint. Within the corner solder joint, the maximum shear stress and shear strain are located at the lower right-hand corner.

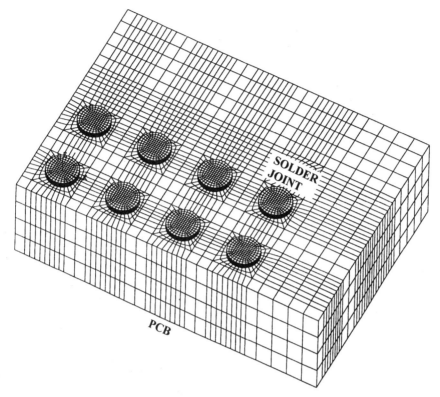

Figure 20.18   Finite element model of ¼ of the assembly (showing the solder joints on PCB).

TABLE 20.2   Material Properties of the NuCSP PCB Assembly

Component	Material	Young's modulus (MPa)	Poisson's ratio	TCE (ppm/°C)
Chip	Silicon	131,000	0.23	2.8
Substrate	BT	19,600	0.18	15
PCB	FR4	22,000	0.28	18
Underfill	Epoxy	7,840	0.35	27
Solder ball	63wt%Sn/37wt%Pb elastoplastic-creep	Temperature-dependent	0.37	22.4

The viscoplastic strain energy density per cycle at the lower right-hand corner of the corner solder joint can be determined by averaging the results of these two cycles, which is 0.26 MPa (38 psi). The typical size of the element area and height of the solder joint are 2×2 mils (0.05×0.05 mm) and 0.44 mil (0.011 mm), respectively. Once we have the average viscoplastic strain energy density per cycle ΔW, the ther-

Material Property of Eutectic Solder

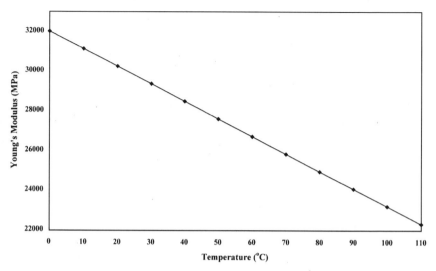

Figure 20.19   Temperature-dependent Young's modulus of 63wt%Sn/37wt%Pb solder.

Stress-Strain Relation for 63Sn/37Pb Solder

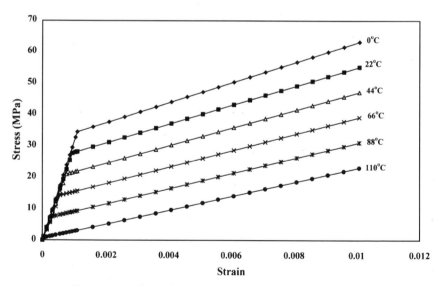

Figure 20.20   Temperature-dependent stress-strain curves of 63wt%Sn/37wt%Pb solder.

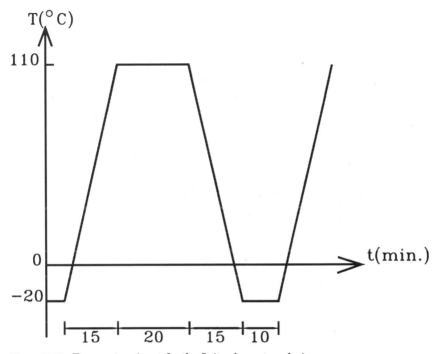

Figure 20.21 Temperature input for the finite element analysis.

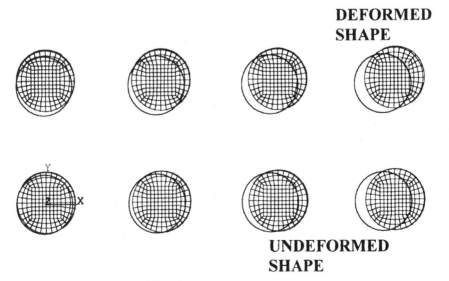

DEFORMED
SHAPE

UNDEFORMED
SHAPE

Figure 20.22 Deformation of the solder joints.

mal fatigue crack initiation life N_o can be estimated from Eq. (13.35) of [6], i.e.,

$$N_o = 7860\Delta W^{-1}$$

and the thermal fatigue crack propagation life N based on the linear fatigue crack growth rate theory can be estimated from Eq. (13.36) of [6], i.e.,

$$da/dN = 4.96 \times 10^{-8}\Delta W^{1.13}$$

or

$$N = N_o + (a_f - a_o)/(4.96 \times 10^{-8}\Delta W^{1.13})$$

where a is the crack length of the solder joint; a_o is the initial crack length, which is assumed to be zero; and a_f is the final crack length, which is 0.0124 in (the diameter of the solder joint). The total thermal fatigue life of the solder joint is $N_o + N$. For our case, N_o, N, and total thermal fatigue life are, respectively, 207 cycles, 4307 cycles, and 4514 cycles for the maximum-strained corner solder joint. Thus, the NuCSP solder joints are reliable in most operating conditions.

20.8 Applications and Advantages

NuCSP is designed for low-pin-count and low-power memory IC and ASIC devices. Because of its low profile through solder-bumped flip-chip interconnect technology, it can be used for miniaturized products.

NuCSP is designed for high-volume production. Because it is simple and utilizes the conventional PCB manufacturing process, the production cost of NuCSP is low.

20.9 Summary and Concluding Remarks

NuCSP is a low-cost solder-bumped flip-chip LGA chip scale package. The key steps in the NuCSP assembly process, such as wafer bumping, substrate design, substrate fabrication, fluxing, pick and place, solder reflow, underfill applications, and inspection, have been discussed here. Furthermore, the effects of underfill on the mechanical and electrical performance of NuCSP have been reported. Finally, the solder joint reliability of NuCSP on PCB has been demonstrated by a three-dimensional nonlinear temperature-dependent finite-element method. Some important results are summarized as follows:

- NuCSP is an LGA package.

- NuCSP has a single-core, two-layer structure.

- NuCSP's size is equal to chip size + 1 mm.
- NuCSP's substrate is manufactured by conventional PCB technology.
- NuCSP is an area-array package with 0.5-, 0.75-, 0.8-, and 1-mm pitch.
- NuCSP is very easy to assemble.
- NuCSP is SMT compatible.
- NuCSP has self-alignment characteristics.
- The solder bumps are reliable.
- The solder joints are reliable.
- NuCSP has high performance.
- NuCSP is low-cost.

20.10   Acknowledgments

The author (JL) would like to thank William Deng, Hermen Liu, and C. T. Lin of UMC for their wafers and useful discussions; C. S. Ho of Aptos for bumping the wafers; T. J. Tseng, David Cheng, and Eric Lao of WWEI for their strong support; and Chris Chang, Tong Chen, Ray Chen, and Livia Hu of EPS for their effective help. I learned a lot about packaging from them.

20.11   References

1. P. Lall, G. Gold, B. Miles, K. Banerji, P. Thompson, C. Koehler, and I. Adhihetty, "Reliability Characterization of the SLICC Package," *Proceedings of the IEEE Electronic Components and Technology Conference,* Orlando, Fla., May 1996, pp. 1202–1210.
2. Y. Kunitomo, "Practical Chip Size Package Realized by Ceramic LGA Substrate and SBB Technology," *Proceedings of the SMI Conference,* San Jose, Calif., August 1995, pp. 18–25.
3. T. Chou and J. H. Lau, "A Low-Cost Chip Size Package—NuCSP," *Circuit World,* vol. 24, no. 1, 1997, pp. 34–38.
4. J. H. Lau, "Solder Joint Reliability of a Low Cost Chip Size Package—NuCSP," *Proceedings of the ISHM Microelectronics Symposium,* Philadelphia, Pa., October 1997, pp. 691–696.
5. J. H. Lau, *Chip on Board Technologies for Multichip Modules,* Van Nostrand Reinhold, New York, 1994.
6. J. H. Lau, *Ball Grid Technology,* McGraw-Hill, New York, 1995.
7. J. H. Lau, *Flip Chip Technologies,* McGraw-Hill, New York, 1996.
8. J. H. Lau and Y.-H. Pao, *Solder Joint Reliability of BGA, CSP, Flip Chip, and Fine Pitch SMT Assemblies,* McGraw-Hill, New York, 1997.
9. J. H. Lau, C. P. Wong, J. L. Prince, and W. Nakayama, *Electronic Packaging: Design, Materials, Process, and Reliability,* McGraw-Hill, New York, 1998.

21

IBM's
Ceramic Mini-Ball Grid Array
Package (Mini-BGA)

21.1 Introduction and Overview

The ceramic mini-ball grid array (mini-BGA) package was reported by IBM in 1995 [1]. This package uses IBM's multilayer ceramic substrate as the interposer. The chip-to-substrate interconnects are C4 solder joints. Underfill adhesive is employed to enhance the reliability of the C4 joints. The package is not molded. Instead, an aluminum cap is used for chip protection and thermal management. The second-level interconnects are BGA solder balls. The ball diameter and pitch are 0.25 mm and 0.5 mm, respectively. The package size is 21×21 mm, and the I/O capacity is up to 1521 (39×39) pins. The total package thickness is 3.4 mm. The main application of this ceramic mini-BGA is for high-speed switching devices. In addition to an active silicon chip, 16 decoupling capacitors are assembled on the alumina substrate. The nominal power rating is 10 W. Along with the mini-BGA, a ceramic test socket was also developed. Therefore, the module can be tested at speed before shipping.

21.2 Design Concepts and Package Structure

The ceramic mini-BGA is a chip scale package with rigid substrate. A schematic diagram of the package structure is illustrated in Fig. 21.1. This package is considered a CSP because of its fine-pitch package I/Os instead of the package-to-chip size ratio. Except for the terminal

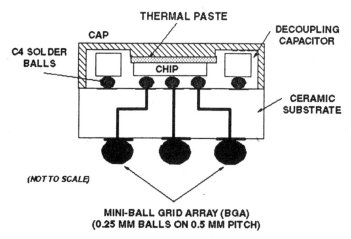

Figure 21.1   Package structure of IBM's ceramic mini-BGA.

Figure 21.2   Silicon chip and decoupling capacitors on the substrate.

pitch and solder-ball material, this package is very similar to IBM's conventional ceramic ball grid array (CBGA) packages [2].

The mini-BGA was developed by IBM mainly for high-speed devices. The silicon die is a 200-MHz switching chip with a size of 10×10 mm. The package interposer is a multilayer ceramic substrate. The size of this substrate is 21×21 mm. Therefore, the package-to-chip size ratio does not conform to the conventional definition for

Figure 21.3 Bottom surface of alumina substrate with mini-ball grid array.

CSP. The first-level interconnects are C4 solder joints with a pitch of 0.23 mm. In addition to the active switching chip, 16 decoupling capacitors are mounted on the same side of the interposer, as shown in Fig. 21.2. In order to improve the thermal fatigue life, underfill adhesive is employed to encapsulate the C4 solder joints.

The package molding is not required for mini-BGA. Instead, a metal cap is adopted for chip protection and thermal management. The cap is nonhermetically sealed. For better heat conduction, a thermal paste is used to connect the backside of the active chip and the bottom of the cap. The package terminals for board-level assembly are BGA solder balls, as shown in Fig. 21.3. The ball diameter is 0.25 mm, and the pitch is 0.5 mm. This package has I/O capacity up to 1521 (39×39) pins. Including the Al cap and the BGA balls, the total package thickness is 3.4 mm. The major attributes of IBM's ceramic mini-BGA are given in Table 21.1.

21.3   Material Issues

In principle, the mini-BGA is a ceramic ball grid array package with finer pitch. The packaging materials include the ceramic substrate, the C4 solder joints, the underfill adhesive, the metal cap, the mounting thermal paste, and the BGA solder balls. The interposer of the mini-BGA is an alumina multilayer ceramic substrate fabricated by IBM, East Fishkill. This interposer is plated with Ni and thin Au on the area-array solder pads at both sides of the substrate. The upper

TABLE 21.1   Major Attributes of IBM's Ceramic Mini-BGA

Chip	
Size	10 mm square
No. of C4 bumps	1195
Speed	200 MHz

Package	
Material	Alumina
Internal metal	Molybdenum
Size	21 mm square
Thickness	3.4 mm
Capacitor	16 (32 nF each)
DCAP type	IBM C4 terminated
Total pkg. I/O	1521 (39×39 array)
Signal I/O	750
Pkg. I/O type	Mini-BGA
Solder type	60/40 Sn/Pb eutectic
I/O pitch	0.25 mm on 0.5 mm
Cap	Al
Power	10 W/12 W (nominal/maximum)
Thermal paste	2.8 W/m · K
Cap seal	Nonhermetic
Card attach	Solder reflow (240–250°C)

side is for the mounting of a C4 solder-bumped flip chip and decoupling capacitors, while the bottom side is for the attachment of mini-BGA solder balls. The C4 pad diameter is 0.12 mm and the pad pitch is 0.23 mm, as shown in Fig. 21.4. On the other hand, at the BGA side, the solder pad pitch is 0.5 mm. The substrate size is 21×21 mm, which is able to accommodate 1521 (39×39) I/Os.

The term *C4* stands for controlled-collapse chip connection, which is a flip-chip bumping technology developed by IBM in the 1960s [3]. This technology uses high-Pb solder as the joint material. For the mini-BGA, the C4 solder bump is 97Pb/3Sn, which has a melting temperature of around 310°C. Since the C4 solder joints are rather small (of the order of 100 μm), the reliability under thermal loading is a major concern. In the mini-BGA package, the CTEs of the silicon chip and the alumina substrate are approximately 3 and 7 ppm/°C, respectively. Therefore, an underfill adhesive is used to encapsulate the C4 solder joints in order to ensure interconnect reliability.

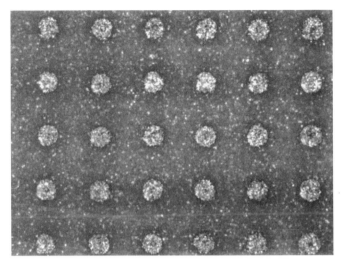

Figure 21.4   Top surface of alumina substrate with C4 pads.

The metal cap of the mini-BGA is made of aluminum. In order to improve the heat-dissipation capability, a thermal paste is employed to connect the backside of the silicon chip and the bottom of the Al cap. The thermal conductivity of this paste is 2.8 W/m · K. For IBM's conventional CBGA, the package I/Os are 90Pb/10Sn solder balls and the ball diameter is 0.875 mm (35 mils). However, for the ceramic mini-BGA, the solder balls are made of 60Sn/40Pb and the diameter is 0.25 mm (10 mils). This can be considered the major difference between these two packages, other than the shrinkage in package size.

21.4   Manufacturing Process

The fabrication of mini-BGA packages is straightforward. The process flow is presented in Fig. 21.5. Once the multilayer ceramic substrate and the C4 solder-bumped silicon chip/capacitors are prepared, the next procedure is to mount the chip/capacitors on the interposer by reflow heating. The reflow temperature is in the range of 350 to 370°C because the C4 joints are high-lead solder (97Pb/3Sn). After electrical testing, the underfill adhesive is dispensed and cured to encapsulate the C4 solder joints. The next few steps are thermal paste dispensing, Al cap installation, curing, and sealing, followed by leak tests. Then the package is ready for BGA ball attachment, as depicted in Fig. 21.6.

Since the solder balls of the ceramic mini-BGA are rather small, a stainless steel mold is needed to make the solder ball preforms while transferring them to the package substrate. The solder mold is fabricated with precision-machined cavities. The geometrical shape and

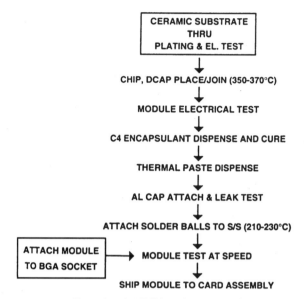

Figure 21.5 Ceramic mini-BGA package assembly process flow.

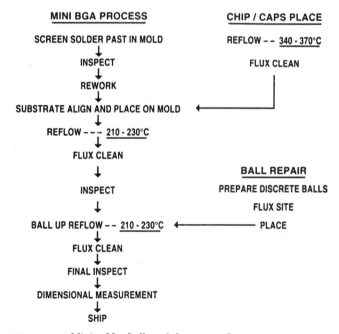

Figure 21.6 Mini solder-ball module process flow.

aspect ratio of the cavities are calculated and modeled to obtain the precise volume of solder. During the development of the mini-BGA, two specifications for solder paste volume, namely, 520 and 800 cubic mils, were designed. The cavities are filled with solder paste by squeegeeing across the mold as shown in Fig. 21.7a. The screened

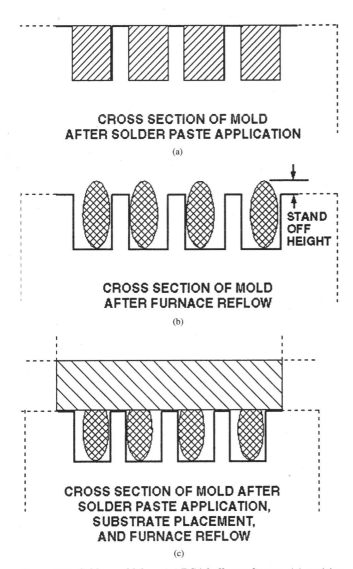

CROSS SECTION OF MOLD
AFTER SOLDER PASTE APPLICATION

(a)

CROSS SECTION OF MOLD
AFTER FURNACE REFLOW

STAND
OFF
HEIGHT

(b)

CROSS SECTION OF MOLD AFTER
SOLDER PASTE APPLICATION,
SUBSTRATE PLACEMENT,
AND FURNACE REFLOW

(c)

Figure 21.7 Solder mold for mini-BGA ball attachment: (a) cavities filled with solder paste; (b) solder reflow without substrate placement; (c) substrate placement over the solder mold followed by the first reflow.

mold is then inspected, and the not-yet-full cavities are refilled and excessive solder paste between the cavities is removed.

Figure 21.7*b* shows the geometry of the solder-ball preforms without the placement of the package substrate. The standoff height above the top surface of the solder mold is critical to the ball-attachment process. The typical height is 2 to 3 mils, which is enough to compensate for the possible warpage of the alumina substrate. For actual solder-ball attachment, the package substrate is placed over the mold with the solder pads aligned with the cavities using a split optical vision device. Then the whole mold is heated to 210 to 230°C to reflow the solder, and the solder-ball preforms are transferred to the bottom of the package substrate, as shown in Fig. 21.7*c*. It should be noted that the stainless steel mold does not interact with the lead-tin solder, and the solder-ball preforms are nonspherical at this moment.

After flux cleaning, inspection is performed for missing or low-volume solder balls. If individual reballing is required, discrete solder-ball preforms can be made as illustrated in Fig. 21.7*b* and then manually placed on the designated solder pads. In order to attach the repaired solder balls and round up all ball preforms, a second reflow at 210 to 230°C is carried out, as shown in Fig. 21.8. The SEM micrograph of the completed mini-BGA solder balls is presented in Fig. 21.9. The measured dimensional characteristics of BGA balls for the two solder-paste volume specifications are given in Tables 21.2 and 21.3.

In addition to the mini-BGA package, IBM also developed a test socket for module testing at speed. The test socket is a multilayer ceramic substrate with through-the-thickness vias connecting pairs of metal pads on both sides of the substrate. The pad diameter is 0.25 mm and the pad pitch is 0.5 mm, which matches the ball pitch of mini-BGA package. These pads are plated with Ni and thin Au. On the bottom side, mini solder balls are attached to the pads using the

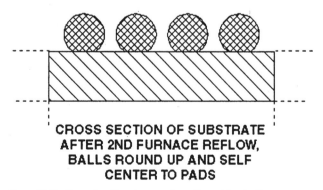

**CROSS SECTION OF SUBSTRATE
AFTER 2ND FURNACE REFLOW,
BALLS ROUND UP AND SELF
CENTER TO PADS**

Figure 21.8 Second reflow to round up the solder balls.

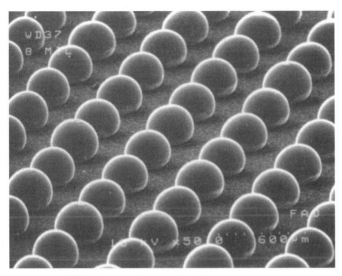

Figure 21.9 SEM picture of min-ball grid array.

TABLE 21.2   Dimensional Characteristics of BGA Balls (520 Cubic Mils of Solder Volume)

	Ball height (mils)			Ball diameter (mils)		
Sample	Minimum	Maximum	Average	Minimum	Maximum	Average
1	7.7	8.4	8.0	8.5	8.8	8.7
2	7.7	8.2	7.96	8.7	9.0	8.9
3	7.9	8.4	8.18	8.7	9.1	8.96
4	7.8	8.4	8.14	8.7	9.2	8.9
5	7.8	8.3	8.05	8.8	9.1	8.96

TABLE 21.3   Dimensional Characteristics of BGA Balls (800 Cubic Mils of Solder Volume)

	Ball height (mils)			Ball diameter (mils)		
Sample	Minimum	Maximum	Average	Minimum	Maximum	Average
1	8.22	11.38	9.46	9.42	10.44	9.84
2	7.95	11.77	9.70	9.77	10.58	10.03
3	8.43	11.29	9.77	9.68	10.44	9.89
4	8.36	10.43	9.39	9.20	9.85	9.53
5	8.43	11.77	9.92	9.55	10.62	9.96

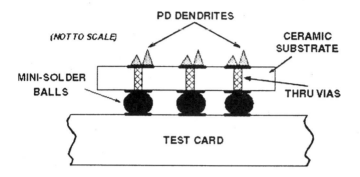

SOCKET SUBSTRATE: 21 MM SQUARE, 1.6 MM THICK

Figure 21.10   Cross section of test socket for IBM's ceramic mini-BGA.

procedures described in the previous section. On the top side, Pd dendrites are deposited on the pads by an electroplating process developed by IBM, Endicott, N.Y. The dendrites consist of 200 to 500 needlelike protrusions per square millimeter. The typical height is 0.2 to 0.3 mm. With standard reflow heating, the ceramic socket is mounted on the test card as illustrated in Fig. 21.10.

The package terminals of the ceramic mini-BGA are mated with the dendritic pads on the top of the test socket. The needlelike dendrites can penetrate the surface oxides on the mini-BGA solder balls and provide reliable contact during the electrical testing. A temporary heat sink is clamped to the top of the package under test in order to improve the cooling and provide dead-weight loading for tighter contact. Since all mini-BGA packages are pretested at speed before shipping, very little rework is expected after the board-level assembly.

21.5   Performance and Reliability

IBM's mini-BGA is a ceramic package with an Al cap. An additional Al heat sink may be installed on top of the Al cap by thermal tape or epoxy. Therefore, good thermal performance is anticipated. The nominal power rating of the ceramic mini-BGA package is 10 W. The maximum allowed power consumption is 12 W.

For most ceramic packages, board-level solder joint reliability is a major concern because of the substantial thermal mismatch between the ceramic substrate and the organic PCB [4]. In the development of the mini-BGA, IBM assembled these packages to a card made of Teflon. The CTE of the Teflon card is 10 to 12 ppm/°C, while that of the ceramic substrate of the mini-BGA is 7 ppm/°C. Since the CTE mismatch is not as large as in the conventional cases, the solder joints may have a prolonged fatigue life. It has been reported that the mini-BGA assembly can sustain 1200 on/off cycles [1]. It should be noted that since the mini-BGA balls are rather small, control of the solder volume is critical to the package-to-card assembly. In addition, according to IBM, a coplanarity of 4 mils or less among the solder balls is required.

21.6 Applications and Advantages

The ceramic mini-BGA package was developed by IBM mainly for high-speed switching devices. In addition to an active silicon chip, 16 decoupling capacitors are mounted on the same package interposer. This package was designed to handle signal switching between processor cards in a symmetric multiprocessor (SMP) configuration. Up to 12 mini-BGA modules and one clock module may be assembled on a Teflon switch card. This card was designed to manage switching among multiple processor cards at a speed of up to 200 MHz.

Besides being designed for high-speed switching applications, the mini-BGA has the major advantages of high power rating and high I/O pin count. The nominal and maximum power ratings are 10 and 12 W, respectively. This package is able to accommodate up to 1521 I/Os. With this capacity, the mini-BGA can be used to package devices with up to 750 signal I/Os and provide a 1:1 ratio for signal and power/ground connections.

21.7 Summary and Concluding Remarks

The ceramic mini-BGA is a chip scale package with rigid substrate. This package is considered a CSP because of its fine-pitch package I/Os rather than its package-to-chip size ratio. Except for the terminal pitch and solder-ball material, this package is very similar to conventional CBGA packages.

The interposer of the mini-BGA is a multilayer alumina substrate fabricated by IBM, East Fishkill. The chip-to-substrate interconnects are C4 joints with 97Pb/3Sn solder. Underfill adhesive is employed to enhance the C4 joint reliability. The package is nonhermetically sealed by an Al cap for chip protection and heat dissipation. The nominal power rating is 10 W. The package terminals are BGA balls with 60Sn/40Pb solder. The ball diameter is 0.25 mm, and the pitch is 0.5 mm. The package size is 21×21 mm, which can accommodate up to 1521 (39×39) I/Os. Besides the mini-BGA package, a ceramic test socket was developed as well. Therefore, the module can be tested at speed before shipping.

The main application of the mini-BGA package is for high-speed switching devices. In addition to an active silicon chip, 16 decoupling capacitors are assembled on the interposer of a single package. Up to 12 mini-BGA modules and one clock module may be assembled on a Teflon switch card. This card was designed to manage switching among multiple processor cards at a speed of up to 200 MHz. Since the CTE mismatch between the ceramic substrate and the Teflon card is not very large, the mini-BGA assembly can sustain up to 1200 power cycles. However, because of the small BGA balls, control of solder volume and ball coplanarity are critical to the package-to-card assembly of ceramic mini-BGA.

21.8 References

1. R. N. Master, R. Jackson, S. K. Ray, and A. Ingraham, "Ceramic Mini-Ball Grid Array Package for High Speed Device," *Proceedings of the 45th ECTC*, Las Vegas, Nev., May 1995, pp. 46–50.
2. T. Caulfield et al., "Surface Mount Array Interconnection for High I/O MCM-C to Card Assemblies," *Proceedings of the ICEMM*, ISHM/MCM, 1993, pp. 320–323.
3. J. H. Lau, *Flip Chip Technologies*, McGraw-Hill, New York, 1995.
4. J. H. Lau and Y. H. Pao, *Solder Joint Reliability of BGA, CSP, Flip Chip, and Fine Pitch SMY Assemblies*, McGraw-Hill, New York, 1997.

IBM's Flip Chip–Plastic Ball Grid Array Package (FC/PBGA)

22.1 Introduction and Overview

IBM reported the preliminary results of technical evaluation of its near-chip-size package in 1998 [1]. The basic configuration of this newly developed package is a flip chip–plastic ball grid array (FC/PBGA). The LSI is a 12×14 mm die with 700 controlled-collapse chip connections (C4). The array pitch of C4 solder bumps is 230 µm. The package interposer is a Driclad laminate with buildup layers. This substrate has an outline of 21×21 mm, which conforms to the JEDEC standard. The package I/Os are 255 BGA solder balls with an array pitch of 1.27 mm. The core technology to implement this FC/PBGA is the buildup surface laminar circuit (SLC) with microvias on the chip carrier. The targeted application of this package is microprocessors. The package reliability of this FC/PBGA has been comprehensively investigated. The results of technical evaluation indicated that this novel package is manufacturable and can satisfy standard reliability requirements.

22.2 Design Concepts and Package Structure

IBM's FC/PBGA is a near-chip-size package with rigid interposer. The cross section of the package structure is illustrated in Fig. 22.1. The LSI is a 12×14 mm flip-chip die with 700 C4 solder bumps. The bump pitch is 230 µm. The flip chip is mounted on a rigid substrate by C4 solder joints and underfill adhesive. The interposer consists of a Driclad laminate and IBM's buildup SLC with microvias. This chip carrier has an outline of 21×21 mm, which conforms to the JEDEC

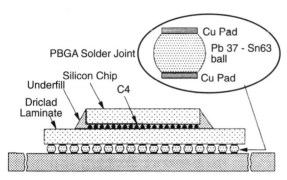

Figure 22.1 Cross section of package structure of IBM's FC/PBGA.

Figure 22.2 Top and bottom views of the rigid interposer of the FC/PBGA.

standard. The second-level interconnects of this package are 255 BGA solder balls. The array pitch is 1.27 mm. The package-to-chip size (or area) ratio of this FC/PBGA is 1.4 (or 2.6). Therefore, IBM considered this newly developed module to be a near-chip-size package instead of a true CSP [1].

The most significant constituent of IBM's FC/PBGA is the multilayer chip carrier. The top and bottom views of this rigid substrate are shown in Fig. 22.2. This interposer provides four signal and two power planes (4S/2P). The core layers are a 2S/2P Driclad laminate. The circuitry is formed by filled plated through holes (PTH) and Cu traces. The diameter of the PTHs is 10 mils, and the design rule for the traces is 3-mil line/5-mil space. The two external signal planes are provided by two SLC layers at each side of the core laminate. A 2-mil line/4.5-mil space design rule is adopted for the traces of the SLC. Microvias with 4-mil diameter are formed by photolithography in the buildup layers for circuit interconnections. A unique feature of the FC/PBGA is that the microvias of the buildup layers directly land on the top of the filled PTHs in the core laminate. In addition, the microvias may exist in the capture pads for C4 joints and BGA solder balls. As a result, the dogbone design is not required, and, hence, the density of interconnects can be substantially increased.

22.3 Material Issues

The FC/PBGA developed by IBM is a near-chip-size package with rigid interposer. The whole module is composed of the silicon die, the C4 solder joints, the underfill encapsulant, the organic substrate, and the BGA solder balls. The interconnects of IBM's FC/PBGA employ conventional C4 and BGA technologies which use high-lead and eutectic Pb/Sn solder materials, respectively. During the development of this package, two commercial underfill materials were used for evaluation. The major difference between these two encapsulants is the glass transition temperature T_g. The comparison in performance will be discussed in Sec. 22.5.

The interposer of the FC/PBGA has several constituents. The main body is made of Driclad dielectric layers together with Cu traces and PTHs. The PTHs are filled with Driclad resin-based paste which contains Cu fillers. At each side of the core laminate, there is an SLC layer. These two buildup layers are made from Advanced Solder Mask's (ASM) photoimagable dielectric material. Microvias are patterned in the buildup layers by photolithography. An additive Cu plating process is employed to form the circuitry for the SLC. On top of the SLC layers, a solder mask is coated for protection and insulation. Two kinds of solder mask were evaluated at the prototyping stage. One solder mask is a liquid photoimagable material, while the other is a photosensitive dry film. The capture pads for the C4 solder bumps and the BGA solder balls are finished with eutectic Pb/Sn solder. Two solder deposition methods were investigated. The comparison will be discussed in the ensuing sections.

22.4 Manufacturing Process

The fabrication of IBM's FC/PBGA consists of two parts. The first part is the preparation of the multilayer interposer. The second part is the assembly of the flip chip and the attachment of BGA solder balls.

Because of the fine pitch and high I/O requirements, the fabrication of the FC/PBGA's substrate is a tedious process. The center core is a Driclad dielectric layer with Cu plating on both sides to provide two power planes. Then additional Driclad dielectric layers are laminated with the center core to establish two signal planes. The circuitry is formed by standard sequential through-hole drilling, full-panel copper plating, and subtractive etching. It should be noted that the PTHs are filled with Driclad resin–based paste which contains Cu fillers.

Figure 22.3   Microvia on the filled PTH.

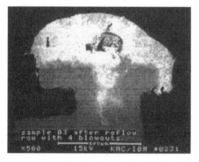

Figure 22.4   Void in the electroplated solder on the microvia.

The aforementioned Driclad laminate provides a 2S/2P circuitry. Once these core layers are established, one SLC needs to be deposited on each side of the laminate to provide additional two signal planes. The buildup layer of the SLC is made of ASM's photoimagable dielectric material. Microvias are patterned in the buildup layer by photolithography. The circuitry of the SLC is formed by an additive Cu plating process. Since the PTHs in the core laminate are filled with conductive paste, the microvias of the SLC are able to land directly on top of the PTHs, as shown in Fig. 22.3. Therefore, no dogbone design is required for the interconnection. In addition, the microvias may be included in the capture pads for the attachment of C4 solder bumps and BGA solder balls. With this feature, the I/O density of the interposer can be substantially increased.

After the SLC layers are completed, a solder mask needs to be coated on the outer surfaces for protection and insulation. Finally, the capture pads on the top and bottom of the interposer are finished with eutectic Pb/Sn solder. During the development of the FC/PBGA,

Figure 22.5 Stencil-printed solder on the microvia (no void).

two methods for the solder finishing were evaluated. The first method was to deposit the solder directly on the Cu pads by electroplating. Although this process is straightforward, voids may be introduced in the solder if a microvia exists in the pad, as shown in Fig. 22.4. The second method was to deposit the solder by stencil printing. The cross-sectional picture in Fig. 22.5 shows no voids in the solder, even with the presence of a microvia. However, in the second method, electroless-plated Ni together with immersion Au is required as the under-bump metallurgy (UBM). For both methods, reflow heating is needed after the deposition of solder, and a coining process is performed to flatten the solder afterwards.

Once the interposer is prepared, the assembly process shown in Fig. 22.6 is conducted to mount the flip chip on the substrate. The last step is the attachment of BGA solder balls to the bottom of the interposer. A graphite mold is employed for the ball placement. The solder balls are attached to the pads by a conventional reflow process. Figure 22.7 presents the cross-sectional picture of a completed FC/PBGA, showing the assembled C4 joints, the PTH in the interposer, and the BGA solder ball.

22.5   Qualifications and Reliability

A comprehensive qualification testing program as shown in Table 22.1 has been carried out to investigate the package reliability of FC/PBGA. The most significant test was thermal cycling (TC). This qualification test was conducted in three different ways. The first one followed the standard specifications of JEDEC A104-A, which uses a temperature range between -55 and $125°C$. Since the T_g of the dielectric material used to build up the SLC is about $115°C$, a modified temperature profile of -25 to $115°C$ was proposed to test some of the specimens under TC. Among this category, 29 specimens were stand-alone packages and 25 specimens were FC/PBGA-PCB assemblies. The ac-

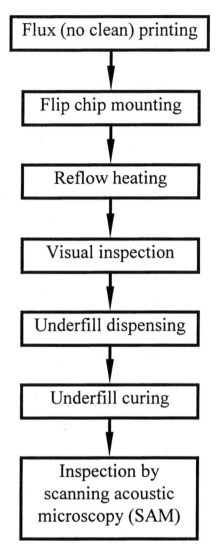

Figure 22.6 Assembly process for the FC/PBGA package.

celerated thermal cycling (ATC) was a standard test to evaluate the fatigue life of assembled packages. The testing temperature range was between 0 and 125°C. The power cycling (PX) tests were conducted to investigate the package reliability under simulated power-on/off conditions. The test chips had built-in heaters and temperature sensors so that the junction temperature could be properly controlled. To study the T_g effect of the SLC (115°C), two temperature conditions (25 to

Figure 22.7 Cross-sectional view of C4
joints, filled PTH, and BGA solder ball.

TABLE 22.1 Summary of Qualification Testing Program

Test	Conditions	Duration	Sample size
TC (JEDEC)	−55 to + 125°C	2350 cycles	43 modules
TC (JEDEC)	−25 to + 115°C	5000 cycles	29 modules
TC (JEDEC)	−25 to + 115°C	5000 cycles	25 modules on cards
ATC	0 to 125°C	4037 cycles	25 modules on cards
PX	25 to 110°C	13,451 cycles	10 modules on cards
PX	25 to 125°C	13,451 cycles	9 modules on cards
HAST	110°C/85% RH/5.5 V	200 h	10 modules on cards
THB	85°C/85% RH/5.5 V	1077 h	20 modules on cards
PCT	121°C/100% RH/2 atm	96 h	14 modules
HTS	150°C	1240 h	10 modules

110°C and 25 to 125°C) were used for testing. The aforementioned
three tests involved temperature cycling. All samples were tested for
the number of cycles specified in Table 22.1. The testing results
showed that most failures were attributed to manufacturing defects
during fabrication. In particular, the major damage mode appeared in
C4 solder joints as a result of voids in underfill. Insufficient plating of
microvias on the PTHs also caused early failure.

According to JEDEC A110, the highly accelerated stress test (HAST) should be conducted under 130°C/85 percent RH for 100 h. However, because of the low T_g of the SLC, a lower temperature (110°C) but longer period (200 h) was used for testing IBM's FC/PBGA. The test results showed that the dielectric integrity of SLC held up throughout the HAST. However, some plated fibers of the laminate may cause shorts during the HAST. Efforts need to be made to eliminate this situation. The temperature/humidity/bias (THB) test was performed according to JEDEC A108. Measurements were taken at regular intervals with the specimens in the testing chamber under the specified temperature and humidity but with the bias voltage being turned off. No failure was detected in this test, indicating that this package could hold up to the normal field operation conditions.

The unbiased autoclave or pressure cooker test (PCT) was conducted to evaluate the resistance of the package to intensive heat, moisture, and pressure. The testing specifications were according to JEDEC A102-B. All specimens which satisfied the manufacturing specifications could pass the PCT without any detectable failure. The high-temperature storage (HTS) test was performed following JEDEC A103. Certain specimens showed a slight increase in the electrical resistance (200 to 400 mΩ). Such symptoms were attributed to defective chip circuitry.

For both the TC and PCT qualification tests, all specimens were subjected to certain high-temperature/humidity preconditioning. Most of them were preconditioned under JEDEC level 3, with a small portion preconditioned under level 2. No detectable failures were observed for all cases. However, since the sample size for level 2 was relatively small, IBM considered the FC/PBGA qualified for JEDEC level 3 of moisture resistance. In addition, it should be noted that during the development of the FC/PBGA, various types of solder masks and underfill encapsulants were evaluated. From the results of qualification tests, it was concluded that there was no observable difference between the two solder masks. However, for the underfill encapsulants, in the one with lower T_g the fillers seemed to have settled at the bottom. This phenomenon might cause a 5- to 15-μm-wide separation between the die and the underfill in certain test items and, hence, reduce the package reliability. Therefore, the underfill with higher T_g was selected for the production of FC/PBGA.

22.6   Summary and Concluding Remarks

IBM's FC/PBGA is a near-chip-size package with the flip chip mounted on a rigid organic substrate. The LSI is a 12×14 mm die which has 700 C4 solder bumps with an array pitch of 230 μm. Underfill encapsulant is required for the assembly of the package. The main feature of IBM's FC/PBGA is the SLC on the external surfaces of the chip carrier. These buildup layers together with the Driclad core laminate can provide a 4S/2P circuitry for I/O distribution. The interposer has an outline of 21×21 mm, which conforms to the JEDEC standard. The package terminals are 255 BGA solder balls with an array pitch of 1.27 mm. The major advantage of IBM's FC/PBGA is the relatively large number of package I/Os. Therefore, the potential application of this package is microprocessors.

A comprehensive qualification testing program has been conducted to investigate the reliability of IBM's FC/PBGA. The moisture resistance of this package was qualified for JEDEC level 3. During the development of the FC/PBGA, various packaging materials were evaluated. It was found that the package reliability is insensitive to the choice of solder mask. On the other hand, a higher-T_g underfill encapsulant is preferred in order to suppress delamination. The results of technical evaluation indicated that this novel package can be fabricated by the existing production facilities. If the produced packages can meet the manufacturing specifications, then the standard reliability requirements can be satisfied.

22.7   References

1. M. Jimarez, L. Li, C. Tytran, C. Loveland, and J. Obrzut, "Technical Evaluation of a Near Chip Scale Size Flip Chip/Plastic Ball Grid Array Package," *Proceedings of the 48th ECTC*, Bellevue, Wash., May 1997, pp. 495–502.

23

Matsushita's MN-PAC

23.1 Introduction and Overview

The MN-PAC was developed by Matsushita Electronics Corporation (MEC) in 1995 [1]. This CSP uses a multilayered ceramic substrate as the interposer and features a proprietary interconnect technology, namely, stud-bump bonding (SBB) [2]. The silicon chip is connected to the interposer by Au stud bumps together with conductive adhesive. Underfill is employed to enhance the package reliability. The package terminals are land grid array (LGA). The current land pitches are 1.0 mm and 0.8 mm. The I/O pin count ranges from 77 to 437. Finer pitch and higher pin count are under development. The package thickness is less than 1.0 mm. Good electrical and thermal performance can be achieved, and the MN-PAC has excellent package reliability. Because of the small form factors, this package is very suitable for portable equipment. The applications include ASICs, DRAMs, DSPs in cellular phones, digital camcorders, and PDAs. The current production volume for the MN-PAC is 3 million per month. It is expected that this may reach 10 million per month by the year 2000.

23.2 Design Concepts and Package Structure

Matsushita's MN-PAC is a CSP with a rigid interposer and a flip-chip configuration. The package structure is illustrated in Fig. 23.1. The chip-level interconnection is established by stud-bump bonding (SBB). The stud bumps are formed on the Al die pads from conventional Au ball bonds by a modified wire bonder. Conductive adhesive is employed to mount the stud bumps on the metallized lands on the top of the interposer. Underfill is used to encapsulate the jointed stud

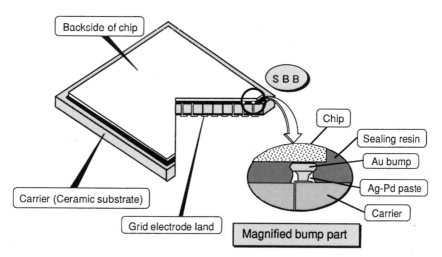

Figure 23.1   Package structure of Matsushita's MN-PAC.

bumps and to enhance the package reliability of the MN-PAC. Currently the SBB technology can be implemented in mass production for chip bond pads with a pitch of 120 μm. With the continuous trend of die shrink, it is expected that finer pitch—up to 80 μm—may be required in the future. It should be noted that the SBB technology has very flexible configuration. Therefore, any kind of bond pad pattern (peripheral or center) on the silicon chip can be accommodated.

The substrate of the MN-PAC is a multilayered ceramic carrier. In order to match the die bond pad pitch, the design rule for metallized traces on the interposer is 60/60 μm for line/space width. In practice, a 50/50-μm specification can be achieved if necessary. For accurate mounting of stud bumps, the tolerance in bonding land position is limited to less than 20 μm. In addition, since the stud bumps are rather small, a coplanarity within 10 μm is required for the top surface of the interposer. The package terminals of the MN-PAC are LGA. The I/Os on the top and bottom faces of the ceramic substrate are connected by metallized through vias. The diameter of the solder pads on the bottom surface of the interposer is 0.6 mm. The tolerance in pad position is 0.1 mm. In addition, the boundaries of these solder pads must be kept at least 0.1 mm away from the edges of the substrate. The picture of a typical 15×15 mm MN-PAC is presented in Fig. 23.2. The underfill and the LGA pattern can be clearly observed.

The MN-PAC was designed for applications with low to medium pin counts. The LGA usually has a full-grid-array pattern. For better assembly reliability, the solder joints at the four corners are used as mechanical support only. The current land pitches are 1.0 mm and 0.8

TOP VIEW

Figure 23.2 Top and bottom
views of a 15×15 mm MN-PAC.

BOTTOM VIEW

TABLE 23.1   Matsushita's MN-PAC Family

	Size (mm × mm)				
Pitch*	17 × 17	15 × 15	13 × 13	11 × 11	9 × 9
1.0 mm	285 pins	221 pins	165 pins	117 pins	77 pins
0.8 mm	437 pins	320 pins	252 pins	165 pins	117 pins
0.5 mm	1085 pins	837 pins	621 pins	285 pins	165 pins

*The 0.8- and 1-mm pitches are in mass production. The 0.5-mm pitch is under development.

mm. A finer pitch, 0.5 mm, is under development. The lineup for the
MN-PAC is presented in Table 23.1. The definitions for the package
outline are given in Fig. 23.3. The overhang value is 0.5 mm.
Including the silicon chip, the SBB, and the ceramic substrate, the
total package thickness is less than 1.0 mm.

23.3   Material Issues

The MN-PAC is a ceramic chip-size package with a flip-chip configuration. The whole package is composed of the silicon die, the stud
bumps, the conductive adhesive, the underfill, and the ceramic inter-

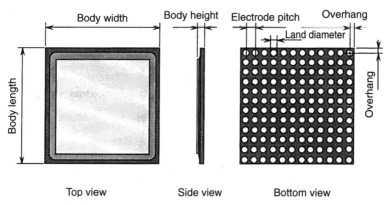

Top view   Side view   Bottom view

Figure 23.3   Outline definitions of the MN-PAC.

poser. Because of the flexibility of SBB technology, the silicon chip in the MN-PAC package may have any bond pad configuration (peripheral or center). The bond pad surface has a conventional Al finish. No additional under-bump metallurgy (UBM) is required. The stud bumps are formed on the Al die pads from the conventional Au ball bonds by a modified wire bonder.

In order to attach the stud bumps to the interposer, conductive adhesive is employed. This material is an epoxy resin with Ag-Pd powder fillers. The adhesive is originally in paste form. After curing, the paste is solidified to bond stud bumps on the metallized lands of the interposer. To improve the SBB joint reliability, the conductive adhesive is relatively compliant so that the deformation due to thermal mismatch may be absorbed. The underfill is made of epoxy resin. In addition to protecting the chip from the environment, this adhesive also provides compressive force to maintain the contact of SBB joints.

The interposer of the MN-PAC is a ceramic substrate. Dimensional stability is a major concern. There are three ways to fabricate this package carrier. The first method is to cofire the substrate with the top face metallization. This approach provides fine-line capability and small-diameter pads. However, the shrinkage error may affect the accuracy of chip connection. Therefore, it is essential to control the substrate shrinkage in order to maintain good chip mounting yield. The second method is to cofire the substrate first and then print and fire the top face metallization on the interposer. This approach provides very accurate bond pad position. However, this technique requires large-diameter metallized lands for via hole joining. Consequently, the wiring capability of the interposer is severely limited. The third approach is to use the newly developed zero-X-Y-shrinkage processing technology [3]. This method provides the most precise dimensions for both the geometry of the package and the position of the top face metallization. However, unlike the previous two approaches, which may use the conventional Al_2O_3 alumina, the third method can use only low-temperature glass ceramic material.

23.4   Manufacturing Process

The package structure of Matsushita's MN-PAC is rather simple, and the manufacturing process is straightforward. The first step is wafer bumping. The stud bumps are formed on the Al bond pads on the wafer by a modified wire bonder. The procedures are illustrated in Fig. 23.4. A loop path is required for the capillary tube to break the wire on the top of the Au ball bond. After the wafer bumping, a mechanical leveling for the coplanarity of stud bumps is performed as shown in Fig. 23.5.

Following the dicing, the silicon chip is flipped over and placed in a shallow bath of Ag-Pd paste, as presented in Fig. 23.6. Once the conductive adhesive paste is transferred to the stud bumps, the chip is mounted on the top of the interposer, and this is followed by curing. Then the package is placed on an inclined hot plate, as shown in Fig. 23.7, for underfill dispensing. The MN-PAC is completed after the underfill is fully cured.

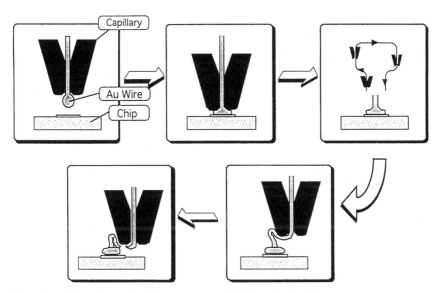

Figure 23.4   Formation of stud bumps by modified Au wire bonding.

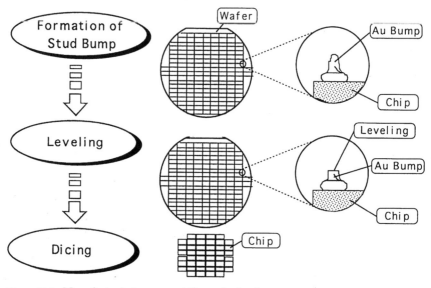

Figure 23.5 Manufacturing process at the wafer level.

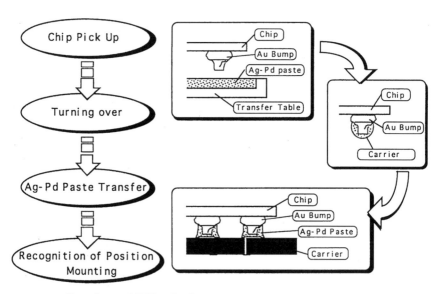

Figure 23.6 Procedures of SBB technology.

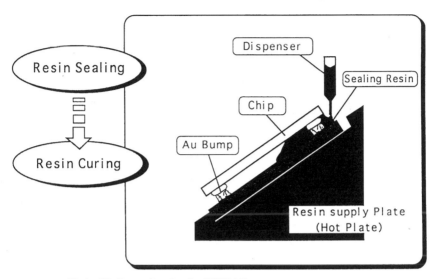

Figure 23.7 Underfill dispensing for the MN-PAC.

TABLE 23.2 Comparison of Electrical Performance between the MN-PAC and QFP

	CSP 13 × 13 (1.0)	QFP 84 pins (0.8)
Lead inductance	0.2 nH	4.2 nH
Switching noise	1/4	1
Crosstalk	1/6	1

23.5   Electrical and Thermal Performance

The wiring path of the MN-PAC is rather short. For a 256-pin MN-PAC, the typical wire length from the die bond pad to the PCB is about 3.0 mm. A comparable QFP has an average wire length of 20 mm. This difference gives the MN-PAC superior electrical characteristics. A numerical comparison between the electrical performance of the MN-PAC and that of the QFP is given in Table 23.2. Detailed graphical comparisons are shown in Fig. 23.8 as well. It is obvious that the MN-PAC outperforms the QFP in lead inductance, switching noise, and crosstalk. Further comparisons between the MN-PAC and the DIP are presented in Figs. 23.9 and 23.10. Substantial differences in electrical switching and address access time can be identified.

The thermal performance of the MN-PAC package was characterized and compared to that of a comparable QFP, as shown in

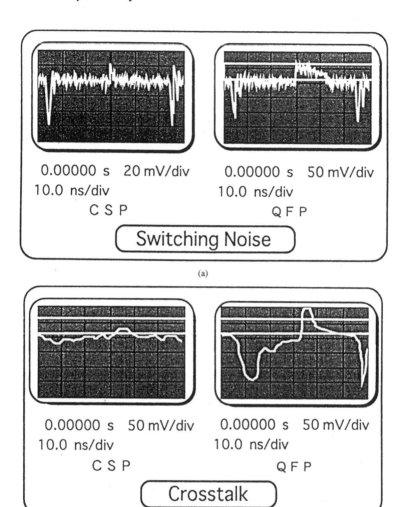

Figure 23.8 Comparison of electrical performance between the MN-PAC and QFP: (a) switching noise, (b) crosstalk.

Fig. 23.11. The thermal resistance θ_{ja} was measured for packages with and without heat sink (Al fin). The testing results are given in Fig. 23.12. It is obvious that the heat-dissipation capability of the MN-PAC is much greater than that of the QFP. This is reasonable because the chip of the MN-PAC is not molded. Further analysis indicated that the major heat path of the MN-PAC is via solder joints to the PCB.

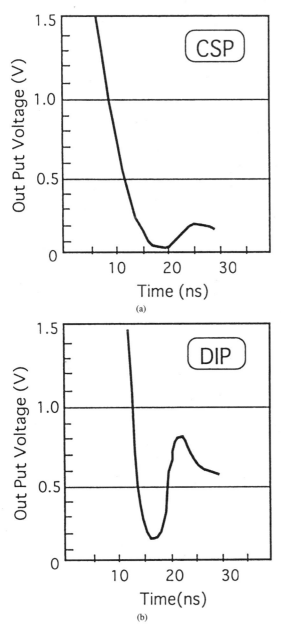

Figure 23.9 Comparison of electrical switching performance: (a) MN-PAC, (b) DIP.

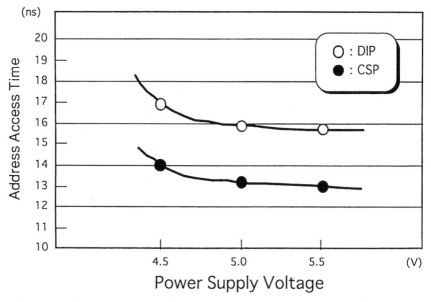

Figure 23.10   Comparison of address access time between the MN-PAC and DIP.

23.6   Qualifications and Reliability

The package reliability of the MN-PAC was extensively investigated. This CSP passed all assigned qualification tests, as shown in Table 23.3. This outstanding performance is due to the moisture-insensitive ceramic substrate and the strengthening provided by the underfill adhesive.

The board-level interconnects of the MN-PAC are LGA. This type of package terminal is easy to handle in SMT, but the self-alignment capability may be a concern because of the relatively small amount of solder. The mountability of MN-PAC packages was investigated, as shown in Table 23.4. It was found that the self-alignment capability is actually quite good. Further evaluation was conducted as presented in Fig. 23.13. The results indicate that the mounting may not be reliable when the offset is close to 0.5 mm or larger. In practice, the pick-and-place tolerance is limited to 0.3 mm to ensure high assembly yield in mass production.

The standoff height of LGA solder joints is very short. Therefore, the board-level solder joint reliability of the MN-PAC could be a critical issue [4]. A thermal cycling test between −40 and 100°C was performed. The package size of tested specimens was 17×17 mm, and the chip size was 16×16 mm. The LGA land pitch was 1.0 mm, and the PCB thickness was 1.0 mm. Various ceramic substrate thicknesses were evaluated. The testing results, presented in Fig. 23.14, indi-

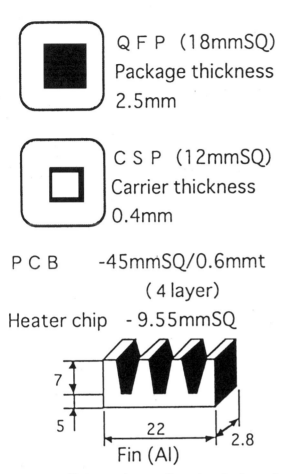

QFP (18mmSQ)
Package thickness
2.5mm

CSP (12mmSQ)
Carrier thickness
0.4mm

PCB   -45mmSQ/0.6mmt
(4 layer)
Heater chip   - 9.55mmSQ

7

5   22
2.8
Fin (Al)

Figure 23.11   Various package configurations for thermal performance evaluation.

cate that, in general, a thinner substrate could improve the thermal fatigue life of solder joints. This is because the thinner substrate allows more bending deformation, which can relieve the thermal stress, as shown in Fig. 23.15. The results also reveal that the LGA solder joints of the MN-PAC assembly may last for almost 1000 cycles (with 50 percent accumulated failure rate) if a 0.4-mm-thick interposer is used. This performance is acceptable for the designated applications of MN-PAC packages.

Another experimental study was conducted to investigate the effect of substrate rigidity on the solder joint reliability of the MN-PAC assembly. The package size was 17×17 mm, and the thickness of the PCB was 0.65 mm. The specimens were thermal cycled between −65

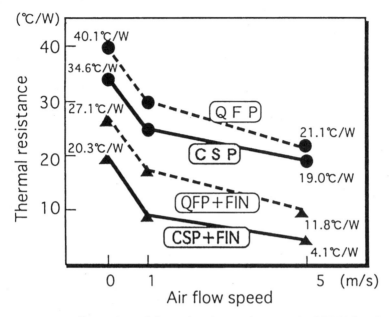

Figure 23.12 Comparison of thermal resistance between the MN-PAC and QFP.

TABLE 23.3   Package Qualification Results for MN-PAC

Test items	Evaluation condition	Target	Judgment
High-temperature bias	125°C/V_{dd} (typical)	1000 h	OK
THB	85°C/85% RH/V_{dd} (typical)	1000 h	OK
High temperature	150°C	1000 h	OK
Temperature shock	−65°C/5 min to 150°C/5 min	50 cycles	OK
Temperature cycle	−65°C/30 min to 150°C/30 min	100 cycles	OK
PCT	134°C/100% RH/3 atm	50 h	OK
Reflow crack	85°C/85% RH/168 h IR reflow (peak temperature 235°C/10 s)	2 times	OK
Vibration	20–200Hz/20G	4 min/sweep	OK
Drop test	Maple board 75 cm	3 times	OK

and 125°C. The Young's modulus of ceramic substrate material B was only half that of material A. The results shown in Fig. 23.16 indicate that the substrate with lower rigidity can substantially improve the solder joint fatigue life. In summary, thinner and softer ceramic substrates can certainly enhance the LGA solder joint reliability of the MN-PAC assembly.

TABLE 23.4   Self-Alignment Features during Board Assembly

	Shift XY 0mm	Shift XY 0.25mm	Shift XY 0.30mm	Shift XY 0.40mm
After mount				● Land of PCB ◉ Land of CSP
After reflow	X-ray test NG 0/24 Electrical test NG 0/24	X-ray test NG 0/8 Electrical test NG 0/8	X-ray test NG 0/8 Electrical test NG 0/8	X-ray test NG 0/8 Electrical test NG 0/8

C S P Size : 11mm×11mm／13mm×13mm   Land pitch 1.0mm

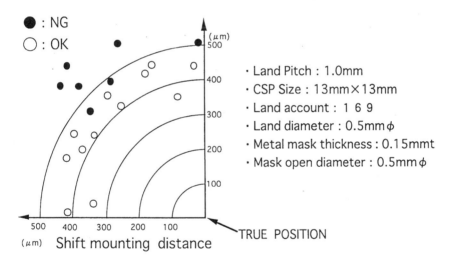

● : NG
○ : OK

(μm)
500
400
300
200
100

· Land Pitch : 1.0mm
· CSP Size : 13mm×13mm
· Land account : 1 6 9
· Land diameter : 0.5mmφ
· Metal mask thickness : 0.15mmt
· Mask open diameter : 0.5mmφ

500   400   300   200   100
(μm)   Shift mounting distance   TRUE POSITION

Figure 23.13   Evaluation of board-level mountability for the MN-PAC.

23.7   Applications and Advantages

The MN-PAC was designed by Matsushita for IC devices with low to medium pin counts. The current lineup includes I/Os from 77 to 437 with terminal pitches of 1.0 mm and 0.8 mm. Finer pitch and higher pin count are under development. The typical applications of MN-PAC packages include ASICs, DRAMs, DSPs in cellular phones, digi-

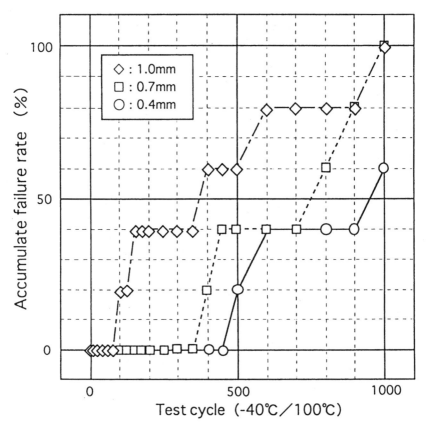

Figure 23.14   Thermal fatigue life of MN-PAC solder joints.

tal camcorders, and PDAs. The production volume in 1998 is 3 million per month. It is expected that this may reach 10 million per month by the year 2000. At present, the cost of the MN-PAC is about 1.2 to 1.5 times that of the QFP with similar pin count. With the growth in production volume, the package price will eventually become the same [5].

The major advantages of the MN-PAC are simple structure, flexible configuration, and small form factors. The package thickness is less than 1.0 mm. Comparisons of the dimensions between various packages are given in Table 23.5 and Fig. 23.17. Because of their low profile and compact size, MN-PAC packages are very suitable for portable equipment. In addition, good electrical and thermal performance can be achieved. Because of the ceramic substrate and the underfill adhesive, the package reliability of MN-PAC packages is outstanding as well.

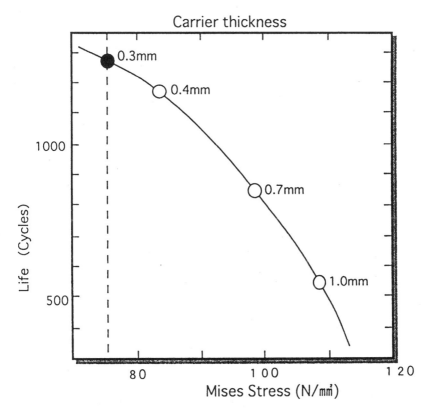

Figure 23.15   Effect of substrate thickness on solder joint reliability.

23.8   Summary and Concluding Remarks

The MN-PAC is a CSP with a rigid interposer. The substrate is a multilayered ceramic carrier. This package adopts a flip-chip configuration. The first-level interconnection is implemented by the SBB technology. The Au stud bumps are formed on the die bond pads by a modified wire bonder and attached to the interposer by conductive adhesive. Underfill is employed to encapsulate the SBB joints and to enhance the package reliability. The package I/Os are LGA. The current land pitches are 1.0 mm and 0.8 mm. Finer pitch and higher pin count are under development. The package thickness is less than 1.0 mm. Because of the package structure, good electrical and thermal performance can be achieved. Also, the package reliability of MN-PAC is excellent and the solder joint reliability is acceptable.

Since the SBB technology is flexible, the MN-PAC can accommodate any conventional ICs with either peripheral or center bond pads. Because of its low profile and compact size, this CSP is very suitable

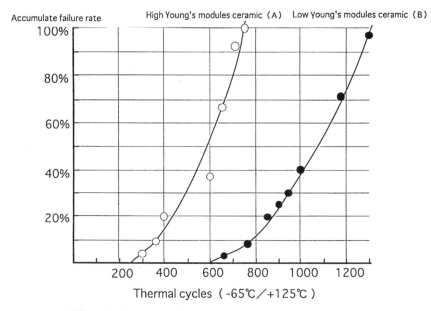

Figure 23.16   Effect of substrate rigidity on solder joint reliability.

TABLE 23.5   Comparison of Form Factors for Various Packages

Pitch (mm)	0.65	0.5	0.4	0.3	1.0	—
Package type	QFP	QFP	LQFP	TQFP	CSP	Bare chip
Mounting area ratio	1	0.73	0.41	0.27	0.17	0.12
Size (mm × mm)						
Body size	28	24	18	14	13	11
Mounting size	31.2	26.6	20	16	13	11
Package thickness (mm)	3.5	3.3	1.4	1.0	0.85	0.4
Pin counts	160	176	164	168	169	—

for portable equipment. Current applications include ASICs, DRAMs, DSPs in cellular phones, digital camcorders, and PDAs. The MN-PAC is already in mass production. The production volume in 1998 was 3 million per month. By the year 2000, this is expected to reach 10 million per month. With the growth in production volume, the package cost of MN-PAC will be substantially reduced and will eventually become comparable to that of other conventional packages.

Figure 23.17 Comparison of package size between the MN-PAC and QFP.

23.9 References

1. Y. Kunitomo, "Practical Chip Size Package Realized by Ceramic LGA Substrate and SBB Technology," *Proceedings of the Surface Mount International Conference,* San Jose, Calif., September 1995, pp. 18–25.
2. Y. Besso, Y. Tomura, Y. Hakotani, M. Tsukamoto, T. Ishida, and K. Omoya, "A Stud Bump Bonding Technique for High Density Multi-Chip Modules," *Proceedings of the IEMTS,* Kanazawa, Japan, June 1993, pp. 362–365.
3. H. Nishikawa et al., "Development of Zero X-Y Shrinkage Multilayered Ceramic Substrate," *Proceedings of the Japan IEMTS,* Kanazawa, Japan, 1993, pp. 238–241.
4. J. H. Lau and Y. H. Pao, *Solder Joint Reliability of BGA, CSP, Flip Chip, and Fine Pitch SMT Assemblies,* McGraw-Hill, New York, 1997.
5. *Nikkei Microdevices,* June 1997.

24

Motorola's SLICC and JACS-Pak

24.1 Introduction and Overview

The slightly larger than IC carrier (SLICC) package was developed by Motorola in 1991 [1]. The basic configuration of this CSP is a flip-chip ball grid array (FC-BGA) package. The first-level interconnects are C4 solder joints, a technology licensed from IBM. The interposer of the SLICC is an organic rigid substrate. Underfill is required for the assembly of the flip chip to the interposer. The package terminals are conventional BGA solder balls with an array pitch of 0.8 mm. Because of the lack of PCB technology for such a fine pitch at that time, a modified version of SLICC, namely, the much larger than IC carrier (MLICC) package, with 1.5-mm-pitch BGA terminals, was developed for the purpose of testing and qualification. Since the SLICC and MLICC packages require costly fabrication processes, they were never put into mass production. To make its CSPs more cost-competitive, Motorola made a large effort to improve the wafer bumping and substrate technology. As a result, a new CSP called the "just about chip size package" (JACS-Pak) was developed in 1997 [2]. The JACS-Pak adopted the FC-BGA configuration as well. A proprietary semi-rigid or flexible substrate is used in this low-profile CSP, making the package thickness less than 1.0 mm. In addition, low-cost eutectic solder bumps are employed to replace the high-lead C4 joints. The JACS-Pak is designed to accommodate both fan-in and fan-in/fan-out I/Os. The pitch of package terminals may be 0.5 or 0.8 mm. This CSP is intended for applications with less than 150 I/Os. The production volume of JACS-Pak is expected to exceed 200,000 units per week in 1998.

24.2   Design Concepts and Package Structure

The SLICC (Fig. 24.1) and the JACS-Pak (Fig. 24.2) are chip scale packages developed by Motorola. The basic configuration of these CSPs is FC-BGA, and both packages use an organic substrate as the interposer. Historically, the SLICC package was considered the first CSP with flip-chip die. However, because of the costly wafer-bumping process and the expensive substrate, the SLICC package was never

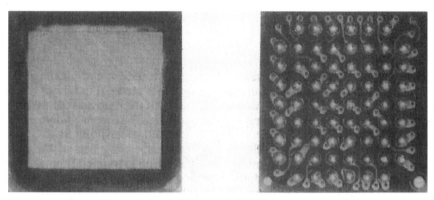

Figure 24.1 Top and bottom views of Motorola's SLICC.

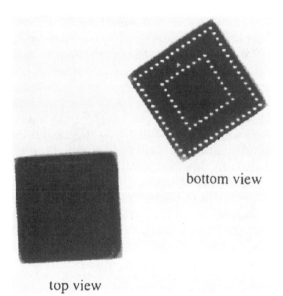

bottom view

top view

Figure 24.2 Top and bottom views of Motorola's JACS-Pak.

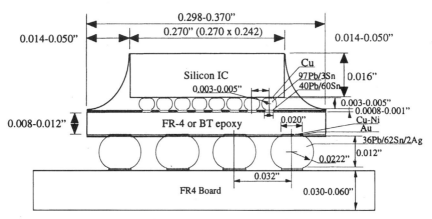

Figure 24.3   Dimensions and relevant features of the SLICC.

commercialized. The JACS-Pak is an improved version of the SLICC with competitive cost-effectiveness. This new CSP is expected to be in mass production in 1998.

The detailed package description for the SLICC is given in Fig. 24.3 [3]. The LSI is a flip-chip die with C4 solder bumps. The bump height is 76 to 127 μm (3 to 5 mils), and the minimum bump pitch is 250 μm (10 mils). The C4 bumps are made of 97Pb/3Sn solder. The high-lead solder bumps are mounted to the substrate by 60Sn/40Pb solder clad on the bond pads. Underfill encapsulation is required to improve the assembly reliability. The interposer of the SLICC is an organic substrate made of glass fiber and FR-4 or BT epoxy. The thickness of the substrate is 0.2 to 0.3 mm (8 to 12 mils), and the size is about 0.7 to 2.5 mm (28 to 100 mils) larger than the chip size. This interposer has two layers of routing traces at the top and the bottom, respectively. The line/space width follows the 4/4 (mil) design rule. The two layers of circuitry are connected by Cu-plugged PTHs which are made by mechanical drilling. The hole and the pad diameter are 0.2 and 0.4 mm (8 and 16 mils), respectively. The metallization of solder-ball bond pads at the bottom of the interposer is Au flash over Cu/Ni. The pad diameter is 0.5 mm (20 mils), and the pitch is 0.8 mm (32 mils). The package terminals of the SLICC are BGA solder balls. The material is 62Sn/2Ag/36Pb, and the ball diameter is 0.56 mm (22.2 mils). Motorola termed its BGA package interconnects C5, which stands for controlled-collapse chip carrier connection. The cross-sectional picture of a SLICC package is presented in Fig. 24.4.

A cost analysis for the SLICC revealed that the package substrate and the wafer bumping contributed approximately 70 percent of the overall packaging cost. Therefore, the modified version, the JACS-

Chip

Underfill
C4 bumps
Interposer

C5 bumps

Figure 24.4 Cross-sectional picture of the SLICC.

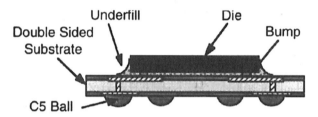

Underfill   Die
Double Sided   Bump
Substrate

C5 Ball

Figure 24.5 Cross-sectional view of the JACS-Pak.

Pak, focused on cost reduction in these two areas. The package structure of the JACS-Pak is illustrated in Fig. 24.5 [2]. The basic configurations are similar to those of the SLICC. However, the costly evaporative high-lead C4 joints are replaced by eutectic solder bumps. In
addition, a proprietary semirigid laminate (glass/epoxy) or flexible
film (polyimide) is substituted for the original rigid substrate. Since
the new substrate is relatively thin, the total package thickness of the
JACS-Pak is less than 1.0 mm, making this CSP a very low-profile
package. The interposer of the JACS-Pak has double-sided routing
traces which can accommodate both fan-in and fan-in/fan-out circuitry. Because the flip chip has eutectic solder bumps, the top bond pads
of this new substrate do not require clad solder. The bottom bond
pads for the C5 solder balls may have a pitch of 0.5 or 0.8 mm. The
family of JACS-Pak CSPs is shown in Table 24.1. Although high pin
counts may be achieved by fully populated BGA balls, the JACS-Pak
is intended for applications with less than 150 I/Os. Therefore, in general, a two-row perimeter array with 0.5-mm pitch or a four-row
perimeter array with 0.8-mm pitch is recommended by Motorola.

TABLE 24.1   Motorola's CSP Family

Body size (terminal count)	Population	0.8-mm pitch	0.5-mm pitch
6 × 6 mm	2 rows		72
(0.5 mm—11 × 11)	3 rows		96
	4 rows		112
	Fully populated		121
8 × 8 mm	2 rows	56	104
(0.8 mm—9 × 9)	3 rows	72	144
(0.5 mm—15 × 15)	4 rows	80	176
	Fully populated	81	225
10 × 10 mm	2 rows	80	136
(0.8 mm—12 × 12)	3 rows	108	192
(0.5 mm—19 × 19)	4 rows	128	240
	Fully populated	144	361
12 × 12 mm	2 rows	96	168
(0.8 mm—14 × 14)	3 rows	132	240
(0.5 mm—23 × 23)	4 rows	160	304
	Fully populated	196	529
14 × 14 mm	2 rows	120	200
(0.8 mm—17 × 17)	3 rows	168	288
(0.5 mm—27 × 27)	4 rows	208	368
	Fully populated	289	729

24.3 Material Issues

Both the SLICC and the JACS-Pak are BGA packages with a flip-chip die. The package constituents in addition to the silion chip are the C4 solder bumps, the underfill encapsulant, the organic substrate, and the C5 solder balls. For the SLICC package, the wafer bumping is implemented by IBM's C4 process with evaporative 97Pb/3Sn. Since this high-lead solder is not reflowed during the assembly of the flip chip, a low-melting-point solder (60Sn/40Pb) needs to be clad on the top bond pads of the interposer. Such flip-chip joints are relatively expensive. Therefore, in order to reduce the packaging cost, the JACS-Pak employs eutectic solder to replace the high-lead C4 bumps. On the other hand, the BGA terminals of the SLICC and the JACS-Pak use the same solder material. The C5 balls may be made of either 63Sn/37Pb or 62Sn/2Ag/36Pb.

During the development of SLICC packages, six underfill encapsulants were considered, as shown in Table 24.2. After evaluation, a

TABLE 24.2   Curing Conditions for Alternative Underfill Materials

Sample	Chemistry	Cure condition
A (baseline)	Bis-A novolac, anhydride cure	20 min @ 120°C 60 min @ 150°C
B	Bis-A novolac, anhydride cure	60 min @ 165°C
C	Elastomer-modified epoxy	30 min @ 150°C
D	Bis-F novolac	20 min @ 120°C 60 min @ 165°C
E	Cyanate ester	20 min @ 120°C 60 min @ 165°C
F	Imidazole cure epoxy	30 min @ 150°C

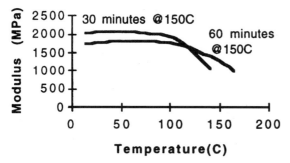

Figure 24.6   Temperature-dependent elastic modulus of underfill.

highly filled, anhydride-cured novolac epoxy was selected. The CTE of this material is 20 ppm/°C. Further tests were performed to evaluate the temperature-dependent elastic modulus of the chosen underfill encapsulant under different curing conditions. The results of these tests are given in Fig. 24.6.

The interposer of the SLICC is a two-layer printed circuit board made of glass-fiber-reinforced FR-4 or BT laminate. The metallization of the bond pads is Au flash over Cu/Ni. Since the die bumps are high-lead solder, the top bond pads are clad with 60Sn/40Pb for flip-chip mounting. Mechanically drilled PTHs are made in the substrate to connect the circuitry at both sides. These PTHs are plugged with Cu to improve the package performance and reliability. Because of the requirements for solder-clad bond pads and small-diameter PTHs by mechanical drilling, the substrate of the SLICC is rather costly. Therefore, one of the emphases of the JACS-Pak was to reduce the cost of the package interposer. As a result, a proprietary interposer made of semirigid (glass/epoxy) laminate or flexible (polyimide) film was developed to replace the original rigid substrate. This interposer has double-sided routing traces which can accommodate both fan-in and fan-in/fan-out circuitry. Because the flip-chip die of the JACS-Pak has eutectic solder bumps, the top bond pads of this new substrate do not require clad solder.

24.4 Manufacturing Process

The fabrication of Motorola's CSP is straightforward. The first step is wafer bumping. For SLICC packages, the standard C4 process, which is a technology licensed from IBM, is followed. For JACS-Pak packages, Motorola has developed a low-cost eutectic-solder-bumping process to reduce the production cost. Once the solder-bumped dies are singulated from the wafer, the CSPs are assembled following the process flow shown in Fig. 24.7 [4].

For both the SLICC and the JACS-Pak, the flip-chip mounting requires only flux printing on the substrate. Since the former uses high-lead solder in wafer bumping, the die bumps do not melt during flip-chip mounting. Instead, the clad solder on the interposer is reflowed and wicks up to form the joints. On the other hand, because the JACS-Pak uses eutectic solder for the flip chip, the whole bump is reflowed during die mounting. The process next to solder reflow is underfill dispensing. To ensure full encapsulation of solder joints underneath the chip, Motorola developed a technique to force the flow of underfill material. A cyclic vacuum and air pressure is applied for 2 to 3 cycles to drive the encapsulant. The scanning acoustic microscopic (SAM) image of the SLICC in Fig. 24.8 shows good-quality underfill encapsulation when this method is used. Even if some voids are trapped at the center, the jeopardy to the integrity of the package should be minimal. The curing of the underfill is performed right after the dispensing process is completed. There are two different curing conditions for the selected underfill. One is 30 min at 150°C, and the other is 15 min waiting plus 60 min at 155°C. Although the former can save a substantial amount of processing time, the latter cur-

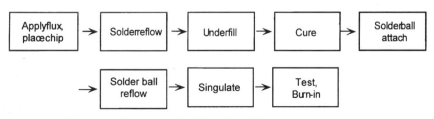

Figure 24.7 Manufacturing process for Motorola's CSPs.

Figure 24.8 Acoustic microscopic inspection after the underfill process.

ing condition is recommended because of concern about package reliability.

Once the underfill has been dispensed and cured, the package is ready for the attachment of C5 solder balls. This process involves flux printing, ball placement, and solder reflow. The conventional practice for BGA packages is followed. At this stage, the CSPs are still in a panel form. After the C5 solder balls are attached, each individual package is singulated from the strip. Subsequently, the CSPs are subjected to burn-in and various tests.

24.5 Electrical and Thermal Performance

Because of the flip-chip configuration and the submillimeter ball-array pitch, Motorola's CSPs have electrical parasitics superior to those of conventional QFPs and BGAs. It was reported that the SLICC and the JACS-Pak can achieve a 4X reduction in inductance and capacitance over standard PQFPs with similar pin counts. The reduction in mutual inductance and mutual capacitance can minimize the crosstalk in digital products and improve the lead-to-lead isolation in RF applications. Also, with a maximum lead length of 3 mm, the package delay may be kept to a minimum.

A comprehensive numerical study was conducted to investigate the thermal performance of the JACS-Pak [5]. The thermal resistance θ_{ja} under various conditions was evaluated for comparison. Both natural and forced convective cooling were considered. Figures 24.9 and 24.10

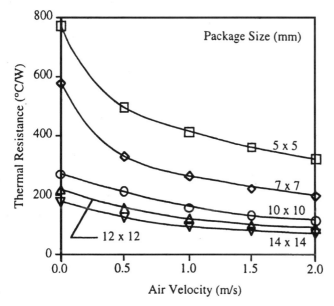

Figure 24.9   Adiabatic PCB thermal resistance as a function of package sizes and air flow rates.

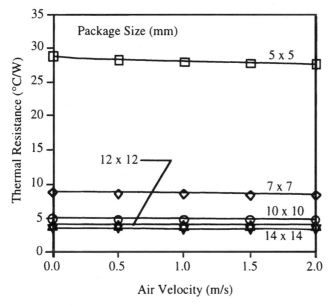

Figure 24.10   Cold-plate thermal resistance as a function of package sizes and air flow rates.

present the values of thermal resistance with an adiabatic PCB and a cold-plate PCB, respectively. The former corresponds to the extreme case with air as the only path for heat dissipation, while the latter represents another extreme case with the PCB as an infinite heat sink remaining at the ambient temperature. In both cases, the results of the simulation indicate that the thermal resistance decreases with respect to the increase in package size and air velocity. Such a trend is reasonable because, in general, more heat can be dissipated with larger area and faster flow. The major difference between Figs. 24.9 and 24.10 is the magnitude of the thermal resistance. From the comparison, it can be concluded that thermal conduction via the PCB is the main path for heat dissipation. Figure 24.11 shows the thermal performance of the JACS-Pak with a regular PCB. The values of θ_{ja} fall between those of the two previous extreme cases and represent a more realistic situation.

Figures 24.12 to 24.15 present the results of a series of parametric studies of the thermal performance of a 10×10 mm JACS-Pak assembled on a regular PCB. The general trend shows that more I/Os, larger die size, larger PCB size, and lower power density can lead to smaller thermal resistance. It should be noted that in Fig. 24.14, no improvement in heat dissipation can be observed after the PCB size

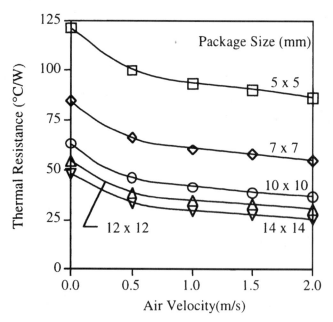

Figure 24.11   Thermal performance of Motorola's CSPs on a regular PCB.

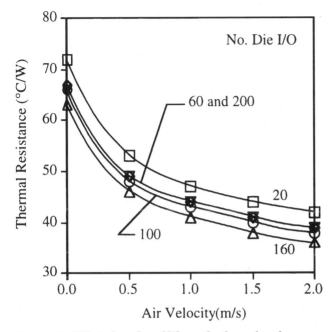

Figure 24.12 Effect of number of I/Os on the thermal performance.

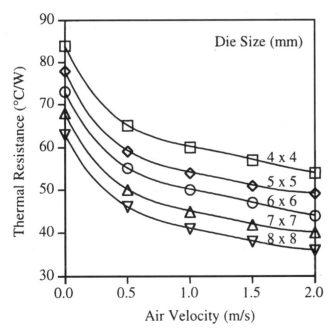

Figure 24.13 Effect of die size on the thermal performance.

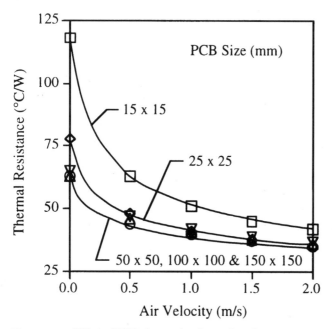

Figure 24.14   Effect of PCB size on the thermal performance.

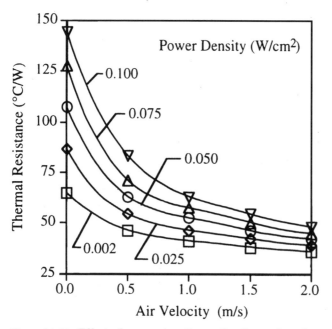

Figure 24.15   Effect of power density on the thermal performance.

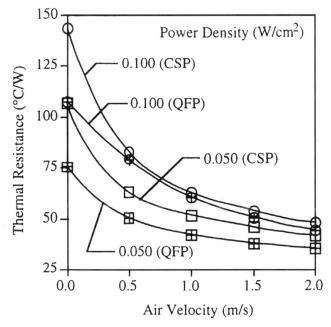

Figure 24.16 Comparison of thermal performance of the JACS-Pak and PQFP.

becomes 5 times the package size. This indicates that the dimensions and layout of the PCB may be optimized with respect to the package size of the CSP. Figure 24.16 shows the comparison between the JACS-Pak and the PQFP for two fixed power densities (0.1 W/cm² and 0.05 W/cm²). It is obvious that thermal performance is better with the latter than with the former. However, it should be noted that although both packages have 160 I/Os and an 8×8 mm die, the package size of the PQFP (28×28 mm) is almost 3 times that of the JACS-Pak (10×10 mm). In view of the advantage in form factor, the deficiency in thermal performance may not be considered a fatal drawback for the CSP.

24.6 Qualifications and Reliability

The package reliability of the SLICC has been investigated by several qualification tests, as shown in Table 24.3. In general, this CSP could pass all testing items and meet the desired goals [6]. To demonstrate the reliability performance of the SLICC, the cumulative failure percentages under temperature/humidity test and condition B thermal shock are presented in Fig. 24.17. In addition to experimental testing, numerical simulation was performed to evaluate the package reliability

TABLE 24.3   Qualification Tests for the SLICC

Test	Preconditioning	Conditions	Sample size	Pass criterion
Liquid–liquid thermal shock, condition B	JEDEC level 5	500 cycles	30	No electrical fails
Air–air temperature cycling, condition B	JEDEC level 5	500 cycles	30	No electrical fails
Autoclave	JEDEC level 5	288 h	30	No electrical fails
85°C, 85% RH	JEDEC level 5	1008 h	30	No electrical fails
Popcorn	JEDEC level 3	2 min @ 220°C	30	No delamination

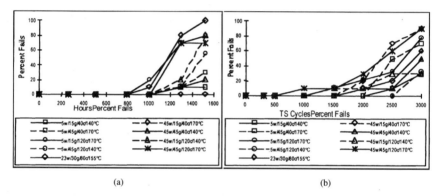

(a)   (b)

Figure 24.17   Cumulative failure percentage of the SLICC under qualification tests: (*a*) temperature/humidity test (85°C/85 percent RH); (*b*) thermal shock (condition B).

of the SLICC [3]. Figure 24.18 presents the stresses which may cause die cracking and delamination in a SLICC package under thermal shock loading. The corresponding stresses in a DCA flip chip are also evaluated for benchmarking. From the comparison, it can be seen that the stresses in the SLICC are always lower, resulting in better package reliability. Similar qualification tests were also performed to investigate the reliability of the JACS-Pak [7]. The qualification data are given in Table 24.4. In general, the reliability level is satisfactory.

The solder joint reliability of the SLICC has been investigated comprehensively by numerical simulation. The stress analysis was performed using a finite-element slice model, as shown in Fig. 24.19. The typical stress contours in the solder joints are presented in Fig. 24.20. These stresses were induced by mechanical three-point bending, as illustrated in Fig. 24.21*a*. The corresponding deformation by numerical simulation is given in Fig. 24.21*b*. More results of stress analysis are

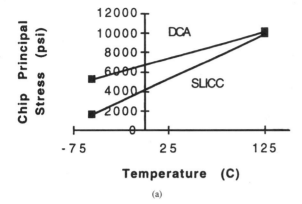

(a)

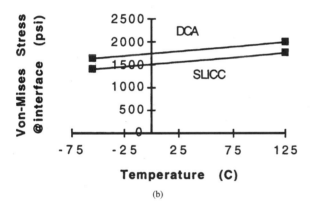

(b)

Figure 24.18 Comparison of stresses between the SLICC and DCA under thermal shock loading: (*a*) principal stress in the chip; (*b*) von Mises stress at the chip-underfill interface.

presented in Fig. 24.22. It can be seen that the load-deflection curve obtained by simulation agrees with the experimental data quite well. With such model verification, the plastic work density obtained from the stress analysis may be used as an index to evaluate the solder joint reliability of the SLICC assembly.

The solder joint reliability of the JACS-Pak assembly was investigated by numerical analysis and experimental testing [7]. Figure 24.23 shows the comparison between finite-element modeling and moiré measurement. The latter is used to verify the accuracy of the former. Once the computational model has been validated, the inelastic strain

TABLE 24.4   Qualification Data for the JACS-Pak

Test	Pass criteria	1% failure probability (characteristic life)	Notes
Liquid–liquid shock, −55 to 125°C, 6 cycles/h	<1% failure @ 400 cycles	Flex 600 (2000) FR4 800 (4212)	C5 fails only
Temperature cycle, −40 to 125°C, 1 cycle/h	<1% failure @ 300 cycles	Flex 600 (1573) FR4 N/A (4443)	C5 fails only
85°C/85% RH, 5.5-V bias	No fails to 168 h	No fails to 1000 h	All leakage readings greater than 1E6 Ω
Autoclave, 121°C, 100% RH	48 h	No fails to 196 h	All leakage readings greater than 1E6 Ω
Aging 125°C	1000 h	2000 h	None

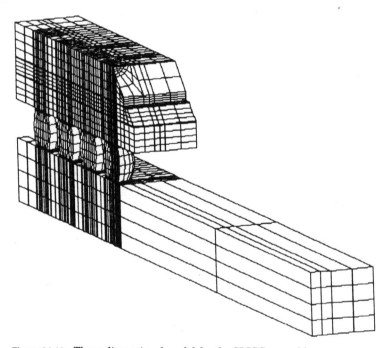

Figure 24.19   Three-dimensional model for the SLICC assembly.

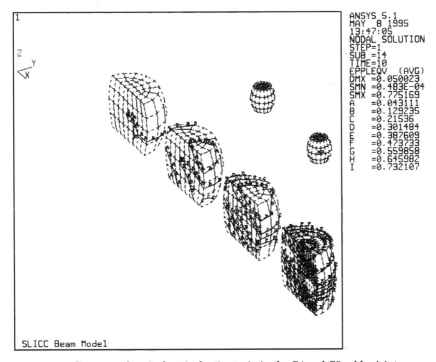

Figure 24.20 Contours of equivalent inelastic strain in the C4 and C5 solder joints.

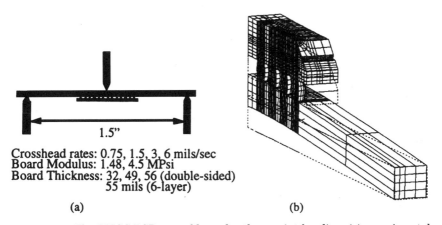

Crosshead rates: 0.75, 1.5, 3, 6 mils/sec
Board Modulus: 1.48, 4.5 MPsi
Board Thickness: 32, 49, 56 (double-sided)
          55 mils (6-layer)

(a)   (b)

Figure 24.21 The SLICC-PCB assembly under three-point bending: (a) experimental setup; (b) deformation by numerical simulation.

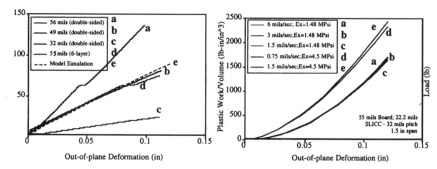

Figure 24.22   The SLICC-PCB assembly under three-point bending: (*a*) load-deflection curves; (*b*) plastic work density versus deflection.

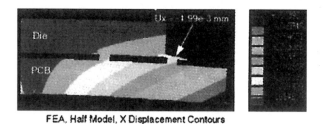

FEA, Half Model, X Displacement Contours

(a)

Experimental Result, Moire Fringes of X Displacement

(b)

Figure 24.23   Comparison of deformation by finite-element analysis and moiré interferometry (JACS-Pak).

obtained from the stress analysis together with the Coffin-Manson equation may be used to evaluate the thermal fatigue life of solder joints. In addition to modeling and analysis, actual thermal cycling tests were performed to characterize the board-level reliability of the JACS-Pak assembly. The test vehicle and the test board are shown in Figs. 24.24 and 24.25, respectively. The specimens were monitored at regular intervals to detect the failure of interconnects. Once a certain sign of damage was detected, a cross section was performed to identify the failure mechanism, as shown in Fig. 24.26. A typical Weibull plot

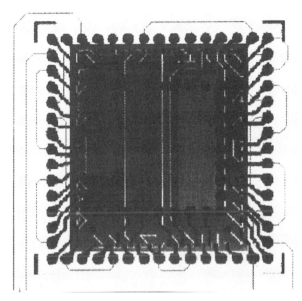

Figure 24.24   Layout of test vehicle for the JACS-Pak.

Figure 24.25   Test board for the
JACS-Pak.

for the solder joint life of the JACS-Pak assembly under thermal cycling tests is given in Fig. 24.27. The comparison between analysis and
testing is summarized in Table 24.5.

24.7   Summary and Concluding Remarks

The SLICC is a chip scale package developed by Motorola. The basic
configuration of this CSP is a flip-chip ball grid array (FC-BGA) package. The first-level interconnects are C4 solder joints. The interposer

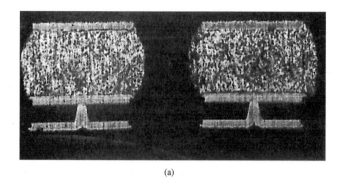

(a)

(b)

Figure 24.26 Cross section of the JACS-Pak's interconnects after thermal cycling tests: (a) cracks appeared at the top of the C5 joints; (b) the C4 joints remained intact.

TABLE 24.5   Summary of Solder Joint Thermal Fatigue Life of JACS-Pak by Analysis (Moiré Measurement + Coffin-Manson) and Cycling Tests

Lot (test)	Substrate	C5 ε_{ave} ($\Delta T = 100$) (moiré)	Number of cycles (Coffin-Manson)	Number of cycles (experimental)	Characteristic field life (years)
2011 (TC)	Flex	1.56%	X (1574)	1574	>24
2011 (TS)	Flex		X (2000)	2000	>20
2502 (TC)	Semirigid	1.12%	1.88X (2954)	4212	>65
2502 (TS)	Semirigid		1.88X (3753)	4443	>45

TC = Thermal Cycling, TS = Thermal Shock

of the SLICC is an organic rigid substrate. Underfill is required for the assembly of the flip chip to the interposer. The package terminals are conventional BGA solder balls with an array pitch of 0.8 mm. Because of the lack of PCB technology for such a fine pitch at that time, a modified version of the SLICC, namely, the much larger than

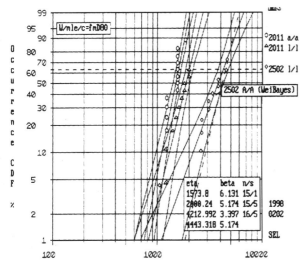

Thermal Cycles (L/L or A/A)

Figure 24.27   Weibull plot for the solder joint life of the JACS-Pak assembly.

IC carrier (MLICC) package, with 1.5-mm-pitch BGA terminals, was developed for the purpose of testing and qualification. Since the SLICC and MLICC packages require costly fabrication processes, they were never commercialized.

To make its CSPs more cost-competitive, Motorola made a great effort to improve the wafer-bumping and substrate technology. As a result, a new CSP called the JACS-Pak was developed. The JACS-Pak also adopted the FC-BGA configuration. A proprietary semirigid or flexible substrate is used in this low-profile CSP, making the package thickness less than 1.0 mm. In addition, low-cost eutectic solder bumps are employed to replace the high-lead C4 joints. The JACS-Pak is designed to accommodate both fan-in and fan-in/fan-out I/Os. The pitch of package terminals may be 0.5 or 0.8 mm. This CSP is intended for applications with less than 150 I/Os. The JACS-Pak is to be in mass production in 1998.

24.8   References

1. K. Banerji and P. Lall, "Development of the SLICC Package," *Proceedings of NEPCON East '95,* Boston, Mass., June 1995, pp. 441–451.
2. J. Aday, C. Koehler, and T. Tessier, "A Laminate-Based Flip-Chip Chip Scale Package," *Proceedings of the Surface Mount International Conference,* San Jose, Calif., September 1997, pp. 12–16.

3. P. Lall, G. Gold, B. Miles, K. Banerji, P. Thompson, C. Koehler, and I. Adhihetty, "Reliability Characterization of the SLICC Package," *Proceedings of the 46th ECTC,* Orlando, Fla., May 1996, pp. 1202–1210.

4. P. Thompson, L. Fischer, G. Hayes, C. Koehler, J. H. Lee, and J. Poarch, "Flip-Chip BGA Process Verification and Transfer," *Proceedings of NEPCON-West,* Anaheim, Calif., February 1997, pp. 1055–1060.

5. B. Chambers, T-Y.T. Lee, and W. Blood, "Thermal Analysis of a Chip Scale Package Technology," *Proceedings of the 48th ECTC,* Bellevue, Wash., May 1998, pp. 1407–1412.

6. P. Thompson, C. Koehler, M. Petras, and C. Solis, "Flip-Chip BGA Assembly Process and Reliability improvements," *Proceedings of the IEMTS,* Austin, Tex., October 1996, pp. 84–90.

7. S. E. Lindsey, J. Aday, B. Blood, Y. Guo, B. Hemann, J. Kellar, C. Koehler, J. Liu, V. Sarihan, T. Tessier, L. Thompson, and B. Yeung, "JACS-Pak Flip-Chip Chip Scale Package Development and Characterization," *Proceedings of the 48th ECTC,* Bellevue, Wash., May 1998, pp. 511–517.

25

National Semiconductor's Plastic Chip Carrier (PCC)

25.1 Introduction and Overview

The plastic chip carrier (PCC) package was developed by National Semiconductor in 1996 [1]. This CSP uses a rigid organic substrate as the interposer. The chip-level interconnects are conventional wire bonds. Although package molding is not required, glob-top encapsulation is employed to protect the die and the bonding wires. The package I/Os are plated flat lands (LCC type). The minimum terminal pitch is 0.65 mm. The bottom lands are connected to the metallization on the top of the substrate through unfilled plated through vias. Additional thermal vias may be made underneath the die for better heat dissipation. Therefore, the thermal performance is outstanding. Compared to conventional leaded plastic packages with the same land pattern, the PCC can accommodate a larger silicon chip. The package thickness of the PCC is about 1.0 mm. This CSP is targeted for low-pin-count applications such as modules in wireless devices. Since the fabrication of the PCC fits the existing packaging infrastructure, this package is expected to be a relatively low-cost CSP.

25.2 Design Concepts and Package Structure

National's PCC package is a CSP with rigid interposer [2]. This package is also known as the substrate chip carrier (SCC). The cross section of the package structure is presented in Fig. 25.1. The substrate is an organic carrier. The top side of the interposer is metallized for the die attachment and the chip I/O connection. Conventional wire

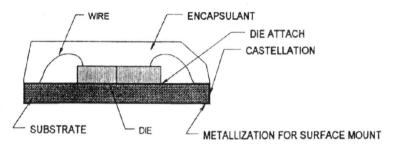

Figure 25.1 Package structure of National's PCC.

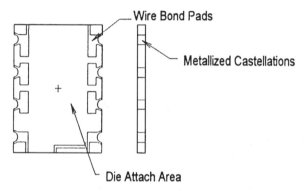

Figure 25.2 Substrate pattern of a typical 8-pin PCC package.

bonds are employed as the first-level interconnects. The package terminals are plated flat lands (LCC type). Because of the limitations of mechanical tooling, the current minimum land pitch is 0.65 mm. The bottom lands are connected to the top metallization of the interposer through punched vias. The via holes are unfilled and plated with metallization for electrical connection. After package singulation, these vias will become castellations, as shown in Fig. 25.2. It should be noted that additional vias may be punched in the substrate underneath the silicon die for thermal enhancement.

The PCC is a plastic package. The package needs to be encapsulated in order to protect the die and the bond wires. Various encapsulation methods were evaluated for the PCC, as shown in Table 25.1. Since cost reduction is the major concern for this CSP, glob-top encapsulation with syringe dispensing was selected. After encapsulation, the thickness of the PCC is about 1.0 mm if the chip thickness is 0.356 mm. As a comparison, for a conventional plastic package such as TSSOP, the silicon chip needs to be ground down to 0.254 mm in order to achieve the same package thickness.

TABLE 25.1   Comparison of Various Encapsulation Methods

Method	Encapsulant type	Advantages	Drawbacks
Conventional molding	B-staged epoxy pellet	Finished product has a uniform shape. Compounds/chemistry well established in industry	Will need dedicated tooling for a package body size/form factor. Will lead to an inefficient use of substrate
Glob top (syringe dispense)	Liquid (moderate viscosity)	Does not require special tooling per package form factor	Flow properties need to be very well understood. May not be suitable for very small geometries
Glob top (screen printing)	Liquid (high shear)	Can be used to print very small geometries	Requires some dedicated tooling (screens) for package form factor

The PCC package outlines are designed to conform to the existing IPC standards for molded packages in body size and lead pitch. The castellations of the PCC can be fabricated with the same diameter as the lead widths specified by EIAJ/JEDEC JC.11, so that the current infrastructure such as land patterns and PCBs may be utilized. It should be noted that with the same lead spacing as SOIC or QFP, the PCC packages would have a smaller footprint.

25.3   Material Issues

The PCC is a leadless plastic chip scale package. In addition to the silicon die, the package materials include the package interposer, the die-attach adhesive, the bonding wires, and the encapsulant. The main feature of the PCC is the design of the package interposer. The chip carrier is an organic substrate such as BT. On top of the interposer, metallization is deposited for die attachment and wire bonding. At the bottom of the substrate, designated lands are plated as well for package-level interconnection. The bottom lands and the wire bond pads are connected through metallized vias made by mechanical punching. The substrates are supplied in a panel form which may be singulated either before testing or after shipping.

National Semiconductor claims that the PCC is a low-cost CSP because this package can fit in the existing infrastructure for electronic packaging. Therefore, the conventional packaging materials such as epoxy resin and Au wires are adopted for die attachment and wire bonding, respectively. Furthermore, in order to minimize the processing cost, glob-top encapsulation is employed instead of molding for chip protection. The encapsulant is in a liquid form and has moderate viscosity. Since the encapsulant is dispensed by a syringe and needle system, no special tooling is required for the encapsulation of PCC packages.

25.4   Manufacturing Process

As shown in Fig. 25.3, the fabrication of the PCC is rather straightforward because of its simple package structure. After wafer sawing, the die is attached to the substrate, and this is followed by conventional wire bonding. The next step is package encapsulation. To avoid the flooding of epoxy resin, a silicone masking is required. The encapsulant is dispensed by a syringe and needle system to reduce the production cost. The silicone masking is removed after the encapsulation is completed.

Package marking is performed once the chip is encapsulated. It should be noted that the substrates are originally in the form of a panel which contains many package units. Each individual package may be inspected after singulation. Alternatively, the packages may be tested while the substrates remain unsingulated. Furthermore, the packages can be supplied in panel form and singulated on site by the end users. To highlight the simplicity of fabrication, a comparison

Substrate Chip Carrier
Saw
|
Die Attach
|
Wire Bond
|
(Silicone Masking)
|
Encapsulation
|
(Silicone Mask Removal)
|
Marking
|
Singulation (Saw)
|
Final Visual

Figure 25.3   Manufacturing process for the PCC package.

between the manufacturing process for conventional plastic packages and the PCC is given in Table 25.2.

25.5 Performance and Reliability

The thermal performance of the PCC has been evaluated in term of θ_{ja} and θ_{jc}. The results were obtained from finite-element simulation and compared to the values for other plastic packages (see Table 25.3). It is obvious that the PCC outperforms conventional plastic packages in

TABLE 25.2 Comparison of Manufacturing Procedures between Conventional Plastic Packages and PCC

Assembly/test operations	Conventional plastic molded package	Plastic chip carrier package (PCC)
Wafer saw	Yes	Yes
Die attach	Yes	Yes
Wire bond	Yes	Yes
Molding	Yes	No*
Dejunk	Yes	No
Plating	Yes	No
Trim/form	Yes	No
Marking	Yes	Yes
Singulation	Yes	Yes
Outgoing/final test	Yes	Yes
Pack/ship	Yes	Yes

*Encapsulation is required.

TABLE 25.3 Comparison of Thermal Performance for Conventional Plastic Packages and PCC

Conventional package	PCC
8 SOIC	PCC
$\theta_{ja} = 180°C/W$	$\theta_{ja} = 125°C/W$
$\theta_{jc} = 35°C/W$	$\theta_{jc} = 7°C/W$
64 TQFP	PCC-64
$\theta_{ja} = 49°C/W$	$\theta_{ja} = 42°C/W$
$\theta_{jc} = 14°C/W$	$\theta_{jc} = 9°C/W$
Mini SO-8	PCC-Mini SO8
$\theta_{ja} = 246°C/W$	$\theta_{ja} = 171°C/W$
$\theta_{jc} = 28°C/W$	$\theta_{jc} = 5°C/W$

TABLE 25.4   Qualification Data for PCC under Thermal Shock Test

Test condition	50 cycles	100 cycles
Thermal shock −65 to 150°C @ 2 cycles/h	0/10	0/10

TABLE 25.5   Qualification Data for PCC under Pressure Cooker Test (PCT)

Test condition	Precondition level	168 h	336 h
PCT (121°C/100% RH)	L1	0/16	0/16
	L3	0/16	0/16

TABLE 25.6   Qualification Data for PCC under
Temperature/Humidity/Bias (THB) Test

Test condition	Precondition level	500 h
THB (85°C/85% RH)	L1	0/22
	L3	0/22

heat-dissipation capacity. This outstanding thermal performance may be attributed to the thermal vias underneath the die-attach area.

A series of qualification tests have been conducted to investigate the package reliability of PCC. These tests include thermal shock (TS), pressure cooker test (PCT), and temperature/humidity/bias (THB) test. For the latter two items, the specimens were preconditioned to both JEDEC level 1 and level 3 specifications. The testing conditions and the external inspection results are given in Tables 25.4, 25.5, and 25.6, respectively. In addition, scanning acoustic microscopy (SAM) was employed to investigate the internal damage after testing. No delamination was detected. Therefore, it may be concluded that the PCC has the same level of reliability as conventional plastic surface-mount components.

25.6   Applications and Advantages

The PCC is a compact-size, low-profile CSP. This package was developed for IC devices with low pin counts and was targeted to replace conventional packages such as TSSOP. A typical application of the PCC is for electronic modules in wireless equipment.

The form factors are the major advantages of PCC. The package thickness is 1.0 mm with a 0.356-mm-thick silicon chip. For a conventional molded package to achieve such a thickness, the die must be ground down to 0.254 mm. In addition, the PCC can accommodate a larger die than conventional plastic packages with the same land pat-

TABLE 25.7   Comparison of Maximum Die Size with the Same
Land Pattern

Land Pattern	Conventional molded packages	Equivalent plastic chip carrier (PCC)
8 SOIC	2.1 × 2.3 mm	2.6 × 4 mm
16 SOIC	2.1 × 6.5 mm	2.6 × 8.5 mm
48 TQFP	4.4 × 4.4 mm	5.7 × 5.7 mm
64 TQFP	7.1 × 7.1 mm	8.8 × 8.8 mm

TABLE 25.8   Comparison of Mounting Area with the Same Package Size

Package type	Conventional mounting area (mm²)	PCC mounting area (mm²)	% area savings using PCC
8 SOIC	29.5	19.5	34
16 SOIC	92.2	73	21
48 TQFP	81	49	40
8 TSSOP	19.2	13.2	31
64 TQFP	144	100	31

tern, as shown in Table 25.7. Furthermore, because of the leadless design, the mounting area of the PCC is much smaller than that of leaded competitors with the same package size. The comparison in package mounting area and the corresponding percentage savings in area
with the PCC are given in Table 25.8.

National Semiconductor claims that in addition to having superior
form factors, the PCC is a relatively low-cost CSP. This is because the
fabrication of this package employs existing technologies in the current
packaging infrastructure. Also, instead of conventional molding, the
package is protected by glob-top encapsulation with syringe dispensing. Therefore, the production cost of PCC could be very economical.

25.7   Summary and Concluding Remarks

The PCC is a leadless plastic package with a rigid interposer. This
CSP uses a metallized organic substrate as the chip carrier. The first-
level interconnects are conventional wire bonds. Although package
molding is not required, glob-top encapsulation is employed to protect
the die and the bonding wires. Including the encapsulation, the package thickness is about 1.0 mm. The second-level interconnects are
plated flat lands (LCC type). The minimum land pitch is 0.65 mm.
The bottom lands are connected to the metallization on the top of the
interposer through punched vias. The unfilled via holes are metal-

lized for electrical connection. Additional vias may be punched underneath the die-attach area to enhance the thermal performance. As a result, the thermal performance of the PCC is superior to that of conventional molded packages. A series of qualification tests have been conducted for the PCC. It can be concluded that this CSP has the same level of package reliability as the other surface-mount components.

Compared to conventional plastic packages with the same land pattern, the PCC can accommodate a larger silicon chip. In addition, because of the leadless design, the mounting area of the PCC is much smaller than that of leaded competitors with the same package size. Therefore, the form factors are the major advantages of PCC packages. Furthermore, since the fabrication of the PCC fits the existing packaging infrastructure, this package is expected to be a relatively low-cost CSP. The PCC package was developed for low-pin-count applications such as modules in wireless devices. This CSP is intended to replace TSSOP in the future.

25.8   References

1. R. Joshi and B. J. Shanker, "Plastic Chip Carrier Package," *Proceedings of the 46th ECTC,* Orlando, Fla., June 1996, pp. 772–776.
2. R. Joshi and A. Chen, "A New Plastic Chip Carrier (PCC) Package," *Proceedings of the Surface Mount International Conference,* San Jose, Calif., September 1996, pp. 231–235.

26

NEC's Three-Dimensional Memory Module (3DM) and CSP

26.1 Introduction and Overview

NEC's 3DM was developed in 1996 for stacked three-dimensional memory modules [1]. This multichip module (MCM) has a flip-chip configuration and uses several rigid substrates with circuitry as the package interposer. This package has almost the same footprint as conventional memory modules such as the TSOP but is equipped with 4 times the memory capacity. The first-level interconnects are either Au stud bumps or solder bumps. Underfill is required to enhance the reliability of the package. After the single-chip elementary module is completed, a die-thinning process is performed to reduce the package thickness to 0.3 mm. The interconnects between stacked elementary modules and the package I/Os from the 3DM to the PCB are peripheral solder balls. With four elementary modules stacked, the 3DM has a total thickness equivalent to that of the TSOP package. More recently NEC implemented a single-chip CSP with high-density wiring capability [2]. This CSP and four 3DMs are mounted on a newly developed D/L substrate to form a MCM. This MCM has superior performance and is used in the application of reduced instruction set computing (RISC).

26.2 Design Concepts and Package Structure

The 3DM is a multichip technology developed by NEC for stacked memory modules [1]. This package consists of several (typically four, as illustrated in Fig. 26.1) elementary modules, each composed of a silicon chip and an interposer. In each elementary module there is a

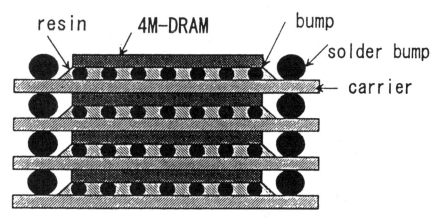

Figure 26.1 Package structure of NEC's 3DM.

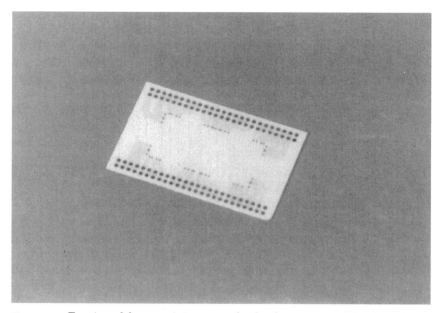

Figure 26.2 Top view of the ceramic interposer for the elementary module of the 3DM.

rigid substrate which is a metallized chip carrier made of ceramic or glass epoxy film. There are two rows of metal pads along the two longitudinal sides of the interposer, as shown in Fig. 26.2. The inner two rows are for chip connection, while the outer two are for the interconnection between elementary modules. The detailed design rules for the interposer of the 3DM are given in Table 26.1.

The chip-to-interposer interconnects of each elementary module

TABLE 26.1   Design Rules for the Interposer of NEC's 3DM

Signal line width	100 μm
Chip bonding pad (via)	100 μm
Number of bonding pads	48
Stack pad (base bonding)	300 μm
Number of stack pads	100
Size	18.8 × 12 × 0.3 mm³
Camber	<5 μm/10 mm

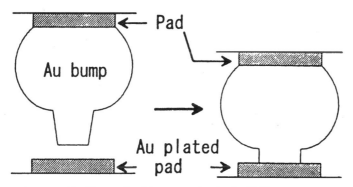

Figure 26.3   Stud bump for chip connection in the 3DM.

in the 3DM are either Au stud bumps or solder bumps, as shown in Figs. 26.3 and 26.4, respectively. The former are joined by thermal compression, and the latter are joined by reflow heating. After assembly, the bump height is about 50 μm. Underfill is required to encapsulate the connected bumps for reliability enhancement. In addition, a die-thinning process is necessary after underfill to reduce the thickness of the elementary modules to 0.3 mm. The whole package is completed by stacking up several elementary modules, as shown in Fig. 26.5. The interconnects between modules are large solder balls. The balls on the top module serve as the package terminals to the next-level substrate. For a four-module stack-up 3DM, the footprint and the package thickness are very close to those of a TSOP. However, the memory capacity of the 3DM is 4 times that of the corresponding TSOP package.

Based on the 3DM technology, NEC developed a single-chip CSP with high-density wiring capability [2]. The structure of the interposer for this CSP is illustrated in Fig. 26.6, and a picture of the top view and the bottom view of this interposer is presented in Fig. 26.7. The base substrate is a glass ceramic laminate with punched vias filled with metallization. On top of this rigid substrate, a thin organic buildup

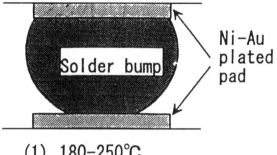

(1) 180-250°C

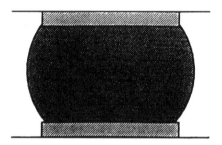

(2) 350°C

Figure 26.4 Solder bump for chip connection in the 3DM.

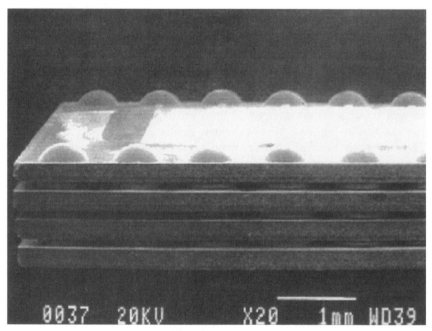

Figure 26.5 SEM picture of the 3DM package.

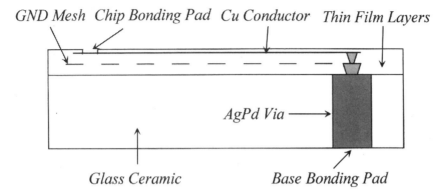

Figure 26.6   Cross section of the interposer for NEC's CSP.

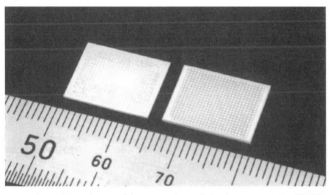

Figure 26.7   Top view and bottom view of the interposer for NEC's CSP.

layer is deposited for circuit redistribution. A picture of the bond pads for chip connection is given in Fig. 26.8. More detailed design rules for the interposer for the CSP are given in Table 26.2. This CSP has a flip-chip configuration, and the first-level interconnects are the same as those of the elementary module in the 3DM package. From the cross section of the interposer shown in Fig. 26.6, it is obvious that the package I/Os of this CSP are land grid array (LGA). As given in Table 26.2, the land diameter is 0.3 mm and the land pitch is 0.5 mm.

Since the aforementioned 3DM and CSP have ceramic-based inter-posers and relatively low-standoff solder joints, their board-level reliability could be a critical issue. Also, if several modules are configured together to form a MCM, a high-density wiring substrate will be needed. Therefore, NEC developed a new D/L base substrate, as shown in Fig. 26.9, for the assembly of the 3DM and CSP [2]. This substrate is so named because it is an organic thin film deposited on a

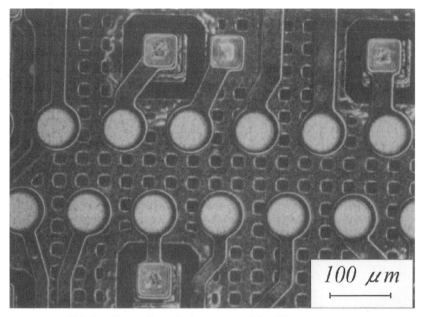

Figure 26.8   Chip bonding pads on the interposer of the CSP.

TABLE 26.2   Design Rules for the Interposer of NEC's CSP

Deposited layer	
Signal line width	25 μm
Signal line pitch	50 μm
Via hole	60 μm
Insulation separation	12 μm
Chip bonding pad	80 μm
Bonding-pad pitch	108 μm
Glass ceramic	
Base bonding pad (via)	300 μm
Bonding-pad pitch	500 μm
Number of I/O pads	525
Size	14 × 11.3 × 0.64 mm³
Camber	<10 μm/10 mm

laminated printed circuit board. The structure of this D/L substrate is illustrated in the cross section presented in Fig. 26.10. The base carrier of this D/L substrate is a PCB with its through-hole vias filled with resin. I/O pins are inserted in certain partially filled holes to serve as the interconnects to the motherboard. As in the configuration of the

Figure 26.9 D/L base substrate for the assembly of NEC's 3DM and CSP.

Chip Bonding Pad   Cu/Resin Thin Film Multilayer

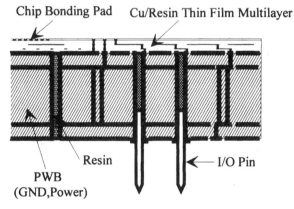

Resin   ← I/O Pin

PWB
(GND,Power)

Figure 26.10 Cross section showing the structure of the D/L base substrate.

CSP interposer, a multilayer thin film containing Cu traces is deposited on the top of the PCB for circuit redistribution. The detailed design rules for this D/L base substrate are given in Table 26.3. From the picture shown in Fig. 26.11, it can be seen that the top buildup layer has a fine-line wiring capability. According to NEC, this newly developed D/L base substrate can be used to carry one CSP and four 3DMs to form a MCM for RISC (reduced instruction set computing) applications. Since there are many chips working together, heat sinks need

TABLE 26.3   Design Rules for the D/L Base Substrate of NEC's MCM

Deposited layer	
Single line width	25 μm
Signal line pitch	50 μm
Via hole	40 μm
Insulation separation	15 μm
PWB	
Via hole	100 μm
Through hole	500 μm
Through-hole pitch	1.27 mm
Number of plate layers	4
Number of I/O pins	179
Size	65 mm × 65mm × 2.0 mm

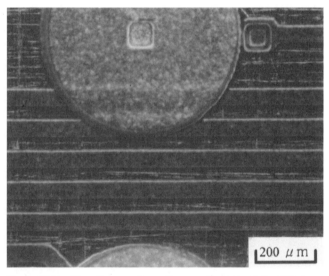

Figure 26.11   Fine-line configuration on the top of the D/L base substrate.

to be installed to enhance the thermal performance. The structure of this RISC MCM is illustrated in Fig. 26.12.

26.3   Material Issues

NEC's 3DM and CSP are packages with a flip chip on a rigid interposer. The whole package consists of the Au stud- or solder-bumped flip chip, the underfill adhesive, and the package substrate. For the 3DM,

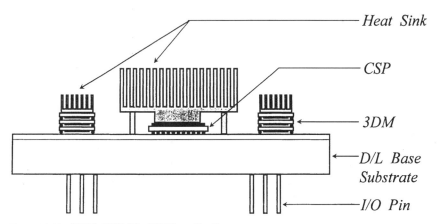

Heat Sink

CSP

3DM

D/L Base Substrate

I/O Pin

Figure 26.12   NEC's MCM for RISC application.

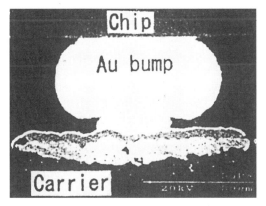

Figure 26.13   SEM micrograph of the cross section.

additional solder balls are needed for the interconnection between elementary modules and the assembly between the whole package and the next-level substrate.

As illustrated in Figs. 26.3 and 26.4, the chip-level interconnects of the 3DM and CSP may be either Au stud bumps or solder bumps. For the former, the stud bumps are deposited on the wire bond pads of the die by a conventional ball bonding method [3]. No extra surface metallization in addition to Al is necessary on the die pads. On the bonding pads of the interposer, a layer of Au plating is required. The stud bumps are mounted on the substrate by thermal compression at 350 to 400°C. The SEM cross-sectional view of a jointed Au stud bump is shown in Fig. 26.13.

For the solder bumps, a high-lead solder (95Pb/5Sn) is used. This solder has a melting point of 310°C, and so the solder bumps for chip connection will not melt during the subsequent reflow process. In order to deposit and join the solder bumps, a special under-bump metallurgy (UBM) is required for both the Al bond pads on the die and the bonding pads on the interposer. The UBM adopted for NEC's 3DM and CSP is electroless Ni/Au plating, as shown in Fig. 26.14. The Ni layer serves as a diffusion barrier, while the Au layer protects the Ni from oxidation and improves the surface wettability for solder. The solder bumps are mounted to the bonding pads on the interposer by a two-stage reflow process. In the prebonding stage, the temperature is elevated to 180 → 250°C to trigger the diffusion of metals. This temperature range was optimized by an experimental study as shown in Fig. 26.15. Then the temperature is raised to 350 → 400°C to melt the solder bumps. The SEM cross-sectional view of a joined solder bump is shown in Fig. 26.16.

The major material used to form the interposer of the 3DM and CSP is glass ceramics. This material has a low dielectric constant and is a low-electrical-resistance cofired conductor compared to the conventional alumina ceramic substrate. The material properties of the glass ceramics employed are given in Table 26.4. It should be noted that for the 3DM package, a glass epoxy film may be used for the interposer as well. Also, the main body of the D/L base substrate for the MCM is a low-cost PCB laminate.

The 3DM is designed for low-pin-count devices such as DRAM modules, and the substrate wiring requirement is relatively low. On the other hand, the CSP and the MCM (for RISC) demand much higher-density wiring capability because of the large number of I/Os. Therefore, a multilayer buildup circuit needs to be deposited on the top surface of the substrates for the CSP and MCM. This buildup circuit is realized by a low-dielectric photosensitive resin which has a relatively low curing temperature of 210°C. The typical properties of

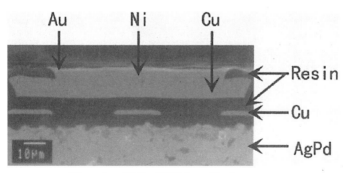

Figure 26.14   Electroless Ni/Au UBM for solder bumps.

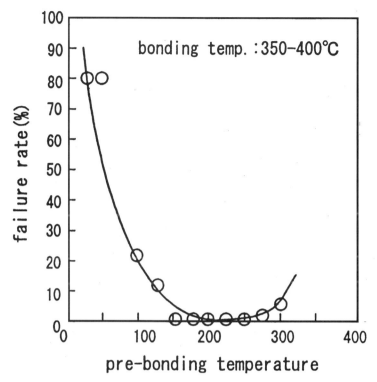

Figure 26.15 Optimization of the prebonding temperature for solder bumps.

Figure 26.16 SEM micrograph of the cross section of a solder bump.

this resin are given in Table 26.5. The functions of this resin are to form via holes in the buildup circuit and to serve as insulation layers. The Cu traces are electroplated between layers of the aforementioned resin. The procedures for fabricating this buildup circuit will be discussed in the following section.

TABLE 26.4   Material Properties of Glass Ceramics for the
Substrates of 3DM and CSP

Dielectric constant (1 MHz)	7.1
Dissipation factor (1 MHz)	0.002
Insulation resistance	$>10^{14}$ $\Omega \cdot$ cm
Coefficient of thermal expansion	5.0 ppm/°C
Flexural strength	280 MPa
Linear shrinkage	13.0 ± 0.3%
Conductor resistivity	2.5–12 $\mu\Omega \cdot$ cm

TABLE 26.5   Material Properties of Photosensitive Resin for the
Buildup Circuit on the Substrate of the CSP and MCM

Dielectric constant (1 MHz)	3.2
Dissipation factor (1 MHz)	0.002
Insulation resistance	$>10^{15}$ $\Omega \cdot$ cm
Coefficient of thermal expansion	70 ppm/°C
Conductor resistivity	1.8 $\mu\Omega \cdot$ cm

26.4   Manufacturing Process

The fabrication of the 3DM begins with wafer bumping and substrate
preparation. The Au stud bumps or 95Pb/5Sn solder bumps are attached to the bond pads on the die at the wafer level. The former do
not need any extra surface metallization in addition to Al, while the
latter require electroless Ni/Au UBM. The Au stud bumps are made
from conventional ball bonds by a modified wire bonder [3]. The solder bumps are formed by depositing solder on the die bond pads followed by reflow heating. The melting point of 95Pb/5Sn is around
310°C. The chip carrier for the elementary module in 3DM is a glass
ceramic substrate which has several rows of bonding pads (for chip
connection and module stackup) and is imprinted with a certain circuit pattern. The Ag-Pd paste and the glass ceramic are cofired to
form the interposer with circuitry. Depending on the type of bumps
selected for chip connection, Au plating (for Au stud bumps) or electroless Ni/Au plating (for solder bumps) is required on the surface of
the chip bonding pads on the interposer.

Once the bumped wafer is diced and the interposer is prepared, the
elementary module of the 3DM is ready to be assembled, as illustrated in Fig. 26.17. The chip is mounted to the substrate by a flip-chip
bonder. For the Au stud bumps, the joints are formed by thermal compression. The required temperature is 350 to 400°C. For the solder

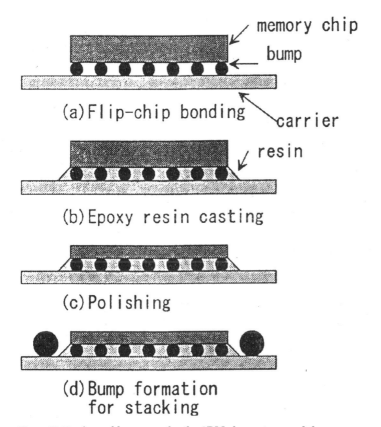

(a) Flip-chip bonding

(b) Epoxy resin casting

(c) Polishing

(d) Bump formation
    for stacking

Figure 26.17   Assembly process for the 3DM elementary module.

bumps, a two-stage bonding process as shown in Fig. 26.4 is performed. At first, the solder bumps and the chip bonding pads are heated to 180 → 250°C to activate the diffusion of metals. Subsequently, the temperature is raised to 350 → 400°C to reflow the solder and form the joints.

Following the chip mounting, the elementary module is placed on an inclined plate and the epoxy resin is dispensed from the top edge into the gap (about 50 μm) between the die and the substrate for underfill. Once the adhesive is cured, a die-thinning process is performed by polishing the back surface of the silicon chip. After die polishing, the total thickness of the elementary module is about 0.3 mm. The next step is the solder-ball attachment for the interconnection between elementary modules. These relatively large solder balls are made of Pb/Sn or Sn/Ag solder, the melting point of which is in the 200 to 240°C range. They are attached to the outer-row bonding pads on the chip carrier by reflow heating at 230 to 250°C. After the formation of these solder bumps, several (three or four) elementary modules are stacked up and reflowed again, as shown in Fig. 26.18, to build the 3DM package. Finally, the eutectic 63Sn/37Pb solder balls are attached to the outer-row bonding pads on the interposer of the top module for the assembly to the PCB.

The package assembly procedures for the CSP and MCM are basically the same as those for the 3DM. However, because of the requirement for high-density wiring capability, the preparation of the substrate is more sophisticated. The main body of the interposer for the CSP is a multilayer glass ceramic substrate. The fabrication process of this substrate is given in Fig. 26.19. The first step is to punch via holes in the glass ceramic green sheets. Then these punched holes are filled with Ag-Pd paste. After several sheets are laminated together, cofiring is performed to make the substrate. Finally, a multilayer Cu/resin buildup circuit film is deposited on the top of the ceramic substrate in order to provide fine-line wiring. This technology is im-

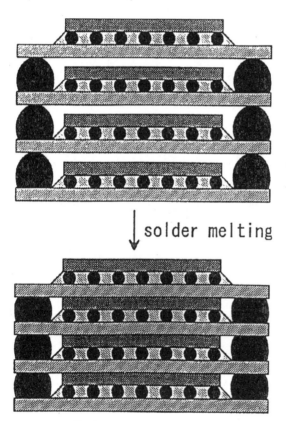

Figure 26.18 Stacking process for the 3DM.

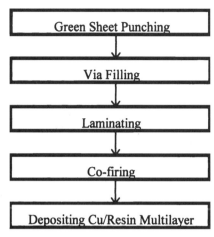

Figure 26.19 Manufacturing process for the substrate for the CSP.

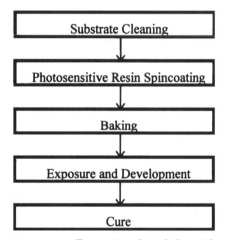

Figure 26.20 Formation of via holes with photosensitive resin.

plemented with a low-dielectric photosensitive resin. The functions of this resin are to form the via holes in the buildup circuit and to serve as insulation layers. The procedures for making via holes and Cu traces in this buildup circuit film are given in Figs. 26.20 and 26.21, respectively.

The formation of via holes begins with plasma cleaning on the substrate surface. Then the photosensitive resin is spin-coated and baked. The exposed films are developed with solvent by the puddle

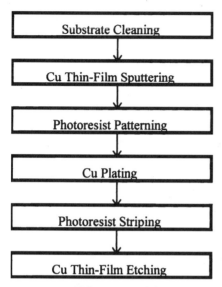

Figure 26.21   Fabrication of Cu traces in the buildup circuit film.

method. After the final curing, the via holes are formed. The fabrication of Cu traces in the buildup circuit also starts with surface cleaning. Afterwards, a Cu thin film is deposited by sputtering. An ordinary photoresist layer is spin-coated on top of the Cu thin film and then patterned by photolithography. Subsequently, the Cu traces are formed by electroplating. Finally, the photoresist layer is stripped and the sputtered Cu thin film is etched away. By the fabrication of alternating layers of photosensitive resin (with via holes) and Cu traces, the multilayer buildup circuit is established.

The manufacturing process for the buildup circuit film on the D/L base substrate for NEC's MCM is similar to that for the CSP. However, the main body of the former is a PCB laminate instead of a multilayer glass ceramic substrate. The through-hole vias in the PCB are filled with resin, as shown in Fig. 26.10. In certain partially filled via holes, I/O pins are inserted to serve as the interconnects to the motherboard. Therefore, unlike those of the 3DM and the CSP, the package terminals of the MCM are pin grid array (PGA).

26.5   Performance and Reliability

For a high-performance electronic device, it is necessary to optimize the impedance between the LSI and the packaging substrate. The characteristic impedance Z_0 depends on the dielectric constant, the geometry of signal lines, and the ground plane. For the CSP interposer and the D/L base substrate of the MCM, the dielectric thickness (insulation separation) was calculated by a three-dimensional static simulation to match Z_0 (50 Ω). Furthermore, the crosstalk coupling noise is a significant issue because the packaging density is restricted by the spacing between parallel lines. The near-end crosstalk K_B and far-end crosstalk K_F in 8-mm parallel lines in the D/L base substrate were simulated by GREENFIELD. The values of K_B and K_F are 1.1 and 0.07 percent, respectively. These results indicate that NEC's MCM should have outstanding electrical performance.

The MCM with the configuration shown in Fig. 26.22 was evaluated for thermal performance. This package consists of one CSP and four 3DMs. A RISC chip in the CSP is rated at 11.5 W, and each SRAM chip in the 3DMs has a power consumption of 0.5 W. Heat sinks were mounted on all modules to enhance the capability for heat dissipation. A computational simulation using FLOTHERM was performed to evaluate the thermal performance of the aforementioned MCM; the results are presented in Fig. 26.23. With an air flow of 0.5 m/s and an ambient temperature of 40°C, the maximum junction temperature $T_{j,\max}$ is under 100°C, which is the allowable temperature at the chip junction. Therefore, the thermal performance of this MCM is considered satisfactory.

A series of qualification tests were conducted to investigate the package reliability of the 3DM. The specimens were 16M DRAMs with Au stud bumps and glass epoxy chip carriers. The tests included temperature/humidity/bias (THB), high-temperature (HT) storage,

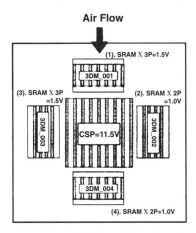

Figure 26.22 MCM configuration for thermal performance characterization.

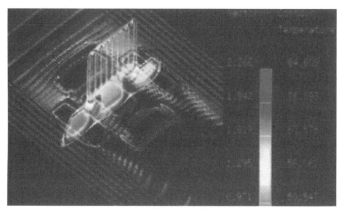

Figure 26.23 Typical results of thermal simulation.

TABLE 26.6   Qualification Test Data for NEC's 3DM Package

Test	Conditions	Single	Stack
		Results (5 pieces/each)	
THB	85°C/85% RH, rated voltage, 1000 h	Passed	Passed
HT	125°C, 1000 h	Passed	Passed
PCT	110°C/85% RH, 1.2 atm, 500 h	Passed	Passed
TC	−40 to 125°C, 1000 cycles	Passed	Passed
LT	−40°C, 1000 h	Passed	Passed

pressure cooker test (PCT), temperature cycling (TC), and low-temperature (LT) storage. The testing conditions and results are given in Table 26.6. No specimens failed any of the qualification tests. Therefore, the 3DM may be considered a reliable package.

26.6   Applications and Advantages

NEC's 3DM was developed mainly for stacked memory module applications. The typical configuration is to stack four 4M DRAMs to make one 16M DRAM. The next generation will be stacking four 64M DRAMs to make one 256M DRAM. The 3DM package is aimed at replacing the conventional TSOP. For a four-stack 3DM (16M DRAM), the footprint and package thickness are about the same as those of a single TSOP (4M DRAM). However, as illustrated in Fig. 26.24, the packaging density of the former (80 Mbits/cm^3) is four times that of the latter (20 Mbits/cm^3). Therefore, the advantage of the 3DM package is obvious.

The CSP developed by NEC is a compact and low-profile package

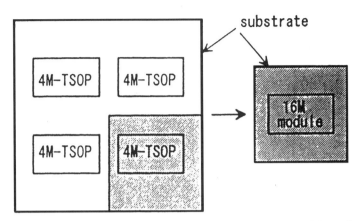

Figure 26.24 Comparison of packaging density between the 3DM and TSOP.

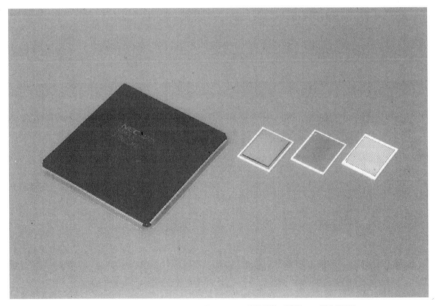

Figure 26.25 Comparison of package size between NEC's CSP and PQFP.

with high-density wiring capability. From the comparison shown in Fig. 26.25, it can be seen that NEC's CSP has form factors much superior to those of the conventional PQFP. This CSP is mainly targeted for RISC applications, as shown in Fig. 26.26. The whole package is an MCM consisting of one D/L base substrate, one CSP, and four 3DMs. Of the 3DMs, two are two-stack modules and the remaining

Figure 26.26   Picture of NEC's MCM for RISC application.

two are three-stack modules. The die in the CSP is a RISC chip with a relatively high power rating (11.5 W). Heat sinks are attached to all modules in order to enhance the thermal performance. According to NEC, this RISC MCM has the advantages of high packaging density, compact size, and good electrical and thermal performance. Although this package is intended to be used in workstations, it can be applied to supercomputers and communication systems as well.

26.7   Summary and Concluding Remarks

The 3DM is a chip scale package for stacked three-dimensional memory modules. This package contains several elementary modules which are composed of one flip chip on a rigid substrate. The chip-level interconnects are either Au stud bumps or solder bumps. Underfill is required to enhance the package reliability. For a four-stack 3DM for DRAM applications, the footprint and the thickness are similar to those of a single TSOP. However, the memory capacity of 3DM is four times as great. Therefore, the packaging density of 3DM is much higher than that of conventional memory modules. A series of qualification tests were conducted to investigate the package reliability of 3DMs. From the testing results, it may be concluded that the 3DM has the same package reliability as conventional surface-mount components.

More recently NEC developed another single-chip CSP with a rigid interposer. The package substrate has a multilayer buildup circuit to provide high-density wiring capability. The other features of this package include compact size and low profile. The major application of this CSP is packaging RISC chips. With a newly developed D/L base substrate, one CSP and four 3DMs are integrated to form a MCM. This RISC MCM has the characteristics of high packaging density, compact size, and good electrical and thermal performance. Although workstations are the main target of NEC's MCM, this package may also find applications in supercomputers and communication systems.

26.8   References

1. N. Takahashi, N. Senba, Y. Shimada, I. Morizaki, and K. Tokuno, "3-Dimensional Memory Module," *Proceedings of the 46th ECTC*, Orlando, Fla., June 1996, pp. 113–118.
2. A. Shibuya, I. Hazeyama, T. Shimoto, N. Takahashi, N. Senba, M. Kimura, Y. Shimada, H. Matsuzawa, and F. Mori, "New MCM Composed of D/L Base Substrate, High-Density-Wiring CSP and 3D Memory Modules," *Proceedings of the 47th ECTC*, San Jose, Calif., May 1997, pp. 491–496.
3. Y. Besso, Y. Tomura, Y. Hakotani, M. Tsukamoto, T. Ishida, and K. Omoya, "A Stud Bump Bonding Technique for High Density Multi-Chip Modules," *Proceedings of the IEMTS*, Kanazawa, Japan, June 1993, pp. 362–365.

27

Sony's Transformed Grid Array Package (TGA)

27.1 Introduction and Overview

The transformed grid array (TGA) package was developed by Sony Corporation in 1996 for mobile equipment [1]. This CSP has a flip-chip configuration and uses an organic substrate as the package interposer. The I/O redistribution may be implemented either on the chip or on the interposer. Solder precoat is the key technology for TGA packages. The bond pads on both sides of the interposer are printed with eutectic solder in advance for chip-level and board-level connection. The flip-chip interconnects are high-lead solder bumps. Underfill adhesive is used to secure the reliability of the package and the flip-chip solder joints. The package I/Os are land grid array (LGA) with a pitch of 0.5 mm. An eight-layer buildup PCB is used for the board assembly of TGA packages. Good package reliability and solder joint fatigue life have been reported. Currently the TGA packages are used in Sony's digital camera VCR "DCR-PC7." With the application of this CSP, a reduction of 37 percent in PCB mounting area can be achieved.

27.2 Design Concepts and Package Structure

Sony's TGA package is a CSP with a rigid interposer. The TGA is so named because of the transformation from the die-pad pitch (150 μm) to the package terminal pitch (0.5 mm). The redistribution of I/Os may be arranged in two ways. One is to deposit a layer of metallization (Cr/Cu/Au) on the active face of the chip to redistribute the peripheral bond pads to an area-array pattern (fan-in only). The other is to use the Cu traces on one side of the interposer for circuit rerouting (fan-in/out). In both cases, a flip-chip configuration is adopted and high-lead solder bumps are used as the chip-to-substrate interconnects.

The interposer of the TGA package is a 0.4-mm-thick organic substrate. The dimension is about 1.0 mm larger than the chip size. Plated through vias are made in grid array to form the package terminal pattern. The diameter of the vias is 150 μm, and the via holes are filled with resin followed by metallization to establish lands for chip-level and board-level connection. The bond pads on both sides of the interposer are precoated with eutectic solder, which is a key technology for the TGA package. If the circuit redistribution is performed at the wafer level, then the flip-chip bumps are just mated with the lands on the upper surface of the interposer, as shown in Fig. 27.1a. Otherwise, as illustrated in Fig. 27.1b, the solder bumps on the die pads are aligned with the peripheral bond pads on the substrate and additional circuit rerouting to the lands of the plated through vias is needed. In both configurations, the flip-chip bumps are surface-mounted to the solder-precoated pads on the interposer by reflow heating. In addition, underfill adhesive is employed to encapsulate these solder joints for better reliability. With the mounting of the flip chip, the total package thickness is 1.0 to 1.2 mm.

The package I/Os of the TGA are land grid array (LGA) on the bottom surface of the interposer. The lands are established at the end of plated through vias and have a perimeter array pattern with four-row procession. The land pitch is 0.5 mm. In order to handle such a fine

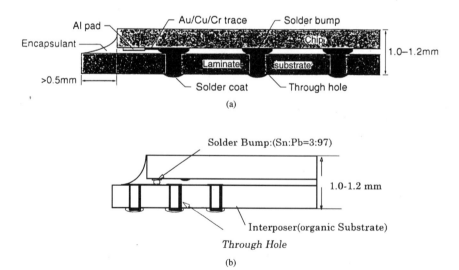

Figure 27.1   Package structure of Sony's TGA package: (*a*) with redistribution on the chip; (*b*) with redistribution on the interposer.

TABLE 27.1   Design Rules for the Multilayer PCB for the Assembly of TGA Packages

Features	Values
Land pitch	500 μm
Land diameter	275 μm
Line/space width	75 μm/75 μm
Line between two lands	1 line
Distribution	2 layers

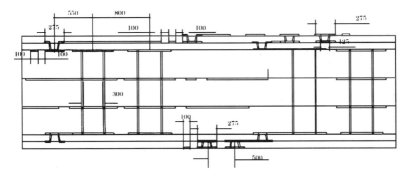

Figure 27.2   Eight-layer buildup PCB for the assembly of TGA packages.

pitch, the board-level assembly of TGA packages requires a multilayer buildup PCB. Since the LGA pattern of this package is a four-row perimeter array with a land pitch of 0.5 mm, the design rules for the corresponding PCB are those given in Table 27.1. The tolerance for the precision of pick-and-place is 50 μm. Because the applications of TGA packages usually require double-side surface mounting for high-density assembly, an eight-layer buildup PCB (four core layers plus two buildup layers on each side of the core) was developed, as shown in Fig. 27.2.

27.3   Material Issues

The TGA package is a CSP with a flip chip on a rigid substrate. The whole package consists of the solder-bumped silicon chip, the underfill adhesive, and the package interposer. The flip-chip bumps of the TGA are made of 97Pb/3Sn solder. The ball limiting metallurgy (BLM) for this high-lead solder is Cr/Cu/Au. Three kinds of underfill materials were evaluated during the development of the TGA. It was reported that the adhesion strength at the interfaces is very critical to the reliability of the package.

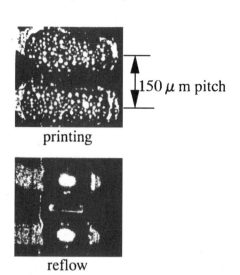

printing

reflow

flattening

Figure 27.3 Solder precoat on the lands of the TGA interposer.

The interposer of the TGA package is an organic substrate. Three types of material were investigated. It was determined that the substrate material should have a low CTE and low moisture absorption capability. For package interconnection, the interposer has plated through vias in a grid array pattern. The via holes are filled with resin and then covered by metallization to form lands on both sides of the substrate. An interposer solder precoat is the proprietary technology for TGA packages. Eutectic solder is screen-printed on the lands, followed by reflow heating and flattening, as shown in Fig. 27.3.

27.4 Manufacturing Process

The manufacturing process for the TGA package is illustrated in Fig. 27.4 [2]. The first stage is substrate preparation. Eutectic solder paste is printed on the pads of the interposer by a conventional screen printer. Reflow heating is required after the solder printing on each side of the substrate. Subsequently, a flattening process is performed

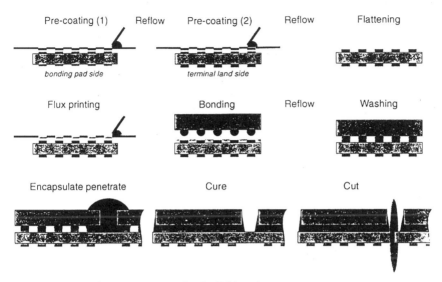

Figure 27.4 Manufacturing process for the TGA package.

to level the solder precoat. Once the substrate has been prepared, flux is printed on the solder precoat, followed by the placement of the solder-bumped flip chip. After reflow heating and cleaning, the silicon chip is surface-mounted on the interposer.

The next step is underfill encapsulation. In fact, the substrate of the TGA is in a strip form and contains several units. The adhesive encapsulant is dispensed in the channel between two flip chips, as shown in Fig. 27.4. The liquid encapsulant is attracted into the gap between the die and the interposer by capillary effect. After curing, the substrate is cut by mechanical sawing for package singulation.

27.5 Qualifications and Reliability

Sony's TGA has been extensively tested for package and solder joint reliability [1]. Since the TGA is in principle a plastic package, popcorning and delamination are the major concern for package reliability. During the development of the TGA, various kinds of substrates, underfills, and chip surface conditions were evaluated for their effects on the package crack resistance. The testing conditions and results are given in Table 27.2. After optimization, the TGA could sustain two rounds of IR reflow heating with preconditioning at 85°C/85 percent RH for 24 h.

Further qualification tests were performed with TGA packages mounted on the PCB. The specimens were preconditioned at 30°C/85

TABLE 27.2 Crack Resistance Testing Results for TGA Packages

Interposer material	Encapsulation resin	Chip surface condition	30°C/85% RH X h × 2 IR reflows				85°C/85% RH X h ×2 IR reflows
			24 h × 2	48 h × 2	96 h × 2	168 h × 2	24 h × 2
Type A	No. 1	Condition A	10/10	—	—	—	
Type C	No. 2	Condition B	0/10	0/10	2/10	8/9	
Type C	No. 2	Condition A	0/10	0/10	0/10	4/10	
Type C	No. 3	Condition A	0/10	0/10	0/10	0/10	0/10

Package size: 10 mm × 10 mm/232 pins/0.5 mm pitch.

TABLE 27.3   Qualification Test Data for TGA Packages

Item	Test condition			Result
T/C	−25 to 125°C, 72 cycles/day	On board	1500 cycles	0/45
THB	85°C/85% RH, 5.0 V	On board	1000 h	0/45
PCT	110°C/85% RH	On board	500 h	0/45
HTB	100°C, 5.0 V	On board	1000 h	0/45

Chip size: 8.69×8.80 mm, memory device.
Package size: 9.69×9.80 mm/100 pins/0.5 mm pitch.

percent RH for 48 h followed by two rounds of IR reflow at 245°C. The testing conditions and results are listed in Table 27.3. From these qualification data, it can be concluded that the TGA has the same package reliability as conventional plastic surface-mount components.

According to Sony, the warranty on solder joints is 10 years. Because of the LGA configuration, the TGA package has relatively small solder joints. Therefore, the board-level interconnect reliability could be a critical issue. The application specification for the TGA requires solder joints to sustain at least 750 cycles of temperature change between −25 and 125°C. A series of temperature cycling tests were conducted to evaluate the thermal fatigue life of TGA solder joints. The cycling frequency was 72 cycles per day. The specimens under test were 10×10 mm TGA packages with 232 pins, and the test board thickness was 0.6 mm. Since the TGA was developed for high-density electronics, double-side surface mount is the standard configuration for board assembly. Various interposer materials, solder joint heights, and interposer thicknesses were evaluated for their effects on the solder joint reliability. The test results are presented in Fig. 27.5. The general trends indicate that a low-CTE substrate, tall solder joint height, and thick interposer could improve the thermal fatigue life of TGA solder joints. In particular, the CTE of Type C substrate is between those of the silicon chip and the PCB. After optimization, the TGA solder joints could sustain 1200 thermal cycles (at a 1 percent failure rate). This result is much better than the required performance (750 cycles). In addition, it was reported that the solder joint reliability is more critical in double-side mounting than in single-side. If the latter configuration were adopted, the thermal fatigue life of solder joints could be doubled.

27.6   Applications and Advantages

Sony developed the TGA package mainly for the ASICs in its digital camera. On the PCB of the model "DCR-PC7" VCR, shown in Fig. 27.6, 20 TGA packages were mounted [1]. More recently, a wire-bonded ver-

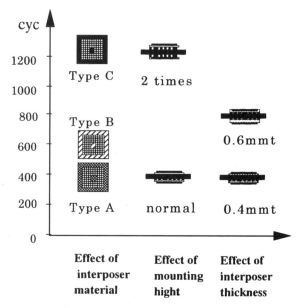

Figure 27.5 Solder joint fatigue life (cycles) for the TGA assembly.

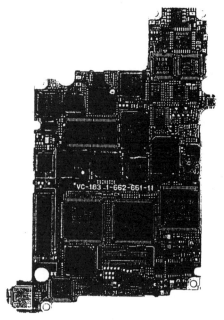

Figure 27.6 Application of TGA packages to Sony's digital camera.

sion of the TGA (W-TGA) was reported. The applications of the W-TGA include CPUs and DSPs in cellular phones. In addition, for both versions of the TGA packages, contract assembly is available [2].

The major advantages of TGA packages are simple structure, low profile, and compact size. According to Sony, with the application of the TGA, the mounting area on the PCB shown in Fig. 27.6 is reduced by 37 percent. Since the package interposer is an organic substrate and the manufacturing process fits in the existing infrastructure, this CSP is relatively cost-effective. Also, good package reliability and solder joint fatigue life were reported. All the above factors make the TGA package an attractive choice for the production of high-volume portable devices.

27.7 Summary and Concluding Remarks

Sony's TGA is a CSP with a rigid organic substrate. The package interposer is an FR-4 laminate with plated through vias. The vias are filled with resin, followed by surface metallization to establish the lands on both sides of the interposer. These lands are precoated with eutectic solder for chip-level and board-level connection. This solder precoat process is the key technology for TGA packages.

The first-level interconnects of the TGA are high-Pb solder bumps. The I/O redistribution may be implemented either on the chip or on the interposer. Underfill adhesive is employed to secure the package reliability and to encapsulate the flip-chip solder joints. The second-level interconnects are LGA. The terminal pattern is a four-row perimeter array with a pitch of 0.5 mm. To handle such a fine pitch, an eight-layer buildup PCB is used for the board assembly of TGA packages. A series of qualification tests were performed. Good package reliability and solder joint fatigue life were reported.

The advantages of TGA packages include simple structure, low profile, and compact size. In addition, this CSP is relatively cost-effective because the package interposer is an organic substrate and the manufacturing process fits in the existing infrastructure. Therefore, the TGA is an attractive package for ASICs in portable electronics. Currently the TGA packages are used in Sony's digital camera VCR "DCR-PC7." With the application of this CSP, a reduction of 37 percent in PCB mounting area can be achieved. A new version of the TGA with wire bonds was developed as well. This package may be applied to CPUs and DSPs in cellular phones. Contract assembly for both versions of the TGA package is available from Sony.

27.8 References

1. K. Kosuga, "CSP (Chip Size Package) Technology for Mobile Apparatus," *Proceedings of the International Symposium on Microelectronics,* Philadelphia, PA, 1997, pp. 244–249.
2. T. Goodman and E. J. Vardaman, *CSP Markets and Applications,* TechSearch International, Austin, TX, 1998, pp. 63–64.

28

Toshiba's Ceramic/Plastic Fine-Pitch BGA Package (C/P-FBGA)

28.1 Introduction and Overview

Toshiba Corporation released a series of fine-pitch ball grid array packages in 1997. The BGAs in this family are considered to be chip scale packages because their package terminal pitch is 0.5 mm or 0.8 mm. The first-level interconnects of these CSPs are either Au wire bonds or flip-chip solder bumps. Although all of them are equipped with a rigid interposer, several kinds of materials are used for the substrate. In addition, various types of encapsulant are used for different substrate materials. The reliability of certain C/P-FBGA packages has been investigated. The qualification testing results indicate that these CSPs have a package reliability similar to that of other surface-mount components. However, the board-level solder joint reliability substantially depends on the material properties of the chip carrier and may be a concern for certain C-FBGA configurations. To resolve this critical issue, a new high-CTE substrate material was developed by Toshiba together with Kyocera. It has been confirmed by thermal cycling tests that the solder joint reliability can be improved by adopting the new substrate material.

28.2 Design Concepts and Package Structure

Toshiba's fine-pitch BGAs are chip scale packages with a rigid interposer. Depending on the substrate material, the CSPs in this family are classified as ceramic or plastic fine-pitch ball grid array (C-FBGA and P-FBGA) packages. For each category, there are wire-bond and flip-chip versions. These packages are considered to be CSPs mainly

Figure 28.1 Top and bottom views of Toshiba's CSP with flip-chip configuration (upper ones are C-FBGA; lower ones are P-FBGA).

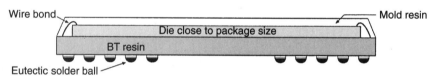

Figure 28.2 Package structure for wire-bonded C/P-FBGA.

because of their fine-pitch (0.5 mm or 0.8 mm) I/Os. The top and bottom views of two typical CSPs with flip-chip configuration are shown in Fig. 28.1.

The cross section of a wire-bonded C/P-FBGA is illustrated in Fig. 28.2. The package structure is very similar to that of the conventional overmolded BGAs, but the outer rim of the substrate does not extend beyond the molded area. For C-FBGA packages, the substrate material may be conventional alumina or high-CTE ceramics, and the potting compound is used for encapsulation. The thickness of the substrate and the encapsulation is 0.4 mm and 0.6 mm, respectively. On the other hand, for P-FBGA packages, the material used for the interposer is BT-832, and the corresponding encapsulant is molding compound. The thickness of the substrate and the encapsulation is 0.49 mm and 0.3 mm, respectively. The package I/Os are area-array solder balls. The pads on the bottom surface of the interposer for solder-ball attachment have a diameter of 0.35 mm. The ball pitch may be either 0.5 mm or 0.8 mm [1].

Figure 28.3 shows the cross section of a typical C/P-FBGA with flip-chip configuration. This category of packages uses eutectic solder bumps as the chip-level interconnects. The bump height is about 50 μm. The minimum bond-pad pitch for peripheral and area-array solder bumps is 200 μm and 250 μm, respectively. The bond-pad size on the die is 50 percent of the pad pitch. Underfill is required to encapsulate the chip-to-interposer solder joints and to enhance the package reliability. The chip carrier may be either ceramic or organic substrate. The package size is about 1.0 mm larger than the die size. The package I/Os are BGA solder balls. The ball pitch may be 0.5 or 0.8 mm. The package thickness excluding BGA ball height is 1.0 to 1.2 mm [2].

28.3 Material Issues

The C/P-FBGA packages are a family of CSPs with a rigid interposer. For the wire-bonded version, the whole package consists of the silicon chip, the die-attach adhesive, the Au bonding wires, the substrate, the encapsulant, and the BGA solder balls. The major material properties are given in Table 28.1. It should be noted that the chip carrier

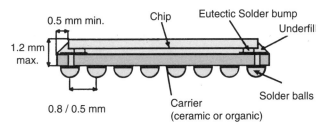

Figure 28.3 Package structure for flip-chip C/P-FBGA.

TABLE 28.1 Properties of Packaging Materials for Wire-Bonded C/P-FBGA

Item	Material	Coefficient of thermal expansion $(10^{-6}/°C)$	Young's modulus (GPa)	Poisson's ratio
Substrate	BT resin (BT-832)	14.6	23.5	0.20
	Glass ceramics	11.5	110.0	0.23
	Al_2O_3	8.0	250.0	0.28
Chip	Si	3.5	170.0	0.28
Encapsulant	Mold resin	7.0	26.0	0.30
	Potting resin	12.4	10.7	0.30
Die attach	Epoxy resin	40.0	5.2	0.30

may be made of organic or ceramic materials. For the former, Toshiba adopted BT-832. The corresponding package encapsulant is a conventional molding compound. For the C-FBGA, either alumina (Al_2O_3) or glass ceramic may be used. In this case, the encapsulation is performed using potting compound instead of molding compound.

The major difference between alumina and glass ceramic is the coefficient of thermal expansion (CTE). That of the former is 8.0 ppm/°C, while that of the latter is 11.5 ppm/°C. A recent numerical study indicated that the board-level solder joint reliability may be optimized if the substrate CTE of the wire-bonded C-FBGA falls in the region of 13 to 14 ppm/°C [3]. As a result, Toshiba, together with Kyocera, developed a new substrate material with a CTE of 13 ppm/°C for the wire-bonded C-FBGA package. The comparison of material properties for three ceramic substrate materials is given in Table 28.2. The properties of the FR-4 PCB are listed as well for reference.

For the flip-chip version of the C/P-FBGA, the packaging materials, in addition to the silicon chip, include the eutectic solder bumps, the ceramic/organic substrate, the underfill adhesive, and the BGA solder balls. The 63Sn/37Pb solder bumps are deposited on the wafer by electroplating. A Ti/Ni/Pd under-bump metallurgy (UBM) is required on the bond pads for the attachment of solder bumps. A uniform bump height of 50 μm with 10 percent deviation can be achieved. The detailed wafer-bumping process will be discussed in the next section.

According to Toshiba, the underfill adhesive is a critical material for the flip-chip C/P-FBGA. Normally the filler content in resin is within the range of 40 to 60 percent. However, the filler size may affect the performance of the underfill. During the development of this package, an experimental study was conducted to evaluate eight different underfill materials with various sizes of fillers. The performance was evaluated in terms of the filling distance in a narrow gap. The gap was generated by putting a glass substrate over a 15×15 mm test chip. The standoff height was maintained at 20 μm. The underfill was dispensed along one side of the gap, and the filling distance from the dispensing side was measured. The test results are presented in Fig. 28.4. It was determined that the filler size in the underfill resin should be smaller than 10 μm.

TABLE 28.2   Comparison of Properties among Various Substrate Materials

Item	Alumina	Glass ceramic	New material	FR-4 PCB
CTE (ppm/°C)	8	11.5	13	12–16
Young's modulus (GPa)	250	110	110	24
Dielectric constant	9.8	5.8	5.3	5.5

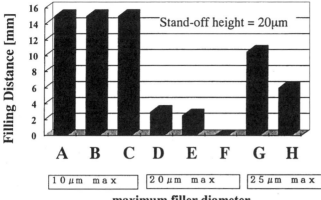

Figure 28.4 Comparison of filling distance performance among different underfills.

28.4 Manufacturing Process

The manufacturing process for the wire-bonded C/P-FBGA is very similar to that for conventional PBGA packages. Therefore, the discussion in this section is focused on the flip-chip version of Toshiba's CSPs.

The fabrication of the flip-chip C/P-FBGA begins with wafer bumping. The detailed procedures are illustrated in Fig. 28.5. The first step is to establish the appropriate UBM. This is implemented by sputtering Ti/Ni/Pd thin films on the whole wafer first. A layer of photoresist is spin-coated on the wafer and then patterned by photolithography to open window access to the die bond pads. Next, 63Sn/37Pb eutectic solder is deposited on the exposed bond pads by electroplating, with the UBM serving as the electrode. After stripping off the photoresist layer, the UBM outside the die bond pads is etched away using plated solder bumps as the etchant resist. At last the wafer is heated in a nitrogen environment to reflow the solder for ball round-up. It is well known that sufficient bump height with good uniformity is critical to the reliability of flip-chip assembly [4]. From an experimental investigation, Toshiba optimized the photoresist thickness and the solder-bump height at 55 μm and 50 μm (as plated), respectively [2]. A typical bump-height distribution is shown in Fig. 28.6. The bump-height variation is less than 10 percent.

The manufacturing process subsequent to wafer bumping is presented in Fig. 28.7. Before the die is mounted on the substrate, a flux transfer step is required. The flux is transferred to the solder bumps by dipping the flip chip in a shallow bath of flux. It was reported by Toshiba that the volume control in flux transfer is rather important [2].

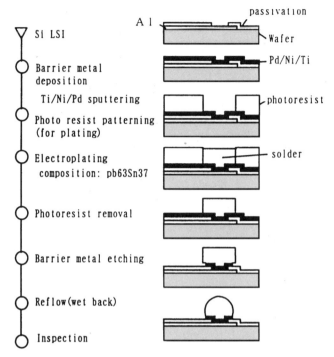

Figure 28.5   Wafer-bumping process for the flip-chip C/P-FBGA.

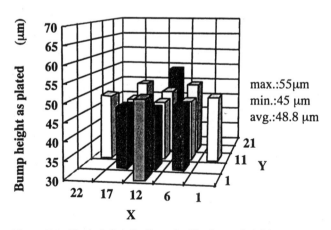

Figure 28.6   Typical distribution of solder-bump height.

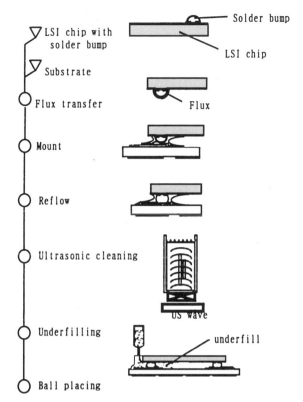

Figure 28.7 Manufacturing process for the flip-chip C/P-FBGA.

Excessive flux may result in a short circuit between adjacent solder bumps after reflow heating. An experimental study was conducted to investigate this issue. From the results shown in Fig. 28.8, it may be concluded that the transferred flux should be limited to less than 1 mg in order to avoid bump short.

After the flux is transferred, the flip chip is mounted on the interposer, with the solder bumps aligned with the corresponding bond pads on the substrate. One of the features of eutectic solder is the good self-alignment capability during reflow. Toshiba also performed a study to characterize the tolerance of misalignment during the pick-and-place process [2]. Two test chips were used for the experiment, as shown in Table 28.3. The test results presented in Fig. 28.9 indicate that for both peripheral and area-array layouts, the flip-chip assembly can tolerate an offset of up to 50 percent of the bump pitch.

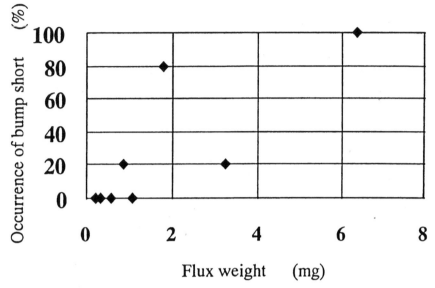

Figure 28.8 Effect of flux weight on the bump short.

TABLE 28.3   Configuration of Test Chips for Flip-Chip C/P-FBGA Packages

	Chip size	Pad pitch	Pad layout	Pin count
Test chip A	10 mm × 10 mm	200 μm	Peripheral	180
Test chip B	6 mm × 6 mm	250 μm	Area array	529

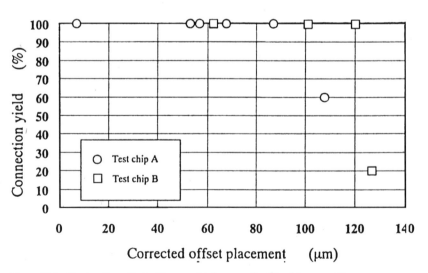

Figure 28.9 Evaluation of misalignment tolerance for flip-chip mounting.

Once the reflow of flip-chip solder bumps is completed, an ultrasonic cleaning is carried out to remove the rosin residue from the flux. The cleaning system developed by Toshiba, namely, Techno Care, is employed for this purpose. The processing conditions for cleaning are given in Table 28.4. It was reported that the use of ultrasound during the cleaning and rinse periods is essential for the removal of rosin residue. Toshiba also indicated that the Techno Care system outperforms the IPA cleaning method and is very effective for both organic and ceramic substrates.

The procedure after flux cleaning is underfill dispensing. There are two major concerns in this process. One is the required flow time to fill the gap between the die and the substrate, and the other is the voids induced in the underfill adhesive. Figure 28.10 presents the results from an experimental investigation of the underfill flow time. It is obvious that, compared to the area-array solder bumps, the peripheral

TABLE 28.4   Conditions for Ultrasonic Cleaning with Toshiba's Techno Care System

Cleaning	15 min, 55°C
Rinse	10 min, R.T.
Drying	3 min, 105°C
Ultrasonic condition	28 kHz, 250 W

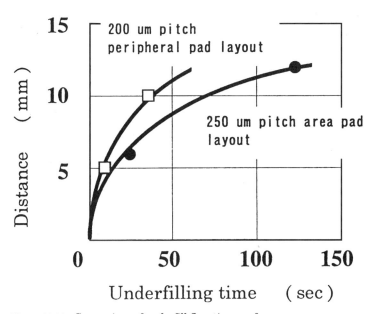

Figure 28.10   Comparison of underfill flow time performance.

layout requires much less time to fill the same gap distance. In addition, from the results obtained in another study, shown in Fig. 28.11, the occurrence of voids inside the underfill depends on the chip size and the standoff height of the gap. In general, use of taller solder bumps and a smaller-size die can suppress the formation of voids during the underfill process.

28.5   Qualifications and Reliability

A number of qualification tests have been performed to investigate the package reliability of a wire-bonded C-FBGA with glass ceramic interposer (CTE = 11.5 ppm/°C). The test items and corresponding conditions are given in Table 28.5. The package reliability is evaluated in terms of the variation in the electrical resistance of the circuit and the degradation of the insulation between two adjacent trace lines on the interposer. The results are presented in Tables 28.6 and 28.7, respectively. After all tests, the change in the aforementioned

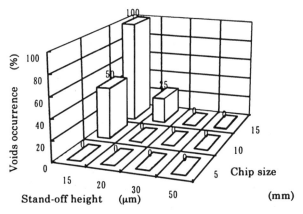

Figure 28.11 Effect of chip size and gap standoff height on the occurrence of voids.

TABLE 28.5   Testing Conditions for Package Qualification

Item	Condition	Number
Thermal shock	−65 to 150°C	20
High temperature/high humidity (1)	85°C/85% RH, no bias	20
High temperature/high humidity (2)	85°C/85% RH, 5.5 V	20
High temperature	150°C	20
Thermal cycling test	−65 to 150°C	20

TABLE 28.6   Change in Circuit Resistance during Qualification Tests

Item	Result
Thermal shock	0 cycles (400 mΩ)
	500 cycles ± 1.1%
High temperature/high humidity (1)	0 h (400 mΩ)
	1000 h ± 3.5%
High temperature/high humidity (2)	0 h (400 mΩ)
	1000 h ± 4.3%
High temperature	0 h (400 mΩ)
	1000 h ± 3.3%
Thermal cycling test	0 cycles (400 mΩ)
	1000 cycles ± 4.1%

TABLE 28.7   Change in Trace Line Insulation during Qualification Tests

Item	Result
Thermal shock	0 cycles $>1 \times 10^{11}\Omega$
	500 cycles $>1 \times 10^{11}\Omega$
High temperature/high humidity (1)	0 h $>1 \times 10^{11}\Omega$
	1000 h $>1 \times 10^{10}\Omega$
High temperature/high humidity (2)	0 h $>1 \times 10^{11}\Omega$
	1000 h $>1 \times 10^{10}\Omega$
High temperature	0 h $>1 \times 10^{11}\Omega$
	1000 h $>1 \times 10^{11}\Omega$
Thermal cycling test	0 cycles $>1 \times 10^{11}\Omega$
	1000 cycles $>1 \times 10^{11}\Omega$

quantities was negligible. Therefore, the wire-bonded C-FBGA is considered to be as reliable as other conventional packages.

The package reliability of flip-chip C-FBGA packages was also investigated with two types of test chips, as shown in Table 28.3. The testing conditions and results are given in Table 28.8. For the specimens with test chip A, no failures were observed in all testing items. For the specimens with test chip B, only two items were confirmed with no failure. The other tests are still ongoing, and the results have not yet been reported.

The thermal fatigue resistance of board-level solder joints is another reliability concern. A computational analysis was performed to investigate the assembly of wire-bonded C/P-FBGA packages [1]. A three-dimensional global-local finite-element model was established, as presented in Fig. 28.12. The thickness dimensions of package con-

TABLE 28.8 Qualification Data for Flip-Chip C-FBGA Packages

Reliability test condition	Results, test chip A	Results, test chip B
TCT (−40 to 125°C) 1500 cycles	0/5	—
1000 cycles	—	0/5
IR reflow (240°C max.), 5 times	0/5	—
TCT after IR reflow 5 times, 1000 cycles	0/5	—
THB (85°C/85% RH/4.2 V) 500 h	0/5	0/5
PCT (130°C/85% RH) 200 h	0/5	—
HTS (125°C) 2000 h	0/5	—
HTS (150°C)1000 h	0/5	—
HTB (125°C/4.8 V) 1000 h	0/5	—

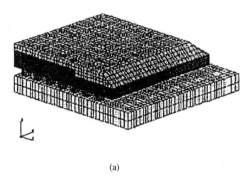
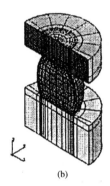

(a) (b)

Figure 28.12 Finite-element model for the analysis of solder joint reliability: (a) global model for the whole assembly; (b) local model for a single solder joint.

stituents are given in Table 28.9. An elastic-creep constitutive relation was adopted to describe the material response of solder joints. The Coffin-Manson equation was employed to estimate the thermal fatigue life of solder joints under cyclic loading between −40 and 125°C. The results of the numerical analysis are given in Table 28.10. It is obvious that the solder joint reliability of the P-FBGA is much better than that of the C-FBGA. This is due to the higher CTE of the BT-832 substrate (14.6 ppm/°C). Similarly, because the CTE of glass ceramic (11.5 ppm/°C) is higher than that of Al_2O_3 (8 ppm/°C), the solder joint thermal fatigue life of C-FBGA is longer with the former substrate than with the latter.

From the aforementioned numerical analysis, it may be concluded that the board-level solder joint reliability depends considerably on the CTE of the package interposer. In order to extend the thermal fa-

TABLE 28.9   Thickness Dimensions Used in Finite-Element Modeling

	Substrate		Encapsulation		Si	Adhesive	Solder ball	PCB
P-FBGA	BT-832	0.490	Molding	0.300				
C-FBGA	Glass ceramics	0.400	Potting	0.600	0.290	0.030	0.435	0.800

All dimensions in millimeters.

TABLE 28.10   Estimation of Solder Joint Thermal Fatigue Life for Wire-Bonded C/P-FBGA Package Assembly

Substrate	Encapsulation	$\Delta\varepsilon_{creep}$ (%)	N_f (cycles)
Al_2O_3	Potting	2.779	150
Glass ceramics	Potting	1.037	1030
BT-832	Molding	0.69	2280

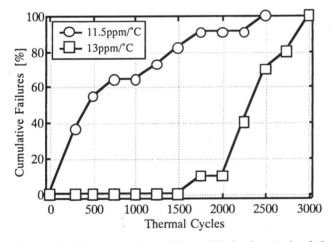

Figure 28.13   Solder joint thermal fatigue life for the wire-bonded C-FBGA with different substrates (temperature cycling between −40 and 125°C).

tigue life of solder joints in the C-FBGA assembly, based on the re-sults of another computational analysis [3], Toshiba, together with Kyocera, developed a new ceramic substrate with a CTE of 13 ppm/°C. This new substrate was applied to C-FBGA assemblies sub-jected to temperature cycling test (−40 to 125°C). The test results for wire-bonded and flip-chip packages are presented in Figs. 28.13 and 28.14, respectively, and are compared with those for C-FBGA pack-ages using a regular glass ceramic substrate. From the presented ex-

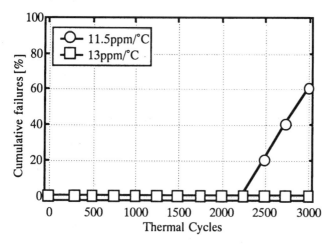

Figure 28.14   Solder joint thermal fatigue life for the flip-chip C-FBGA with different substrates (temperature cycling between −40 and 125°C).

perimental data, it is obvious that the new substrate material outperforms the regular glass ceramic. However, in view of the substantial thermal mismatch between the silicon chip and the new substrate material, Toshiba still recommends the regular glass ceramic for flip-chip C-FBGA packages.

28.6   Applications and Advantages

Toshiba's CSPs are fine-pitch ball grid array packages with ceramic or organic substrate. This family of chip scale packages was developed for IC devices with medium pin counts (100 to 300). The current application of C/P-FBGA is in a standard 2-Mbit SRAM module. The ASICs in wireless systems may be another niche for this series of CSPs.

   The major advantage of Toshiba's C/P-FBGA is the flexibility in package configuration. This family of CSPs can offer wire-bonded and flip-chip versions with either ceramic or organic substrate. In addition, with the flip-chip package, both peripheral and area-array bond-pad layouts can be accommodated. This diversity allows C/P-FBGA packages to fulfill various requirements from different customers.

28.7   Summary and Concluding Remarks

Toshiba's C/P-FBGAs are a family of CSPs with a rigid interposer. This series of packages has wire-bonded and flip-chip versions, and the package substrate may be made of organic or ceramic materials.

The first-level interconnects of the C/P-FBGA are either Au wire bonds or eutectic solder bumps. For the flip-chip version, the bond-pad pitch on the die may be 200 μm for peripheral layout or 250 μm for area-array layout. The package I/Os are BGA solder balls. A pitch of 0.5 mm or 0.8 mm can be achieved. Depending on the substrate material, molding or potting resin compound is used to encapsulate the wire-bonded C/P-FBGA packages. For the flip-chip version, under-fill adhesive is required to encapsulate the eutectic solder joints between the die and the interposer. This series of CSPs was developed for IC devices with medium pin counts (100 to 300). The applications of C/P-FBGA include SRAM modules and ASICs in wireless systems.

The reliability of C/P-FBGA packages has been investigated. The qualification testing results indicate that these CSPs have package reliability similar to that of other surface-mount components. It was reported that the board-level solder joint reliability depends considerably on the CTE of the package substrate and may be a concern for certain C-FBGA configurations. To resolve this critical issue, Toshiba, together with Kyocera, developed a new high-CTE substrate material. From the results of computational analysis and thermal cycling tests, it was confirmed that the thermal fatigue life of solder joints could be substantially extended by adopting the newly developed substrate material.

28.8 References

1. T. Ikemizu, Y. Fukuzawa, J. Nakano, T. Yokoi, K. Miyajima, H. Funakura, and E. Hosomi, "CSP Solder Ball Reliability," *Proceedings of the IEEE/CPMT IEMTS,* Omiya, Japan, October 1997, pp. 447–451.
2. H. Aoki, C. Takubo, T. Nakazawa, S. Honma, K. Doi, M. Miyata, H. Ezawa, and Y. Hiruta, "Eutectic Solder Flip Chip Technology—Bumping and Assembly Process Development for CSP/BGA," *Proceedings of the 47th ECTC,* San Jose, Calif., May 1997, pp. 325–331.
3. H. Yonekura, M. Higashi, N. Hamada, K. Yamaguchi, M. Kokubu, M. Ikemizu, J. Nakano, T. Yokoi, and H. Funakura, "Evaluation of Solder Joint Reliability between PWB and CSP by Using High TCE Ceramic Material," *Proceedings of the 48th ECTC,* Bellevue, Wash., May 1998, pp. 1260–1264.
4. J. Lau, *Flip Chip Technologies,* McGraw-Hill, New York, 1995.

29

Chipscale's Micro SMT Package (MSMT)

29.1 Introduction and Overview

The Micro SMT (MSMT) was released by ChipScale, Inc., in 1994 [1]. This package is classified as a wafer-level CSP because all fabrication procedures are performed on the wafer. The first-level interconnects of the MSMT are thin Au beams which connect the die bond pads with peripheral metallized silicon posts. The package I/Os are the flat lands of silicon posts with surface metallization. Encapsulation is required for package integrity and chip protection. The MSMT has a face-up configuration, and the active face of the IC is covered by encapsulant. An optional silicon cap may be added on top of the encapsulant to improve package coplanarity and heat dissipation. The package thickness is about 0.5 mm with the silicon cap. The MSMT is for packaging IC devices with peripheral bond pads only. The package I/O pin counts range from 2 to 144. The terminal pitch varies from 0.3 to 0.63 mm. The MSMT package may find a wide spectrum of applications for passive and active devices such as diodes, transistors, op amps, and RFICs. Currently Motorola is the major licensee of MSMT technology. Its Micro SMT products are called μSurf and are used mainly to replace existing transistors and SOICs.

29.2 Design Concepts and Package Structure

The MSMT package is known as a wafer-level CSP because in the fabrication process the whole wafer is packaged instead of a single die. This technology has been in use in the military market since 1990 for microwave diodes (two to four leads) [2]. In 1993, ChipScale was founded to refine and develop the MSMT packaging technology for transistors and integrated circuits.

The package structure of the MSMT is quite different from that of other CSPs. A three-dimensional cutaway view of the MSMT is presented in Fig. 29.1 for illustration. The first-level interconnects are thin Au beams which serve as a bridge for electrical connection between the die bond pads and the peripheral package terminals. In fact, each interconnect is a pair of shallow V-shaped double beams, as shown in Fig. 29.2. These structures are configured in this way in

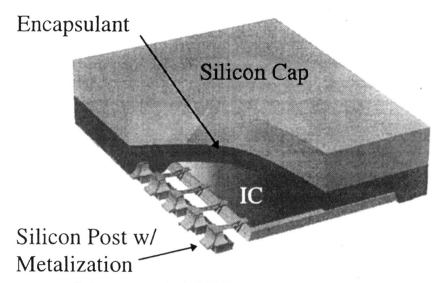

Encapsulant

Silicon Cap

IC

Silicon Post w/ Metalization

Figure 29.1   Package structure for the MSMT.

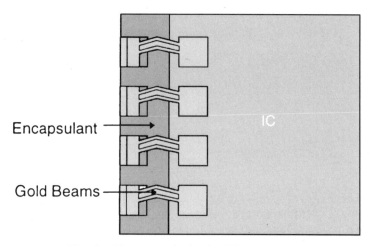

Encapsulant

Gold Beams

IC

Figure 29.2   First-level interconnects for the MSMT.

order to accommodate deformation from thermal expansion. The package I/Os of the MSMT are the bottom lands of metallized silicon posts. These pillars are formed by etching away the silicon from the scribe line area that surrounds the die. For high-density packaging applications, the posts may be formed at the locations which are currently used for die bond pads. A certain metallization is deposited on the surface of silicon posts for electrical connection.

For package integrity and chip protection, encapsulation of the MSMT is required. The encapsulant covers the active face of the die and encloses the Au beams together with the upper portion of the silicon posts. For larger IC devices, a silicon cap should be attached to the top of the encapsulant to maintain the coplanarity and position of the peripheral silicon posts within 6 μm. This cap may also provide thermal enhancement and impedance control. However, for smaller chips, such as diodes and transistors, this cap may be omitted from the package. Depending on the location of the silicon posts, the package-to-chip area ratio of the MSMT varies from 1.0 to 1.2. This form factor complies with the dimensional requirements for chip scale packages. The thickness dimension of MSMT is rather small. Including the silicon cap, the package thickness is about 0.5 mm. The cross-sectional picture of a typical MSMT package is presented in Fig. 29.3 [3].

The MSMT package was developed mainly for low-pin-count devices, although on some occasions it may accommodate more than 100 I/Os. The current lineup ranges from 2 to 144 pins. The lead pitch varies from 0.3 to 0.63 mm. It should be noted that the MSMT can package only silicon chips with peripheral bond pads. Figures 29.4 to 29.6 illustrate some typical MSMT packages with various terminal layout patterns. Because of the lack of a real package substrate, the standardization of wafer-level CSPs is rather difficult. ChipScale has suggested that the MSMT may be standardized based on lead pitch

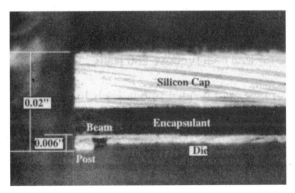

Figure 29.3 SEM micrograph of the cross section of the MSMT.

Figure 29.4 MSMT transistor with 3-pin terminal layout.

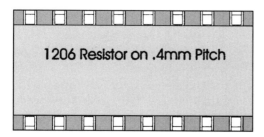

Figure 29.5 MSMT resistor with dual-in-line terminal layout.

Figure 29.6 MSMT IC with perimeter terminal layout.

instead of package outline or footprint [4]. Table 29.1 shows the JEDEC/EIAJ standard for small outline packages. The pin counts for a certain lead pitch may be adopted as an index for the standardization of MSMT packages.

TABLE 29.1   Package Size and Allowable Pin Counts Based on JEDEC/EIAJ Standard

Package designator		Maximum leads per package									
		Lead pitch (mm)									
Metric	English	0.3	0.35	0.4	0.45	0.5	0.55	0.6	0.65	0.8	1.0
1005	0402	4	3	2	N/A	N/A	N/A	N/A	N/A	N/A	N/A
1608	0603	10	8	6	5	4	3	2	N/A	N/A	N/A
2012	0805	16	13	10	8	7	6	5	4	2	2
3216	1206	26	21	18	15	13	11	10	9	6	4
3225	1210	32	27	23	19	17	15	13	12	8	5
4532	1812	45	38	33	28	25	22	20	18	13	9
4564	1825	67	56	49	42	38	34	30	28	21	16
5025	2010	44	37	32	27	24	21	19	17	13	9
5650	2220	65	55	47	41	36	33	29	27	21	15
6032	2512	55	47	40	35	31	27	25	22	17	12
7227	3611	60	51	44	38	34	30	27	24	19	14
7343	2917	71	60	52	46	40	36	33	30	23	17
8530	3412	71	60	52	45	40	36	32	29	23	17

29.3   Material Issues

ChipScale's MSMT is a CSP with a wafer-level fabrication process. The whole package consists of the silicon chip, the Au beam interconnects, the metallized silicon posts, the package encapsulant, and the silicon cap (optional). It should be noted that the silicon posts, which are formed by photolithography and silicon etching, are originally part of the wafer. The surface metallization on the silicon posts, as well as the Au beams, is deposited by sputtering and evaporation. Although silicon is the regular material for the optional cap, glass or ceramic may be used instead.

According to ChipScale, the encapsulant plays a key role in the structure of the MSMT package [5]. In addition to providing a protective cover for the silicon chip, the encapsulant layer gives the entire package a flexible mechanical support. Furthermore, the encapsulant, together with the beam-and-post interconnect structure, serves to relieve thermal stresses due to CTE mismatch. Therefore, the encapsulant material is a critical element of the package reliability of the MSMT. To achieve the aforementioned objectives, the encapsulant should have the following features: low CTE or Young's modulus, good adhesion strength and chemical resistance, small-size filler, and minimum void contents.

TABLE 29.2   Material Properties of Acceptable Encapsulants for the
MSMT Package

	CTE (in/in/°C)	Flexural modulus (psi)
Encapsulant A	15×10^{-6}	Not available
Encapsulant B	20×10^{-6}	Not available
Encapsulant C	Not available	2.21×10^6

During the development of the MSMT packages, 20 different encapsulants were evaluated. The materials were epoxy, polyimide, and Teflon. The specimens were coated and cured on 4-in wafers with a thickness of 20 mils. Then a thinning process was performed on the back side of the wafers until the thickness became 6 mils. If the warpage of the wafer exceeded 3 mils, the corresponding encapsulant was disqualified. For those specimens that passed the warpage test, chemical and temperature resistance tests were conducted. No material degradation or interfacial delamination was allowed. At the end of the testing program, only three encapsulants, shown in Table 29.2, were considered acceptable. Of these, encapsulant C was selected for the MSMT because of its superior performance. In addition, the filler size of this material is 1.9 μm, which allows the encapsulant to flow through the Au beams and fill the trenches easily.

29.4   Manufacturing Process

The fabrication of the MSMT involves common semiconductor technologies such as photomasking, silicon etching, and metal deposition. Instead of each individual die being packaged, the whole wafer is packaged at the same time. The manufacturing process is illustrated in Fig. 29.7. Since most of the procedures involved are existing and well-proven IC fabrication techniques, the MSMT packages are expected to be cost-effective and reliable.

The making of the MSMT begins with the deposition of metal bridges at the scribe area between dies on the wafer. The patterned metallization is implemented by photolithography and sputtering. The next two steps are to shield the IC areas by photomasking and then to etch the silicon underneath the patterned metal beams. Subsequently, the whole wafer is encapsulated to fill the cavities created during silicon etching and to cover the active face of the ICs. For large-size dies, an optional cap wafer may be attached to the top of the encapsulant. Afterward, the active wafer is etched from the back side to form the silicon posts. Finally, the surface of the silicon posts is metallized by evaporation and the dies are singulated from the wafer.

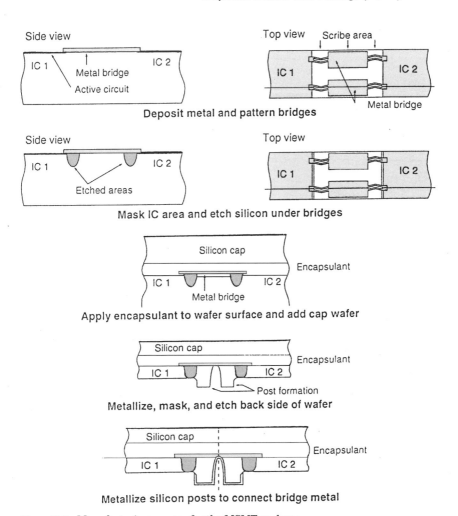

Figure 29.7   Manufacturing process for the MSMT package.

29.5   Performance and Reliability

The MSMT is an ultra-thin package with very short lead distance. Therefore, good electrical and thermal performance is anticipated. According to ChipScale, the inductance and capacitance of the MSMT are 0.1 to 0.2 nH and 0.02 to 0.03 pF, respectively. These figures are much better than those of most other CSPs. The thermal resistance of the MSMT is about 50°C/W. This superior heat-dissipation capability is attributed to the thin package structure and the heat-sink-like silicon cap. Since the MSMT has a face-up configuration and the back side of the IC is exposed to the outside, the bottom of the silicon chip may be placed in direct contact with the PCB. It has been reported that this may substantially enhance the thermal performance [2]. The MSMT is a plastic-encapsulated package. However, the qualification data are not available in the literature. The assembly of MSMT packages to the PCB is performed with conventional SMT. The second-level interconnects are metallized flat lands together with eutectic solder joints. No underfill adhesive is required. The separate silicon posts, the curved Au beam leads, and the compliant encapsulant provide a flexible structure to absorb the thermal stress induced by CTE mismatch. Also, most MSMT packages are relatively small. Therefore, the board-level solder joint reliability is not a major concern for MSMT assembly.

29.6 Applications and Advantages

The Micro SMT technology has been used by M-pulse Microwave for making microwave diodes since 1990. Since establishment of ChipScale in 1993, this technology has been refined for packaging transistors and integrated circuits. The MSMT was developed mainly for low-pin-count devices, as demonstrated in Figs. 29.8 and 29.9. However, this package may also be used for device with more I/Os, as shown in Fig. 29.10. The applications of the MSMT include diodes, transistors, MOSFETs, power regulators, op amps, resistor networks, and RFICs. In addition, the MSMT can also be tested at the wafer level in order to provide known good dies (KGD) for multichip module (MCM) applications [6]. Currently Motorola is the major licensee of MSMT technology. Its Micro SMT products are called μSurf and are used mainly to replace existing transistors and SOICs.

The major advantages of MSMT packages are their miniature form factors and superior electrical performance. From the comparisons shown in Figs. 29.8 and 29.9, it is obvious that the MSMT is much more compact than the corresponding conventional packages. Table 29.3 presents a comparison among various chip scale packages (including flip-chip and wire-bond chip-on-board). It can be seen that the MSMT package is superior to the μBGA in form factors and is comparable to flip chip in electrical performance. In addition, no underfill is required for the assembly of the MSMT. All these features make the MSMT an attractive option to the packaging industry.

Another advantage of MSMT technology is the ability to adopt existing and well-proven manufacturing processes from the semiconductor industry. As a result, the MSMT packages are expected to be cost-effective and reliable. Also, instead of each individual die being packaged, the whole wafer is packaged at the same time. This feature

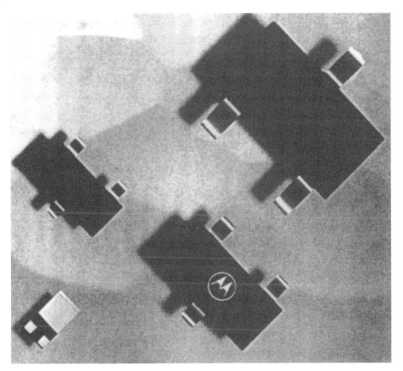

Figure 29.8 Comparison between Motorola's μSurf (0402 MSMT) and its
conventional transistor packages.

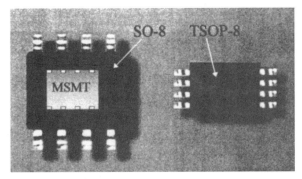

Figure 29.9 Comparison between the MSMT and a small-
outline package.

can minimize the extra packaging expense when die shrink is implemented or a larger wafer is used. With more units obtained from a
single wafer, die shrinks and larger wafers actually lead to lower
packaging cost per module for the MSMT packages.

Figure 29.10 Bottom view of a 100-pin/0.3-mm-pitch MSMT package.

TABLE 29.3 Comparison of Outline and Performance among Various Chip Scale Packages

Package parameter	COB (with wire bonds and pads)	Flip-chip	μBGA	MSMT
Pitch, mm	0.15	0.25	0.5	0.3
Footprint area, mm²	125	120	150	110
Package/chip area	1	1.2	1.5	1.1
Height, mm	0.4–0.6	0.5–0.7	0.84	0.5 with cap, 0.3 without cap
Inductance, nH	1–2 (0.75-mm wire)	0.1–0.2 (0.5-mm bump and trace)	0.5–2.1 (1-mm bump and trace)	0.1–0.2 (0.5-mm beam and post)
Capacitance, pF	0.02	0.3	0.05–0.2	0.02–0.03
Primary applications	General-purpose	General-purpose	ICs above 200 leads	ICs and discretes under 144 leads
PCB attachment	Wire bond and epoxy cover	Solder and underfill epoxy	Solder	Solder

29.7 Summary and Concluding Remarks

The MSMT is a wafer-level CSP with all fabrication procedures performed on the wafer. The package was developed for IC devices with peripheral bond pads only. The first-level interconnects are thin Au beams deposited by photolithography and sputtering. These electrical bridges connect the die bond pads with peripheral metallized silicon posts, which are formed by etching away the silicon from the scribe line area surrounding the die. The package I/Os are the flat lands of silicon posts with surface metallization. Encapsulation is required for package integrity and chip protection. An optional silicon cap may be attached to the top of the encapsulant to enhance package coplanarity and heat dissipation. The package thickness is about 0.5 mm with the silicon cap.

The MSMT was developed mainly for low-pin-count applications, although higher-I/O devices may be accommodated. The current line-up ranges from 2 to 144 pins, and the lead pitch varies from 0.3 to 0.63 mm. The major advantages of MSMT packages are their miniature form factors and superior electrical performance. Another feature of MSMT technology is its adoption of existing and well-proven manufacturing processes from semiconductor industry. Therefore, these packages are expected to be cost-effective and reliable. MSMT packages may find a wide spectrum of applications in passive and active devices such as diodes, transistors, MOSFETs, power regulators, op amps, resistor networks, and RFICs. In addition, the MSMT can also be tested at the wafer level in order to provide KGDs for MCMs. Since the whole wafer is packaged at the same time, extra packaging expenses associated with die shrink or wafer expand can be minimized. With more modules obtained from a single wafer, lower unit packaging cost can be achieved for the MSMT packages by the implementation of die shrinks and larger wafers.

29.8   References

1. J. Young, "The MSMT Package for Integrated Circuits," *Micro SMT Application Note,* ChipScale, Inc., San Jose, Calif., 1994.
2. J. Young, "Wafer Level Processing: Working Smarter," *Chip Scale Review,* vol. 1, no. 1, 1997, pp. 28–31.
3. P. Marcoux, "Chip Scale Packaging Meets SMT and MCM Needs," *Proceedings of NEPCON-West '95,* Anaheim, Calif., 1995, pp. 228–233.
4. J. Young, "Designing Wafer Level Chip Scale Packages to JEDEC and EIAJ Outlines," *Proceedings of NEPCON-West '97,* Anaheim, Calif., 1997, pp. 1519–1523.
5. P. Marcoux, "Encapsulant Selection for Micro SMT Wafer Level Chip Scale Packaging," *Proceedings of NEPCON-West '97,* Anaheim, Calif., 1997, pp. 360–363.
6. J. Young, "Chip Scale Packaging Provides Known Good Die," *Proceedings of NEPCON-West '95,* Anaheim, Calif., 1995, pp. 52–59.

30

EPIC's Chip Scale Package

30.1 Introduction and Overview

EPIC Technologies, Inc., released a chip scale package in 1997 [1]. For this CSP, circuit redistribution is performed at the wafer level to transform peripheral bond pads to area-array bumps on the die. If a process carrier is fabricated in advance, this packaging technology can also be implemented on separate silicon chips. EPIC's CSP adopts a flip-chip configuration. The first-level interconnects are formed by Cu plating on Al bond pads. A proprietary metallurgy is required as a diffusion barrier between the Cu and Al. The package I/Os are Cu bumps formed by electrolytic plating. A Ni/Au metallization needs to be plated on the surface of the Cu bumps. The nominal bump height is 45 μm. In addition to the metallization, two layers of polymer dielectrics are deposited on the wafer for protection and insulation. Because of the thin package structure and short I/O path, EPIC's CSP has superior electrical and thermal performance. In addition to the package, a corresponding high-density PCB was developed as well. A 25-μm design rule can be achieved for both CSP and PCB. The assembly of this package is fully SMT-compatible. The board-level reliability was investigated. Because of the existence of compliant polymer dielectric layers, the solder joint reliability is not a major concern for this CSP.

30.2 Design Concepts and Package Structure

EPIC's CSP is a package with I/O redistribution implemented at the wafer level. The cross section of the package structure is illustrated in Fig. 30.1. This CSP adopts a flip-chip configuration. Two layers of polymer dielectrics and two layers of Cu metallization are alternately

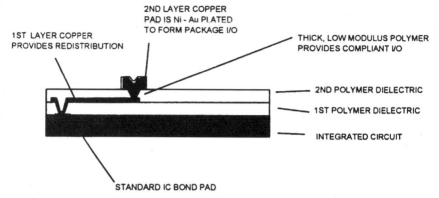

Figure 30.1 Package structure for EPIC's CSP.

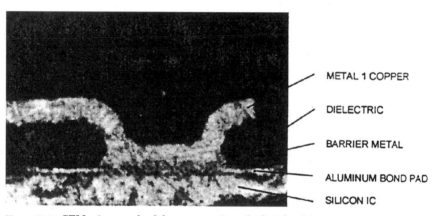

Figure 30.2 SEM micrograph of the cross section of a first-level interconnect.

deposited on the active face of the silicon chip to redistribute the I/Os from peripheral bond pads to area-array bumps. The base layer in direct contact with the passivation of the IC is a 22-μm-thick polymer dielectric film. Photo-defined vias are made along the perimeter of the die to expose the bond pads.

The first-level interconnects are formed by Cu plating on Al bond pads, as shown in Fig. 30.2. A proprietary metallurgy is required as a diffusion barrier between the Cu and Al. The Cu is deposited by electroplating, and its thickness is about 8 μm. This layer is arranged in a desired trace pattern to redistribute the peripheral I/Os into an area-array format. On top of the Cu traces, a second layer of polymer dielectric film is deposited for protection and insulation. The package I/Os are Cu bumps formed by electrolytic plating with a thickness of 40 μm. These bumps are in contact with the aforementioned Cu traces through photo-defined vias on the second polymer dielectric layer. The surface of these Cu bumps is finished with a 5-μm-thick Ni layer and an Au thin film of a few hundred angstroms. Therefore, the total bump height is about 45 μm. The cross section of such a Cu bump is shown in Fig. 30.3. The bottom view of a typical CSP is presented in Fig. 30.4. It can be seen that the peripheral bond pads have been redistributed to area-array bumps.

In addition to the CSP, a high-density PCB for the assembly of the CSP was developed by EPIC. The fabrication of this PCB follows the same procedures used for the CSP. The results of a design rule study for various CSP bump pitches and pin counts are presented in Fig. 30.5 [1]. It was identified that a PCB configuration of 25-μm lines and 60-

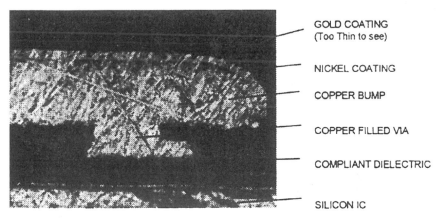

GOLD COATING
(Too Thin to see)

NICKEL COATING

COPPER BUMP

COPPER FILLED VIA

COMPLIANT DIELECTRIC

SILICON IC

Figure 30.3 SEM micrograph of the cross section of EPIC's second-level interconnect.

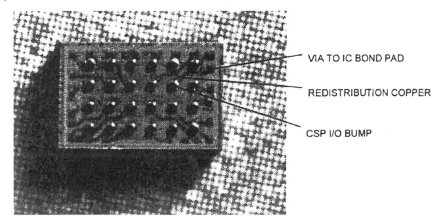

VIA TO IC BOND PAD

REDISTRIBUTION COPPER

CSP I/O BUMP

Figure 30.4 Bottom view of a typical EPIC CSP.

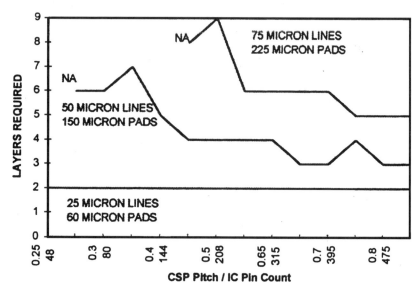

Figure 30.5 Design rules for high-density PCB for EPIC's CSP.

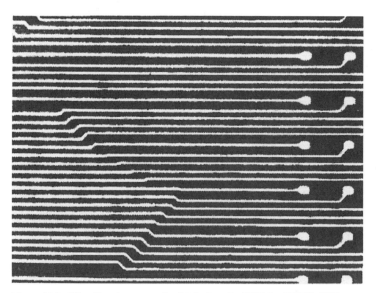

Figure 30.6 Fine-line traces on the high-density PCB for EPIC's CSP.

μm pads could accommodate all CSP pitch/pin count combinations with only two layers of metallization. Therefore, this design rule was adopted for EPIC's CSP assembly board. The trace and pad pattern of a typical high-density PCB is shown in Fig. 30.6.

30.3   Material Issues

In principle, EPIC's CSP is a package with a built-up circuit attached to the active face of the IC for I/O redistribution. In addition to the silicon chip, two layers of polymer dielectrics and two layers of metallization are deposited on the die at the wafer level. The dielectric layers are photo-imagable polymer films with low Young's modulus. This material is originally in liquid form and is spray-coated on the wafer. The curing of the polymer may be performed by UV light source or by baking. The functions of these dielectric layers include protecting the IC from moisture and alpha-particle attack, providing insulation between metallizations, and relieving the thermal stress due to CTE mismatch.

There are two major metal layers in EPIC's CSP. The first layer is deposited by electroplating to form Cu traces for I/O redistribution. The second layer is deposited by electrolytic plating to make Cu bumps for package terminals. Before the deposition of each Cu layer, a proprietary metallurgy needs to be plated, using an electroless process, in order to establish a diffusion barrier. In addition, on the surface of the Cu bumps, a Ni/Au metallization is required in order to prevent oxidation and to improve solderability for board-level assembly.

30.4   Manufacturing Process

The packaging process of EPIC's CSP essentially establishes a two-layer built-up circuit for I/O redistribution on the wafer. The manufacturing process flow is given in Fig. 30.7. The fabrication begins with spray-coating a polymer dielectric layer on the passivation side of the wafer. This photoimagable dielectric is exposed to a collimated UV light source through a mask. The UV light crosslinks the polymer until it is half cured. The unexposed regions are then developed and removed from the wafer. As a result, via holes are made on the die bond pads for subsequent interconnection. The wafer is further baked to fully cure the polymer dielectric layer.

The next step is to treat the surface of the dielectric layer and the exposed bond pads with a special electroless metallization process. This is a wet chemistry process similar to the method used to plate through holes in conventional PCB. The wafer is immersed in a series of chemical baths and rinse tanks. A thin metallization is electrolessly plated on the surface of the dielectric layer and on the die bond pads through the via holes. This metal coating not only serves as a seed layer for the subsequent electroplating, but also forms a diffusion barrier for the first-level interconnection.

After the electroless metallization, a photoresist is coated on the wafer and patterned by photolithography. Cu is then deposited on the

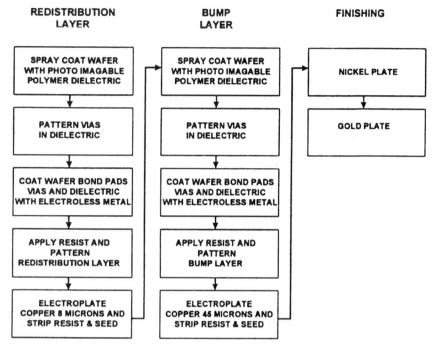

Figure 30.7 Manufacturing process for EPIC's CSP.

exposed seed layer by electroplating. This Cu layer establishes the first-level interconnects on the die bond pads and forms routing traces for I/O redistribution. The next step is to strip the photoresist and to etch the unpatterned thin seed layer away. At this point, the peripheral bond pads have been redistributed to an area-array format. It should be noted that, in addition to silicon wafers with Al bond pads, the aforementioned process can be applied to GaAs wafers with Au bond pads as well.

The subsequent steps to form Cu bumps for package I/Os are basically a repeat of previous procedures to build another two layers of polymer dielectric and Cu. The only difference is that the Cu bumps are deposited by electrolytic plating, so that a larger thickness can be achieved. Finally, an electroless Ni/Au plating on the surface of the Cu bumps is required in order to prevent oxidation and to improve the wettability for SMT assembly.

For certain special applications, the aforementioned manufacturing process may be applied to separate dies instead of to the whole wafer. In this case, a process carrier needs to be fabricated according to the procedures illustrated in Fig. 30.8. At first, the ICs are placed to 12-μm accuracy with their active faces attached to an alignment plate.

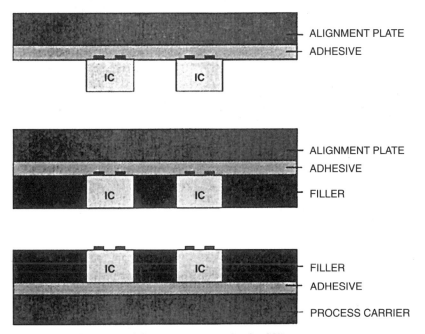

Figure 30.8   Fabrication of the process carrier for the CSP.

On the bottom surface of the alignment plate, a tacky adhesive film is installed in advance. This adhesive is the same material used to lap wafers and is formulated to allow removal from the silicon chip without damaging or contaminating the ICs. Subsequently a liquid filler material is dispensed to encapsulate the dies. After the encapsulant is fully cured, a thinning process is performed by lapping on the back side of the dies. Typically, the silicon chips are lapped to a thickness of 250 μm. At the end, a process carrier is attached to the bottom of the encapsulated dies and the top alignment plate is removed. The manufacturing process shown in Fig. 30.7 is then followed to fabricate the CSPs accordingly.

30.5 Performance and Reliability

It has been reported that EPIC's CSP has outstanding thermal performance. The heat may be dissipated from the back of the IC or through the package I/Os to the PCB. This package has been successfully applied to a GaAs device with a power density of 130 W/cm². The junction-to-ambient temperature change was less than 30°C in this case.

The electrical performance of EPIC's CSP is also superior to that of conventional packages. For a typical CSP as shown in Fig. 30.4, the inductance is 0.5 nH if the package is mounted on a PCB with a return ground plane (10 nH and 2.5 nH for PQFP and PBGA, respectively). This value can be made even lower if more metal layers are added to the CSP. The lead resistance of this CSP is less than 0.1 Ω, which is relatively small compared to that of other packages.

The package reliability of EPIC's CSP has been investigated by computational simulation. The cracking and blistering of polymer dielectric layers may be a concern for this CSP. Currently two qualification tests are ongoing to verify the package reliability. EPIC's target is to survive thermal shock (TS, −55 to 125°C) for 500 cycles and the pressure cooker test (PCT) for 168 h. The testing results are not available yet.

EPIC's CSP is SMT-compatible for board-level assembly. The required placement accuracy is 0.25 mm. The Cu bumps are joined to the PCB by a conventional solder reflow process. No underfill adhesive is needed. EPIC's target for solder joint thermal fatigue life is 1000 cycles under thermal cycling. A preliminary computational analysis indicated that this goal could be achieved for CSPs with a package size less than 20×20 mm. Thermal cycling tests are ongoing for the verification of solder joint reliability.

30.6   Applications and Advantages

EPIC's packaging technology is fully compatible with existing ICs. Since this CSP is still under development, no specific applications have been reported. EPIC claims that its CSP technology is very cost-effective. The projected pricing for this CSP is U.S.$155 per wafer for high-volume production. For a 8-in wafer containing dies with 48 to 368 I/Os, the packaging cost per pin ranges from 0.3 to 0.05 cent. In addition, the typical packaging yield of this CSP is higher than 98 percent. Therefore, the overall cost is much less than the commonly accepted penny-a-pin figure.

In addition to cost-effectiveness, the other merits of EPIC's CSP include low profile, compact size, and good electrical and thermal performance. For modules with the same packaging density, the mounting area of this CSP is about 24 percent of that of TSOP and only 5 percent of that of PQFP. In general, this CSP is very close to a bare flip chip. However, because of the compliant polymer dielectric layers, no underfill is required for the board-level assembly. Good reworkability is another advantage of EPIC's CSP. This feature is a consequence not only of the absence of underfill, but also of the robust Cu bumps which will not stick to the PCB during the removal of CSP.

30.7   Summary and Concluding Remarks

EPIC's package is a CSP with I/O redistribution performed at the wafer level. If a process carrier is fabricated in advance, this packaging technology can also be implemented on separate dies. EPIC's CSP has a flip-chip configuration. Two layers of polymer dielectrics and two layers of Cu metallization are alternately deposited on the active face of the silicon chip to redistribute the I/Os from peripheral bond pads to area-array bumps. The functions of the dielectric layers are protection and insulation. The first-level interconnects are Cu plating on Al bond pads. A proprietary metallurgy between the Cu and Al is required as a diffusion barrier. The package I/Os are Cu bumps formed by electrolytic plating. A Ni/Au metallization needs to be plated on the surface of the Cu bumps in order to avoid oxidation and to improve solderability for board-level assembly.

In addition to the package, a corresponding high-density PCB was developed as well. A 25-μm design rule can be achieved for both CSP and PCB. The assembly of this package is fully SMT-compatible. No underfill adhesive is required. Because of the existence of compliant polymer dielectric layers, the solder joint reliability is not a critical issue for this CSP. Another major advantage of this package is cost-effectiveness. Because of the high production yield and the simplicity of the fabrication process, the packaging cost per pin for EPIC's CSP is far less than the commonly accepted figure. This essential merit, together with superior electrical/thermal performance and good reworkability, should make this CSP attractive to the electronics industry.

30.8   References

1. J. E. Kohl, C. W. Eichelberger, S. K. Phillips, and M. E. Rickley, "Low Cost Chip Scale Packaging and Interconnect Technology," *Proceedings of the CSP Symposium SMI '97,* San Jose, Calif., 1997, pp. 37–43.

Flip Chip Technologies' *Ultra*CSP

31.1 Introduction and Overview

The *Ultra*CSP is a wafer-level chip scale package developed by Flip Chip Technologies (FCT). This CSP is basically a flip chip equipped with a redistribution layer (RDL). FCT's RDL features a thin-film benzocyclobutene (BCB) interlayer dielectric and Al/NiV/Cu under-bump metallurgy (UBM). The first-level interconnects are formed by sputtering the UBM on the Al bond pads. The peripheral I/Os are redistributed to an area-array pattern by traces made of UBM materials. The second-level interconnects are eutectic or high-Pb solder bumps. The bump height is 0.25 to 0.4 mm, and the minimum bump pitch is 0.4 mm. Underfill is optional for the board-level assembly. The *Ultra*CSP has a very low package profile. The total thickness, including solder bumps, is less than 1.0 mm. This CSP is mainly for ICs with low to medium pin counts (<150). The applications aimed at are memory modules (flash, SRAM, DRAM). With the redistribution capability, the *Ultra*CSP can provide a common footprint for various IC devices from different sources.

31.2 Design Concepts and Package Structure

FTC's *Ultra*CSP is a chip scale package based on a technology of I/O redistribution at the wafer level [1–3]. This CSP adopts a flip-chip configuration. As shown in Fig. 31.1, a redistribution layer (RDL) is employed to route the I/Os from peripheral bond pads on the die to area-array package terminals. FCT's RDL consists of two layers of polymer dielectric and one layer of under-bump metallurgy. The first layer of polymer dielectric has a thickness of 4 μm and is di-

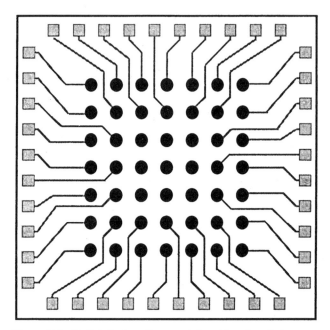

Figure 31.1 Redistribution from peripheral bond pads to area-array I/Os.

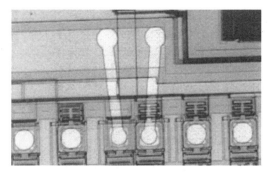

Figure 31.2 First-level interconnects and UBM traces for the *Ultra*CSP.

rectly coated on the active face of the IC for passivation. For the first-level interconnection, windows are opened in this passivation layer on top of the die bond pads. The interconnects are made by depositing the UBM on the polymer dielectric through the open windows. The sputtered UBM layer is also patterned to form desired routing traces (runners), as shown in Fig. 31.2. The line/space width of the traces is currently 38 μm and will be reduced to 25 μm in the future. The second layer of polymer dielectric is coated over the patterned UBM to protect the traces and to define the bonding area of package I/Os.

The second-level interconnects of the *Ultra*CSP are eutectic or high-Pb solder bumps. These solder bumps are attached to the UBM bond pads through windows defined by the second dielectric layer, as illustrated in Fig. 31.3. The bump height is 0.25 to 0.4 mm, and the minimum bump pitch is 0.4 mm. The cross-sectional view of a typical *Ultra*CSP is presented in Fig. 31.4. This CSP has a very low package profile. The total thickness, including solder bumps, is 0.65 to 1.0 mm. This package is developed mainly for ICs with low to medium pin counts (<150). With the redistribution capability, FTC's *Ultra*CSP can provide a common footprint for various IC devices from different sources [2].

31.3 Material Issues

The *Ultra*CSP is basically a flip chip with I/O redistribution at the wafer level. In addition to the silicon chip, this CSP consists of two polymer dielectric layers, a UBM layer, and solder bumps (Fig. 31.3). The functions of the dielectric layers are to protect the circuitry and to define the windows for interconnection. During the development of the *Ultra*CSP, three materials were evaluated for the dielectric layers, namely, polyimide (PI), silicon nitride (SiN), and benzocyclobutene (BCB). After comparison, BCB was chosen because of its superior performance. This material has very low moisture uptake (<0.2 percent at 85 percent RH) and very high resistance to chemical solvents. In addition, because of its hydrophobic chemistry, BCB can substantially reduce the mobility of ions.

The UBM of the *Ultra*CSP is Al/NiV/Cu (Fig. 31.3). The justifications for using this metallization are given in Table 31.1. This UBM is deposited on top of the first dielectric layer by sputtering. Unlike conventional UBM, this thin metal layer not only establishes the bond pads for solder-bump attachment, but also serves as the first-level interconnects and the traces for I/O redistribution. It should be noted that the selected UBM has the ability to heal probing marks on the surface metallization of die bond pads. Therefore, it can be used for the bumping of probed wafers.

The solder bumps of the *Ultra*CSP may be either eutectic (63Sn/37Pb) or high-Pb (95Pb/5Sn) solder. For the former, controlling the alloy composition to within ±2 percent deviation is desired. Otherwise, the melting point of the solder may considerably increase, and the conventional temperature profile cannot reflow such solder

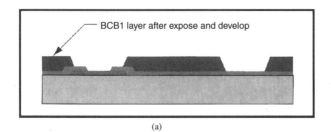

(a)

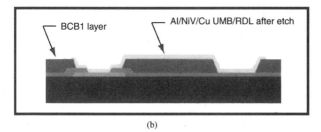

(b)

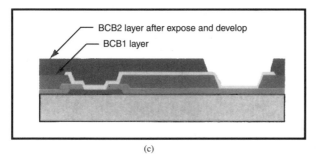

(c)

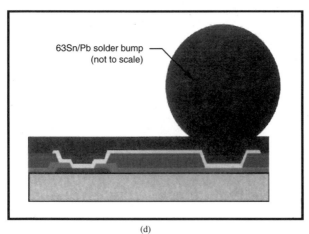

(d)

Figure 31.3 (*a*) BCB1 layer after expose and develop, (*b*) thin film UBM/RDL after etch, (*c*) BCB2 layer after expose and develop, and (*d*) Ultra CSPRDL and solder ball.

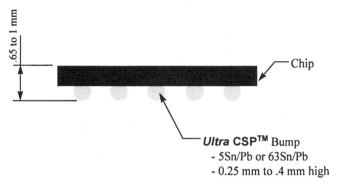

Figure 31.4   Cross-sectional view of FTC's *Ultra*CSP.

TABLE 31.1   Requirements for the UBM of the *Ultra*CSP

Good adhesion to wafer passivation

Good solder wettability

Low ohmic contact to the final metal

Providing robust solder diffusion barrier

Protecting die bond pad metallization from the environment

Minimization of stress on IC

Ability to heal probed wafers

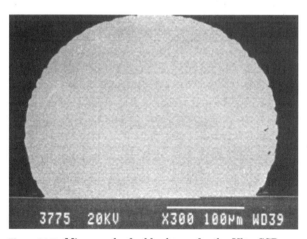

Figure 31.5   Micrograph of solder bump for the *Ultra*CSP.

bumps. Note that the solder is deposited on the UBM in the paste form. The bonding area is defined by the window on the second dielectric layer. After reflow heating, the solder bump is formed as shown in Fig. 31.5. The typical bump height is 0.25 to 0.4 mm.

31.4   Manufacturing Process

The fabrication of the *Ultra*CSP involves common clean-room procedures such as dielectric coating, metal deposition, and photomasking. Instead of each individual die being packaged, the whole wafer is processed at the same time. The manufacturing flow is illustrated in Fig. 31.6. Since most procedures are performed with existing and well-proven IC fabrication technologies, the *Ultra*CSP packages are expected to be relatively reliable.

In addition to high yield, the other general feature of CSPs is cost-effectiveness. These characteristics are especially obvious for wafer-level CSPs. Because the manufacturing process is performed on the whole wafer instead of a single silicon chip, the packaging cost per die or per I/O for the *Ultra*CSP always decreases when die shrink or wafer expand is implemented. This advantage can be illustrated by the examples given in Table 31.2. The figures in this table are estimates based on the assumption of 85 percent production yield. The bumping cost per wafer is $128 and $150 for 150- and 200-mm wafers, respectively. It should be noted that the cost for I/O redistribution has been absorbed in the bumping cost and the underfill cost is optional.

31.5   Performance and Reliability

The *Ultra*CSP is a very thin package without any encapsulation. Therefore, the heat-dissipation capability should be relatively good. Also, since the lead length is rather short, the self-inductance and resistance should be very low. It has been reported that the nominal resistance of UBM and traces is 13 mΩ. Also, because of the relatively thick polymer dielectric layer, the coupling between the RDL and the underlying IC is minimal.

The package reliability of the *Ultra*CSP has been investigated with the THB test (85°C/85 percent RH/5-V bias). This CSP could withstand 1000 h without failure. This good performance is mainly due to the low moisture uptake of the BCB dielectric layers. The adhesive

TABLE 31.2   Cost Analysis for FCT's *Ultra*CSP

Wafer size (mm)	150	200	150	200	200
Die size (mm)	6 × 6	6 × 6	8 × 8	8 × 8	4 × 4
Number of bumps	144	144	196	196	68
Bump cost/die (cents)	37	23	68	42	10
Underfill material cost/die (cents)	4	4	6	6	2
Underfill process cost/die (cents)	6	6	8	8	4
Total cost/die (cents)	47	33	82	56	16
Total cost/bump (cents)	0.33	0.23	0.42	0.29	0.24

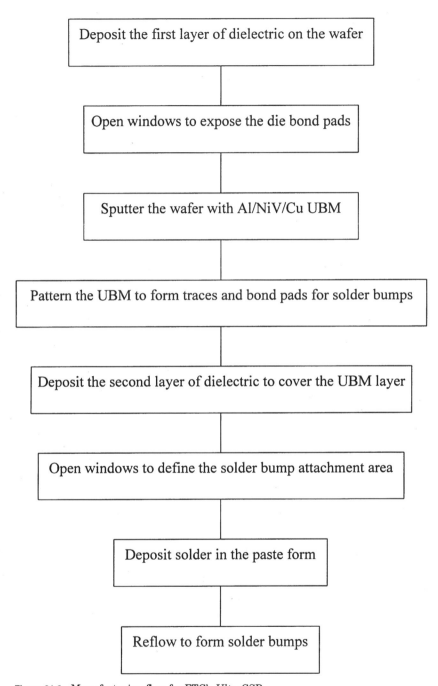

Figure 31.6   Manufacturing flow for FTC's *Ultra*CSP.

strength between the UBM and the solder bumps is another package reliability concern. This property of the *Ultra*CSP was evaluated by the bump shear test. The specimens should be preconditioned with three cycles of reflow before testing. The solder bumps were sheared at a height of 25 μm above the die passivation surface. The ram speed was 175 μm/s. The desired failure mode should be cohesive failure in the solder bump, as shown in Fig. 31.7. Other failure modes such as interfacial separation, delamination in the UBM, and silicon cratering (Fig. 31.8) are not acceptable. Even with cohesive failure, the shear strength of solder bumps should meet a specification of 3.1 mg/μm² (or above). For a typical *Ultra*CSP with good-quality solder bumps, the shear resistance of a single bump ranges from 45 to 50 g.

A typical solder joint of an *Ultra*CSP assembly is shown in Fig. 31.9. Since the standoff height is relatively small, the board-level solder joint reliability could be a concern. Thermal cycling tests have been conducted to investigate this issue. The specimens were monitored

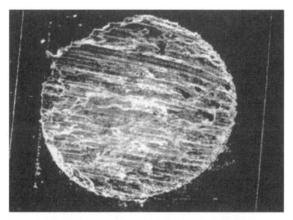

Figure 31.7 Fracture surface in the solder bump by shear test.

Figure 31.8 Cratering on silicon by bump shear test.

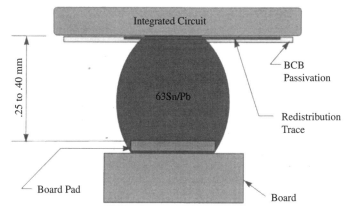

Figure 31.9 Typical board-level solder joint of the *Ultra*CSP.

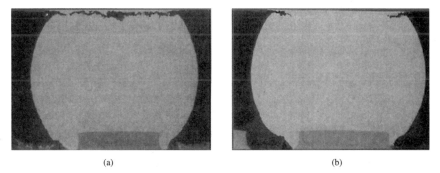

(a)   (b)

Figure 31.10 (a) Thermal fatigue failure of *ULTRA*CSP solder ball, and (b) partial thermal fatigue failure of *Ultra*CSP solder ball.

every 100 cycles. A 20 percent increase in electrical resistance is considered to be failure. One set of *Ultra*CSP specimens was mounted on ceramic substrates without underfill. The testing conditions were −50 to 150°C with a cycle period of 80 min. The experimental results indicated that these solder joints have a thermal fatigue life equivalent to that of the standard FOC bumps. Another set of specimens was assembled to an FR-4 PCB with underfill. The testing conditions were −40 to 125°C with a cycle period of 60 min. The solder joints could last for 1500 cycles. Such a result is considered satisfactory for an *Ultra*CSP assembly. For all the cases, the thermal fatigue failure of the solder is near the chip side as shown in Fig. 31.10.

31.6 Applications and Advantages

The *Ultra*CSP redistributes the package I/Os from fine-pitch peripheral bond pads to coarser-pitch area-array solder bumps. This CSP was developed mainly for ICs with low to medium pin counts (<150).

The targeted applications involve memory modules (flash, SRAM, DRAM). With the redistribution capability, the *Ultra*CSP can provide a common footprint for ICs from different suppliers.

In addition to the I/O redistribution ability, another major advantage of the *Ultra*CSP is that the UBM can heal the probing mark on the die bond pad. Therefore, this technology is very suitable for packaging probed wafers. Furthermore, because the whole wafer is processed at the same time, the extra production expenses for die shrink or wafer expand can be minimized. With more units obtained from a single wafer, the implementation of die shrinks and larger wafers can substantially reduce the packaging cost per die (or per bump) for *Ultra*CSP packages.

31.7   Summary and Concluding Remarks

FCT's *Ultra*CSP is a flip-chip chip scale package with I/O redistribution at the wafer level. The redistribution is achieved with FCT's RDL technology, which features two layers of BCB interlayer dielectric and one layer of Al/NiV/Cu UBM. The first-level interconnects are formed by sputtering UBM on the Al bond pads. The UBM layer is also patterned to form routing traces to redistribute the I/Os from peripheral bond pads to an area-array format. The second-level interconnects are eutectic or high-Pb solder bumps. The bump height is 0.25 to 0.4 mm, and the minimum bump pitch is 0.4 mm. Underfill is optional for the board-level assembly.

The *Ultra*CSP is a true chip size package with a very low package profile. The total thickness, including solder bumps, is less than 1.0 mm. This CSP was developed for packaging ICs with low to medium pin counts (<150). The intended applications are memory modules (flash, SRAM, DRAM). With the redistribution capability, the *Ultra*CSP can provide a common footprint on ICs from different suppliers. In addition to the I/O redistribution ability, another major advantage of the *Ultra*CSP is that the UBM can heal the probing mark on the die bond pad. Therefore, this technology is very suitable for packaging probed wafers. Like other wafer-level CSPs, the *Ultra*CSP can easily cope with die shrink and wafer expand. Consequently, the implementation of die shrinks and larger wafers can substantially reduce the packaging cost per die (or per bump) for *Ultra*CSP packages.

31.8   References

1. P. Elenius, "FC^{2}SP-(Flip Chip-Chip Size Package)," *Proceedings of NEPCON-West '97,* Anaheim, Calif., 1997, pp. 1524–1527.
2. P. Elenius, "Flip Chip Bumping for IC Packaging Contractors," *Proceedings of NEPCON-West '98,* Anaheim, Calif., 1998, pp. 1403–1407.
3. P. Elenius and H. Yang, "The Ultra CSP Wafer-Scale Package," *High-Density Interconnect,* vol. 1, no. 6, October 1998, pp. 36–40.

32

Fujitsu's *Super*CSP (SCSP)

32.1 Introduction and Overview

The *Super*CSP (SCSP) is a new wafer-level chip scale package developed by Fujitsu in 1998 [1, 2]. This CSP adopts the conventional technology for I/O redistribution on the wafer but features a new molding process for encapsulation. A pallet of encapsulant is placed at the center of the wafer. A polymer film is used to cover the whole wafer over the pallet. A thermal compression process is applied to melt the pallet, and the polymer film is peeled off from the encapsulated wafer afterwards. The typical thickness of encapsulation is 90 μm. The package I/Os of Fujitsu's SCSP are eutectic solder balls with a diameter of 0.4 mm and a pitch of 0.75 mm. The solder balls are deposited after the wafer is encapsulated and the dies are singulated. A comprehensive qualification program has been conducted to investigate the reliability of the SCSP. The testing results revealed that the SCSP has good package reliability and is qualified for JEDEC level 3. Like other CSPs in the same category, Fujitsu's SCSP has the advantages of offering common footprints and coping with die shrink or wafer expand easily. The potential applications of SCSP include flash, SRAM, and other memory devices for portable equipment.

32.2 Design Concepts and Package Structure

Fujitsu's SCSP is a molded wafer-level chip scale package. This CSP adopts the conventional technology for I/O redistribution on the wafer but features a new molding process for encapsulation. The cross section of the package structure is illustrated in Fig. 32.1. On the passivation (SiN) of ICs, a polyimide layer is deposited for protection. The

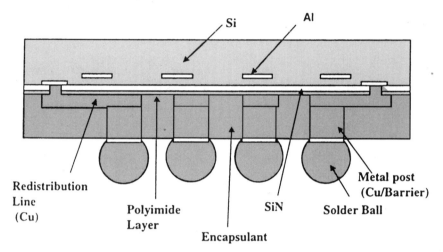

Figure 32.1   Cross section of package structure of Fujitsu's SCSP.

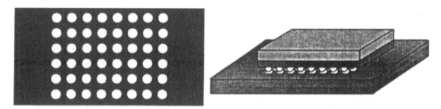

Figure 32.2   Bottom view and the assembly of Fujitsu's SCSP.

first-level interconnection is implemented by sputtering Ti/Ni on the conventional Al bond pads of the dies. Cu traces are patterned by photolithography to redistribute the I/Os from peripheral bond pads to the desired format. On the bonding lands for the second-level interconnects, thicker metal posts which consist of Cu and under-bump metallurgy (UBM) are formed by a plating process. The UBM is composed of plated Ni and Pb/Sn solder.

The main feature of the SCSP is the encapsulation of the wafer by a proprietary molding process. This process involves a covering polymer film which is to be peeled off from the wafer after the encapsulation. The molding compound encapsulates all Cu traces and metal posts. The typical thickness is 90 μm. The package terminals of the SCSP are eutectic BGA solder balls. These solder balls are attached to the aforementioned UBM after the dies are singulated from the wafer. The typical BGA configuration of the SCSP is 0.4-mm-tall balls with an array pitch of 0.75 mm. The bottom view and the assembly of Fujitsu's SCSP are illustrated in Fig. 32.2.

32.3 Material Issues

The SCSP developed by Fujitsu is a wafer-level chip scale package with molding encapsulation. The package consists of the silicon die, the routing traces, the UBM, the molding encapsulant, and the BGA solder balls. The LSI of the SCSP may be a conventional chip without any modification. On the passivation of dies, a thin layer of polyimide is coated for protection and better adhesion. The Cu traces are patterned on the top of polyimide layer to redistribute the I/Os. The first-level interconnection between the Al bond pads on the dies and the Cu traces is formed by sputtered Ti/Ni. The UBM of the bonding lands at the other end of the Cu traces is plated Ni and Pb/Sn solder. The eutectic solder balls are mounted to the UBM as the package terminals. The attachment of BGA solder balls is performed by the conventional SMT reflow process.

The unique feature of Fujitsu's SCSP is the molding process. The encapsulant is silica-filled epoxy molding compound and is in a pallet form before molding. The coefficient of thermal expansion (CTE) is less than 30 ppm/°C. During the molding process, the pallet melts under heating (170°C) and pressure. The viscosity is less than 120 ps and larger than 160 cm spiral flow. In addition to the encapsulant, a covering polymer film is involved in the molding process. The film is a thermoset material with a Young's modulus of 2 GPa at 200°C. The surface roughness of this polymer film is rather important and needs to be maintained at under 300 nm Ra. After encapsulation, the polymer film is removed by mechanical peeling.

32.4 Manufacturing Process

The fabrication of Fujitsu's SCSP consists of three major parts. The premolding part is I/O redistribution on the wafer. The conventional thin-film deposition and photolithography technologies are adopted in this part to form the routing pattern and the UBM. The Cu traces are made by a plating and etching process. The first-level interconnection is implemented by sputtering Ti/Ni on the Al bond pads of the dies, while the UBM for the second-level interconnects is plated Ni and Pb/Sn solder.

The second part is the molding process for the whole wafer, which is a unique feature of the SCSP. As illustrated in Fig. 32.3, the wafer is mounted in a mold and a pallet of encapsulant is placed at the center of the wafer. It should be noted that a thin polymer film is installed at the bottom of the upper mold. Then the upper and lower molds are compressed together to provide heating and pressure. The pallet melts when the mold temperature reaches 170°C. Further clamping flattens the melted encapsulant. After 5 min of clamping,

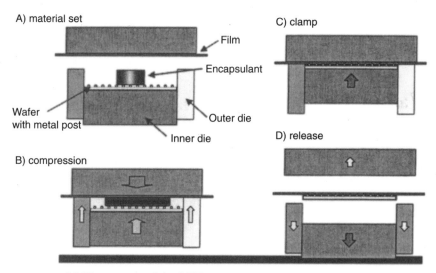

Figure 32.3 Molding process of the SCSP.

the molding compound is fully cured, and then the encapsulated wafer together with the covering polymer film is released from the mold. Subsequently, the polymer film is peeled off from the wafer as shown in Fig. 32.4 and the molding process is completed. It should be noted that the in-plane stiffness and surface roughness of this polymer film must be properly selected in order to successfully remove the film without damaging the UBM on the encapsulated metal posts.

The postmolding part of fabrication involves the attachment of BGA solder balls. The dies are singulated from the encapsulated wafer as presented in Fig. 32.5. The singulated packages are placed on a carrier tray in an array form, as shown in Fig. 32.6. Eutectic solder balls are then aligned to the flat lands on the bottom of SCSP. The attachment of the BGA solder balls is completed by the conventional reflow process.

32.5 Qualifications and Reliability

The package reliability of Fujitsu's SCSP has been investigated. The testing items included temperature cycling (MIL-833, Condition C), pressure cooker test (121°C/100 percent RH, 2 atm), high-temperature storage (150°C), and moisture sensitivity. The qualification data are given in Table 32.1. It can be seen that this package passed the first three testing items and was qualified for JEDEC level 3 for moisture sensitivity. Therefore, Fujitsu's SCSP may be considered to have the same package reliability as the conventional surface-mount plastic packages.

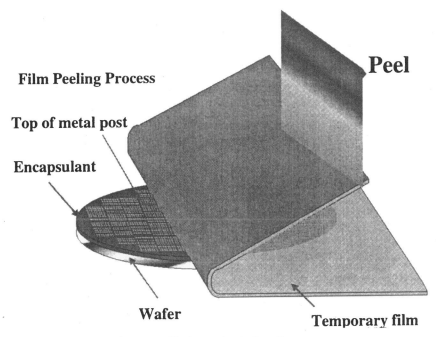

Film Peeling Process

Top of metal post

Encapsulant

Peel

Wafer

Temporary film

Figure 32.4   Removal of covering film by mechanical peeling.

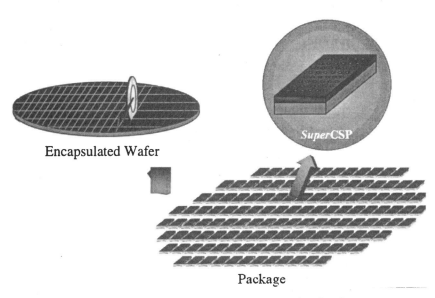

Encapsulated Wafer

*Super*CSP

Package

Figure 32.5   Singulation of SCSP packages from the encapsulated wafer.

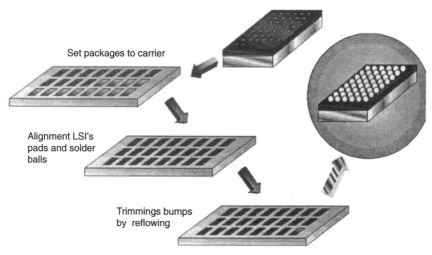

Set packages to carrier

Alignment LSI's
pads and solder
balls

Trimmings bumps
by reflowing

Figure 32.6   Attachment of BGA solder balls to the SCSP packages.

TABLE 32.1   Qualification Data for Fujitsu's SCSP

Test	Test condition	Sample number	Results
Pressure cooker test	121°C/85% RH	10	168 h passed
High-temperature storage	150°C/in the air	10	500 h passed
Moisture sensitivity test	JEDEC level 3	10	Passed

(Precondition: Prebake, 125°C, 24 h, 85°C/85% RH, 24 h, IR Reflow).

TABLE 32.2   Qualification Data for Fujitsu's SCSP on PCB

Test	Test condition	Sample number	Results
Temperature cycle	−55/+125°C (condition B)	10	500 cycles, Passed
Pressure cooker test	121°C/85% RH	10	168 h, Passed
Free fall test (150 g, 1-m height)	Vertical and Horizontal	10	20 times, Passed

In addition to the package reliability, the board-level reliability of SCSP was investigated. The test items included temperature cycling (MIL-833, Condition B), pressure cooker test (121°C/100 percent RH, 2 atm) and free-fall test (150 gm, 1 m height). The test data are given in Table 32.2. It can be seen that the SCSP passed all testing items. In addition, a finite-element analysis was conducted to study the solder joint reliability, as shown in Fig. 32.7. It was found that under the

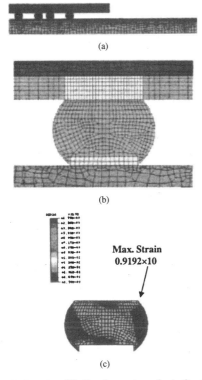

(a)

(b)

Max. Strain
0.9192×10

(c)

Figure 32.7 Finite-element analysis for the solder joint reliability: (a) the global model; (b) detailed meshes around a solder joint; (c) typical strain distribution from the finite-element analysis.

same loading condition, the maximum inelastic strain in the solder joint of the SCSP is less than that in Fujitsu's FBGA [1]. From these results, it may be concluded that the board-level reliability of Fujitsu's SCSP should be acceptable for most applications.

32.6 Summary and Concluding Remarks

Fujitsu's SCSP is a chip scale package with molding encapsulation. The I/O redistribution is implemented on the wafer level using conventional thin-film deposition and photolithography technologies. The molding is performed by melting a pallet of encapsulant on the wafer, using a thermal compression process. The typical thickness of encapsulation is 90 μm. The package terminals of Fujitsu's SCSP are eutectic BGA solder balls. The ball diameter is 0.4 mm, and the array pitch is 0.75 mm. The solder balls are attached after the encapsulated packages are singulated from the wafer.

The reliability of the SCSP has been investigated by a comprehensive qualification program. The test results revealed that the SCSP has good package reliability and is qualified for JEDEC level 3. Like other CSPs in the same category, Fujitsu's SCSP may offer common footprints and can easily cope with die shrink or wafer expand. Because of its compact dimensions, Fujitsu's SCSP should find good applications for flash, SRAM, and other memory devices for portable equipment.

32.7 References

1. M. Hou, "Super CSP: The Wafer Level Package," *Proceedings of Semicon-West,* San Jose, Calif., July 1998, pp. F1–10.
2. M. Hou, "Wafer Level Packaging for CSPs," *Semiconductor International,* vol. 21, no. 8, July 1998, pp. 305–308.

Mitsubishi's Chip Scale Package (CSP)

33.1 Introduction and Overview

Mitsubishi Electric Corp. released its chip scale package (CSP) in 1994 [1]. (Mitsubishi claimed that they were the first to use the term "Chip Scale Package.") This CSP was aimed at low-pin-count applications such as memory modules. Later on, Mitsubishi developed the second generation of molded CSP for high-pin-count applications such as ASICs [2]. Although there are certain differences in the fabrication process, both versions of the CSP use the same wiring technology for I/O redistribution at the wafer level. The basic configuration of Mitsubishi's CSP is a flip-chip die with molding encapsulation. A multilayer metallization is employed to route the center or peripheral die bond pads to the desired package I/O pattern. The first-level interconnects are sputtered TiN on the Al bond pads. A Ni/Au layer is deposited on the top of the TiN and serves as the under-bump metallurgy (UBM). On the UBM are package terminals which consist of inner high-lead solder, transferred copper land, and external eutectic solder bump. Depending on the pin count and the terminal pattern, the bump pitch of Mitsubishi's CSPs ranges from 0.5 to 1.0 mm. The performance and reliability have been investigated comprehensively, and good results have been reported. Mitsubishi's CSP has both "package" features and "bare die" features. With the I/O redistribution capability, the standardization of footprints is very easy to implement. Therefore, this chip scale packaging technology has attracted considerable attention in the electronics industry.

33.2 Design Concepts and Package Structure

Mitsubishi's CSP was one of the earliest chip scale packages reported in the literature. There are two generations of this family of chip scale packages. The first version was released in 1994 [1], and the second version was developed in 1996 [2]. The corresponding package structures are illustrated in Figs. 33.1 and 33.2, respectively. From the schematic diagrams presented, it can be seen that the former is mainly for low-pin-count applications such as memory modules, while the latter is for high-I/O devices such as ASICs. Although there are certain differences in the fabrication process, both versions of the CSP use the same wiring technology for I/O redistribution at the wafer level.

Figure 33.3 shows the cross-sectional view of the package structure for the first generation of the CSP. The chip-level interconnection is formed by depositing a multilayer metallization (total thickness 0.5 μm) on the die bond pads. This metallization is patterned to redistribute I/Os into the desired format, and a 40-μm design rule is adopted. On top of the routing metallization, there is a 7-μm-thick polyimide film which serves as a protection layer and also defines the bonding area of the package terminals. Depending on the terminal pitch, the

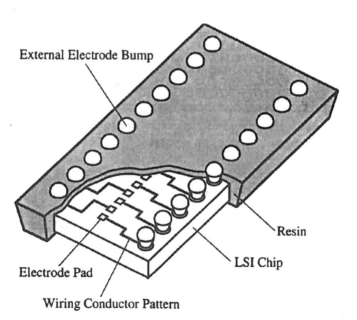

Figure 33.1 Schematic diagram for the package structure of Mitsubishi's low-pin-count CSP.

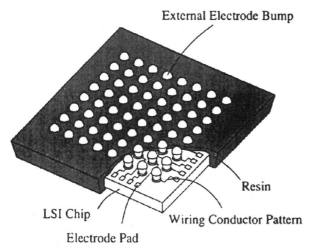

External Electrode Bump

Resin

LSI Chip   Wiring Conductor Pattern

Electrode Pad

Figure 33.2 Schematic diagram for the package structure of Mitsubishi's high-pin-count CSP.

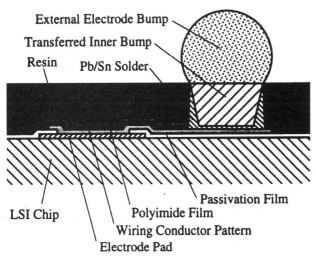

External Electrode Bump

Transferred Inner Bump

Resin   Pb/Sn Solder

Passivation Film

LSI Chip   Polyimide Film

Wiring Conductor Pattern

Electrode Pad

Figure 33.3 Cross-sectional view of the first- and second-level interconnects (the first generation).

openings on the polyimide film may be 400 or 500 μm in diameter. The package I/O consists of three portions. At the bottom is a 40-μm-thick high-lead solder which sits on the UBM provided by the metallization layer. In the middle is a 100-μm-tall Cu bump wrapped around by the inner high-lead solder. On the top is the eutectic solder bump which serves as the external terminal of the package. Unlike other wafer-level CSPs, Mitsubishi's CSP has molding encapsulation covering the whole package. During the development of the first generation of the CSP, Mitsubishi fabricated three kinds of prototype, as shown in Fig. 33.4. All the prototypes had the same package size (6.35×15.24×0.65 mm) and the same chip size (5.95×14.84×0.4 mm). The number of I/Os ranged from 32 to 96, and the terminal bump pitch could be 0.8 or 1.0 mm. For larger bump pitch, the diameter of the external solder bumps was 680 μm. Otherwise, the bump diameter was 550 μm.

The package cross section for the second generation of CSPs is illustrated in Fig. 33.5. Unlike the first generation, in the second gen-

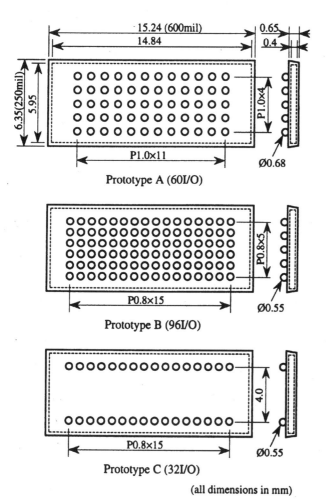

Prototype A (60I/O)

Prototype B (96I/O)

Prototype C (32I/O)

(all dimensions in mm)

Figure 33.4 Package outlines for CSP prototypes (the first generation).

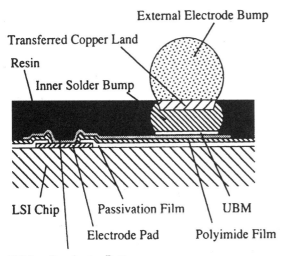

Transferred Copper Land

Resin

Inner Solder Bump

External Electrode Bump

LSI Chip   Passivation Film   UBM

Electrode Pad   Polyimide Film

Wiring Conductor Pattern

Figure 33.5 Cross-sectional view of the first- and second-level interconnects (the second generation).

TABLE 33.1 Specifications for Mitsubishi's Second-Generation CSP

Package body size	Pin count	Bump pitch
9.0 × 9.0 mm	256	0.5 mm
16.6 × 16.6 mm	1024	0.5 mm

eration the polyimide film is underneath the routing metallization. Also, instead of covering the whole metal traces, the UBM is deposited only at the designated pads to define the bonding area of the package terminals. Since this version of CSP is for high-I/O applications, a 25-μm design rule is adopted for the wiring conductor pattern. Also, the diameter of the terminal bond pads is reduced to 250 μm and the external solder-bump pitch becomes 0.5 mm. The outlines of two second-generation prototypes are given in Table 33.1.

33.3 Material Issues

Mitsubishi's CSP is a molded flip-chip die with I/O redistribution at the wafer level. In addition to the silicon chip, the package consists of the routing metallization, the polyimide film, the molding encapsulant, and the package terminals. Although the two generations of CSP have certain variations in fabrication process, the packaging materials are essentially the same.

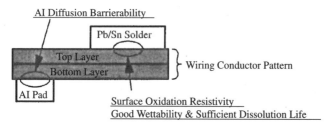

Figure 33.6 Requirements for the routing metallization and UBM.

The most significant element in this CSP is the routing metallization (or wiring conductor, to use Mitsubishi's term). This metallization performs three functions, namely, interconnection to the die bond pads, conducting traces for I/O redistribution, and UBM for package solder bumps. For the first-level interconnection, the metallization should also act as a diffusion barrier for the Al surface finishing on the bond pads. On the other hand, as the UBM for solder bumps, the wiring conductor must be solder-wettable, have sufficient dissolution life, and provide resistance to surface oxidation. In general, no single material can fulfill all the above requirements. Therefore, a multilayer metallization, as shown in Fig. 33.6, is employed as the wiring conductor for the CSP.

During the development of Mitsubishi's CSP, three materials were considered as candidates for the bottom layer of routing metallization. The chosen materials were Cr, TiN, and Ti. In order to evaluate its performance as a diffusion barrier, each metal film (0.1 μm thick) was sputtered on the Al bond pads on a wafer and then subjected to the curing condition for the polyimide film (350°C for 1 h). Auger electron spectroscopy (AES) was used to investigate the interdiffusion behavior between each chosen metal film and the Al bond pads. For comparison, the AES was performed before and after the curing tests with a sputtering rate of 7.5 nm/min. The results of atomic concentration (A.C.) profiles in the depth direction are presented in Fig. 33.7. No observable difference before and after the curing tests could be identified by the AES for all three selected materials. Therefore, it may be concluded that all three candidates are valid.

For the top layer of metallization, Cu/Au, Pd, and Ni/Au material systems were considered. To evaluate the resistance to oxidation, these three metal films were subjected to the curing condition for the polyimide film (350°C for 1 h), and x-ray photoelectron spectroscopy (XPS) was used to measure the surface oxygen concentration before and after the curing tests. From the experimental results shown in Table 33.2, it can be seen that Pd had the best performance. Because

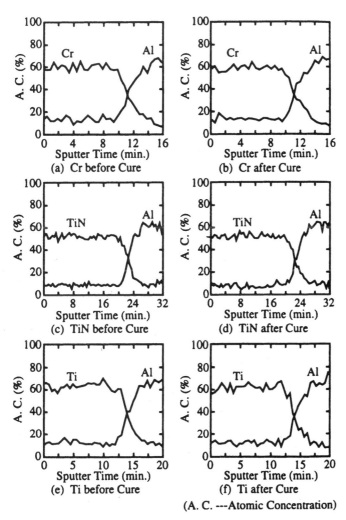

Figure 33.7   Comparison of AES results for various metallizations.

TABLE 33.2   Comparison of Surface Oxidation for Various Metallizations

Metal(s) (thickness)	Surface oxygen concentration (atomic%)	
	Before cure	After cure
Cu/Au (10,000/1000 Å)	7.2	36.9
Pd (3000 Å)	Not detected	24.4
Ni/Au (3000/1000 Å)	5.7	37.4

Au diffused into Cu and Ni under the curing condition, the other two specimens had relatively high oxidation levels on the surface. However, since the atomic concentration of oxygen is not very severe, all three material systems were considered acceptable for resisting surface oxidation.

In general, a metal which has good wettability for Pb/Sn solder also dissolves easily into the melting solder during reflow heating. Therefore, it is important to ensure that the top layer of metallization has sufficient dissolution life while providing a wettable surface to the inner high-lead solder. Since good wettability and sufficient dissolution life would lead to high adhesion strength between the solder and the UBM, the bump shear test as shown in Fig. 33.8 was employed to evaluate the performance of the five cases specified in Table 33.3. TiN was chosen to serve as the bottom layer of metallization for all speci-

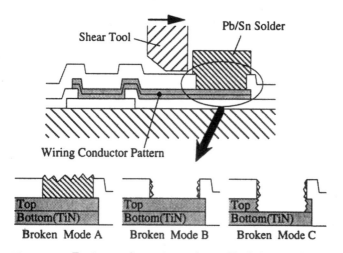

Figure 33.8 Testing configuration and possible failure modes of bump shear test.

TABLE 33.3   Configurations of Specimens for Bump Shear Tests

Sample name	Combination	Thickness (Å)
Top#1	Cu/Au	10,000/1000
Top#2	Pd	6000
Top#3	Pd	9000
Top#4	Ni/Au	1500/1000
Top#5	Ni/Au	3000/1000

Bottom: TiN (1000Å).

mens. As illustrated in Fig. 33.8, three failure modes may result from the bump shear test. Mode A is the shear failure of solder, which indicates good adhesion strength between the solder and the UBM. Mode B corresponds to separation at the solder/UBM interface. This failure mode is due to a poor wetting condition. The bottom layer of metallization is exposed in Mode C, which implies excessive dissolution of the top layer of metallization. The specimens were subjected to 350°C for 150 s. Bump shear tests were performed before and after the heating. From the shear strength and the failure mode, the wettability and the dissolution life could be evaluated. The results are given in Table 33.4. The use of Ni/Au (thicknesses 0.3/0.1 μm) gave good results in terms of both wettability and dissolution life. The final choice for the multilayer wiring conductor was TiN/Ni/Au, with corresponding thicknesses of 0.1/0.3/0.1 μm. It should be noted that TiN was selected in view of its advantage in the fabrication process. The bottom layer of metallization exists underneath the Ni. Since the other two candidates, Cr and Ti, look very similar to Ni, it is difficult to distinguish the quality of each metal deposition using optical microscopy. On the other hand, the optical appearance of TiN is different from that of Ni. Therefore, TiN was chosen for the bottom layer of wiring conductor.

The package terminals of Mitsubishi's CSP have three parts, namely, the inner high-lead solder, the intermediate Cu land, and the external solder bump. The composition of the inner solder is 95Pb/5Sn; this has a melting point of 310°C. This solder serves as an adhesion layer between the UBM and the package I/O. The external solder bumps are made of 63Sn/37Pb eutectic solder. Between the inner solder and the external solder bump is a Cu land transferred from a stainless steel base frame. This Cu land mainly serves two purposes. One is to act as a medium to join the inner solder and the external solder bump together, and the other is to increase the bump height so that the molding encapsulant can be accommodated and the board-level joint reliability may be improved. The Cu lands origi-

TABLE 33.4   Comparison of Wettability and Dissolution Life for Various Metallizations

Item (criteria)	Wettability (no mode B at all)	Dissolution life (no mode C at all)
Top#1	Good	No good
Top#2	Good	No good
Top#3	Good	No good
Top#4	Good	No good
Top#5	Good	Good

nally sit on the steel base frame, as shown in Fig. 33.9. They are transferred to the die of the CSP with the reflow of inner solder followed by mechanical peeling of the base frame. In order to maintain a reliable transfer process, the adhesion strength between the Cu lands and the base frame must be properly controlled. During the development of this CSP, the two configurations for the transfer base frame shown in Fig. 33.10 were evaluated. Bump shear tests were performed to measure the adhesion strength before and after a heat treatment which simulated the reflow condition of the inner solder. The test results are given in Fig. 33.11. Since frame 2 had consistent and relatively low shear strength between Cu lands and the base frame, this configuration was chosen for the fabrication of Mitsubishi's CSP.

Figure 33.9 Stainless steel base frame with Cu lands.

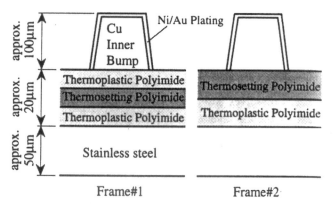

Figure 33.10 Cross-sectional view of the two configurations for the attachment of Cu lands on the base frame.

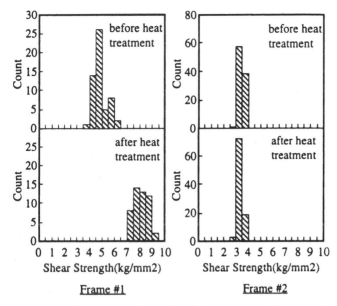

Figure 33.11   Comparison of interfacial shear strength between Cu lands and the base frame.

33.4   Manufacturing Process

Although Mitsubishi's CSP is considered a wafer-level chip scale package, not all fabrication procedures are completed on the wafer. Also, there are certain differences in the manufacturing process between the two generations of CSP. The wafer-level process for the first-generation CSP is presented in Fig. 33.12. At first, the multilayer metallization (TiN/Ni/Au of 0.1/0.3/0.1 μm, respectively) is sequentially deposited on the wafer by sputtering. The wiring pattern is then formed by photolithography and wet etching. A 40-μm design rule is adopted for the trace patterning. The next step is to deposit a 7-μm-thick polyimide film on the routing metallization for protection and insulation. This film is patterned by photolithography to open windows on corresponding pads of the wiring conductor to define the bonding areas of the package terminals. Once the pattern is formed, the wafer is heated up to 350°C for 1 h to cure the polyimide film. Subsequently, a thick photoresist (PR) layer is coated on the wafer, and it is patterned with openings on the corresponding windows in the polyimide film for the deposition of the inner solder. The 40-μm-thick 95Pb/5Sn solder is deposited on the exposed UBM by evaporation followed by liftoff. Afterwards, wafer dicing is performed to singulate the dies for subsequent assembly process. A typical die of prototype A with completed wafer-level process is shown in Fig. 33.13.

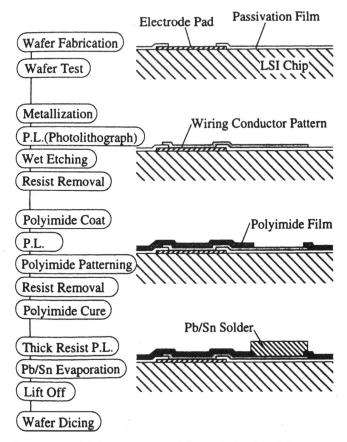

Figure 33.12   Fabrication process at the wafer level for interconnection, I/O redistribution, and UBM deposition (the first generation).

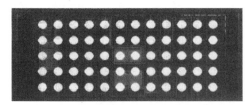

Figure 33.13   Bottom view of the CSP after the wafer-level process (the first generation).

Once the dies are singulated from the wafer, they are ready for the subsequent assembly and packaging process, as shown in Fig. 33.14. First, the dies are flipped over and mounted on the base frame with the inner solders aligned to the corresponding Cu lands. The whole assembly is then heated to 350°C in an $H_2 + N_2$ environment for 150 s.

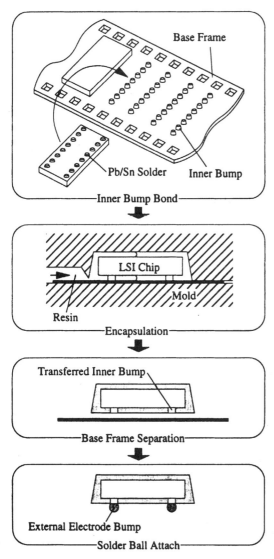

Base Frame

Pb/Sn Solder   Inner Bump

Inner Bump Bond

LSI Chip

Mold

Resin

Encapsulation

Transferred Inner Bump

Base Frame Separation

External Electrode Bump

Solder Ball Attach

Figure 33.14 Molding and bumping process for the CSP (the first generation).

The inner solders are melted to wet the Cu lands. After being joined together, the assembly is encapsulated by transfer molding. Then the stainless steel base frame is peeled off by mechanical means and all Cu lands are transferred to the CSPs. Finally, flux is deposited on the exposed Cu lands and eutectic solder balls are placed through a patterned metal mask. The package is heated up to 230°C for 60 s on a hot plate to reflow the eutectic solder. With the aforementioned process, a minimum bump pitch of 0.8 mm can be achieved. The pictures of completed prototypes are shown in Fig. 33.15.

The second generation of Mitsubishi's CSP is aimed at fine-pitch high-I/O applications such as ASICs. In order to achieve this objective, certain modifications in the manufacturing process are necessary. Figure 33.16 shows the wafer-level process for this CSP. Unlike the procedure for the other version, the polyimide film is coated on the wafer first. Then the bottom layer (TiN) of wiring conductor is de-

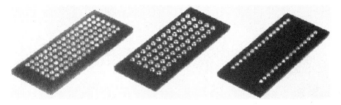

Figure 33.15 Completed prototypes of the CSP (the first generation).

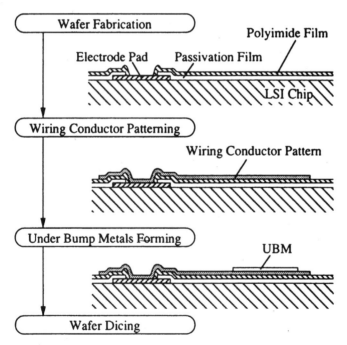

Figure 33.16 Fabrication process at the wafer level for interconnection, I/O redistribution, and UBM deposition (the second generation).

posited and patterned by sputtering, photolithography, and wet etching. Since this CSP is for fine-pitch applications, a 25-μm design rule is adopted for the routing traces. The next step is to deposit the top layer (Ni/Au) of metallization as the UBM for package terminals. It should be noted that, unlike the configuration of the first-generation CSP, the UBM is deposited at the designated locations only. The bonding pads have a diameter of 250 μm and a pitch of 0.5 mm, as shown in Fig. 33.17. After the deposition of the UBM, wafer dicing is performed to singulate the dies for the subsequent packaging process.

The procedures for forming package terminals for this CSP differ from those for the earlier version. As illustrated in Fig. 33.18, instead of being attached to the UBM, the inner high-lead solders are deposited to the Cu lands on the base frame first. Then the singulated dies are flipped over and placed on the inner bumps by a flip-chip bonder. The joints are formed by reflow heating in an $H_2 + N_2$ environment. Note that flux is not used in this mounting process. The next two steps are encapsulation by transfer molding and separation of the base frame by mechanical peeling. It should be noted that for this

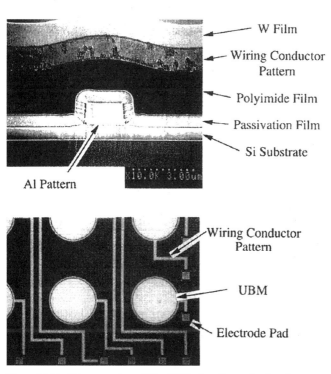

Figure 33.17   Close-up view of the CSP after the wafer-level process (the second generation).

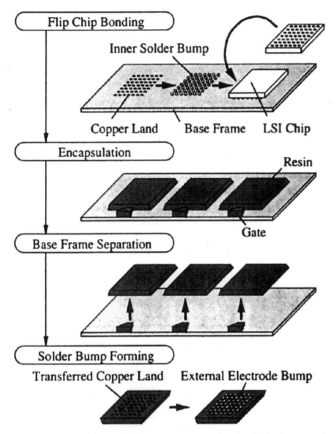

Figure 33.18   Molding and bumping process for the CSP (the second generation).

version of the CSP, the Cu lands are plated directly on the ferroalloy base frame. No adhesive layer is used. Therefore, as indicated by the experimental data shown in Fig. 33.19, the interfacial shear strength is much lower than that of the previous version (Fig. 33.11). This mechanical property is desired for the fine-pitch configuration.

In addition to the variations for inner solders and Cu lands, the formation of external solder bumps is also different from the previous procedure. Instead of solder-ball placement, a stencil printing process is employed to deposit the eutectic solder paste on the Cu lands, as shown in Fig. 33.20. Afterwards, the solders are reflowed by a scanning YAG laser beam. To prevent the bridging of fine-pitch solder bumps, the reflow is performed with the stencil present. Once all the solder bumps are formed, the stencil is removed and the CSP is com-

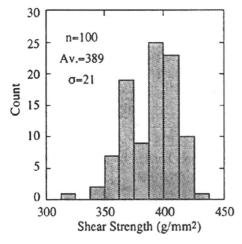

Figure 33.19   Shear strength of plated Cu lands on the ferroalloy base frame.

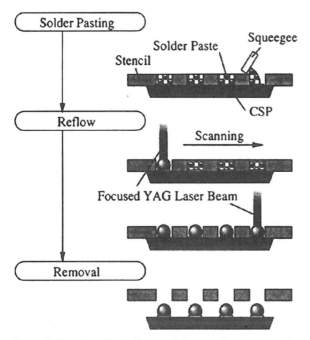

Figure 33.20   Stencil printing and laser reflow process for the external solder bumps (the second generation).

Figure 33.21 Top and bottom view of completed high-pin-count CSP (the second generation).

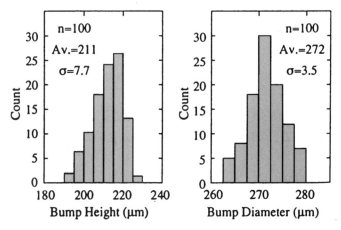

Figure 33.22 Dimensional statistics for the external solder bumps (the second generation).

pleted as shown in Fig. 33.21. From the measurement data given by Mitsubishi in Fig. 33.22, the average height and diameter of the external solder bumps are 211 μm and 272 μm, respectively. Since the standard deviation in bump dimensions is relatively small, the aforementioned solder-bump formation process is considered to be valid and reliable.

33.5 Performance and Reliability

The electrical performance of the second-generation CSP was analyzed by electromagnetic field analysis. The corresponding SPICE model is illustrated in Fig. 33.23. The inductance, capacitance, and

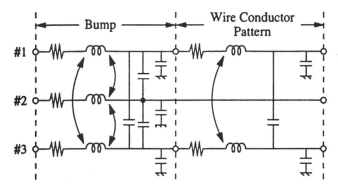

Figure 33.23 SPICE model for the simulation of electrical performance.

TABLE 33.5 Electrical Performance by Numerical Simulation

	Bump	Wire conductor pattern
L_s	0.052 nH	0.95 nH/mm
L_m	0.01–0.02 nH	0.16–0.28 nH/mm
C_o	0.17–0.24 pF	0.20–0.21 pF/mm
C_m	0.01–0.02 pF	0.12–0.65 fF/mm
R	2.9 mΩ	670 mΩ/mm

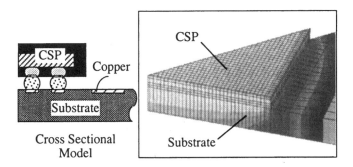

Temperature Distribution with FEM

Figure 33.24 Finite-element model for the simulation of thermal performance.

resistance are given in Table 33.5. These values are relatively small compared to those for other conventional packages.

The thermal resistance θ_{ja} of a 1024-pin CSP was evaluated by a three-dimensional finite-element analysis as shown in Fig. 33.24. The CSP was mounted on a single-layer PCB with a thickness of 1.6 mm.

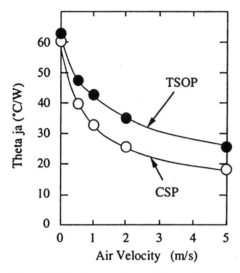

Figure 33.25 Comparison of thermal resistance between Mitsubishi's CSP and TSOP.

Because of the geometrical symmetry, only one-eighth of the assembly needs to be modeled. The results of the analysis are presented in Fig. 33.25. The thermal resistance of a conventional TSOP with similar package size (12.7×22.2×1.2 mm) is also given for comparison. It is obvious that the CSP outperforms the TSOP, especially under forced convection cooling. During the analysis, it was found that the main path for heat dissipation is via package terminals to the PCB. Therefore, it is expected that the thermal resistance of this CSP would further decrease if a multilayer PCB is used.

Since Mitsubishi's CSP has molding encapsulation, the package reliability in a high-temperature and high-humidity environment may be a concern. An experimental study was conducted to investigate the moisture sensitivity of a 1024-pin CSP. From the test results shown in Fig. 33.26, the moisture absorption has become saturated after 48 h. Further reliability tests were performed with various storage periods under the 85°C/85 percent RH condition. The specimens were prebaked at 125°C for 24 h and were subjected to IR reflow at 230 ± 5°C after the high-temperature/humidity storage. Scanning acoustic microscopy (SAM) was employed to inspect the damage inside the CSP after the tests. As indicated by the test results given in Table 33.6, no cracks or delamination could be observed. Therefore, this CSP is considered to have the same level of package reliability as other conventional plastic packages.

The board-level solder joint reliability of the first-generation CSP was investigated by a finite-element analysis. A typical model of a sol-

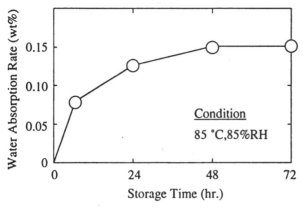

Figure 33.26 Moisture absorption feature of Mitsubishi's molded CSP.

TABLE 33.6 Qualification Testing Results

| | Results (failures/sample size) | |
Storage time (h)	Crack	Delamination
6	0/5	0/5
24	0/5	0/5
48	0/5	0/5

Conditions: (1) Baking: 125°C, 24 h, (2) storage: 85°C/85% RH, (3) IR reflow: 235°C max. (230 ± 5°C).

der joint is shown in Fig. 33.27. Two types of substrate (ceramic and organic) were studied. The assembly was subjected to a thermal loading from −40 to 125°C. The maximum equivalent plastic strain in the solder joints was obtained from the analysis, and the Coffin-Manson equation was employed to estimate the thermal fatigue life cycles. The results are given in Table 33.7. Except for the case with prototype C assembled on the organic substrate, the board-level solder joint reliability of CSP is considered comparable to that of conventional TSOP. In addition to numerical simulation, a thermal cycling test was conducted for the case with prototype B assembled on the organic substrate. The testing temperature profile was −40 to 125°C with a period of 1 h per cycle. As well as the CSP, a bare-die flip chip and an underfilled CSP assembly were tested simultaneously for comparison. The test data are given in Table 33.8. From these results, it may be concluded that the CSP has much better board-level solder joint reliability than the bare-die flip chip and that the underfill can prolong the solder joint fatigue life under thermal cycling.

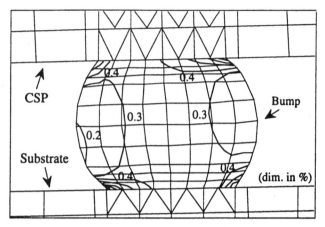

Figure 33.27 Finite-element model for the prediction of solder joint fatigue life.

TABLE 33.7 Simulation Results for the Prediction of Solder Joint Thermal Fatigue Life

Substrate	Al_2O_3		Glass-Epoxy	
CSP	$\Delta\epsilon_t$ (%)	N_f(50) (cycle)	$\Delta\epsilon_t$ (%)	N_f(50) (cycle)
Prototype A	1.478	769	1.628	621
Prototype B	1.492	753	1.832	478
Prototype C	1.666	590	3.310	129

N_f(50): 50% failure cycle number

TABLE 33.8 Solder Joint Thermal Cycling Test Results

	Result of test (failures/sample size)			
Structure	0 cycles	100 cycles	200 cycles	500 cycles
Bare die	0/30	10/30	25/30	30/30
CSP	0/20	0/20	0/20	8/20
CSP with underfilling	0/3	0/3	0/3	0/3

Test condition: -40 to 125°C, 1 cycle per hour.

33.6 Applications and Advantages

Mitsubishi's CSP is a flip-chip die with molding encapsulation. There are two generations in this family of chip scale packages. The earlier version was developed for low-pin-count devices and has package bump pitches of 0.8 and 1.0 mm. The bump pitch of the later version

TABLE 33.9   Comparison of Electrical Performance among
Various Packages

Package	Pin count	Body size (mm^2)	L_s (nH)
CSP	256	9.0	0.15– 3.7
BGA	225	27.0	6.8 –14.4
QFP	240	32.0	12.1 –15.1

is 0.5 mm, and the number of I/Os may exceed 1000. As a result, this series of CSPs may cover a wide range of applications, from memory modules to ASICs.

The major advantages of Mitsubishi's CSP are compact package size and good electrical performance. A comparison among various packages is given in Table 33.9. It is obvious that for similar pin count, the CSP is superior to the BGA and QFP in package size and self-inductance. Also, in a previous comparison, shown in Fig. 33.25, the CSP outperformed the TSOP in thermal resistance. Furthermore, unlike other wafer-level CSPs, Mitsubishi's CSP has molding encapsulation covering the whole package. Although not yet fully verified by qualification tests, it is expected that Mitsubishi's CSP should be more robust than the other wafer-level CSPs. On the other hand, since the molding and bumping processes are performed on the singulated dies, Mitsubishi's CSP may not gain much cost advantage with die shrink or wafer expand.

33.7 Summary and Concluding Remarks

Mitsubishi's CSP is a molded chip scale package with a flip-chip configuration. The I/O redistribution is performed at the wafer level. A multilayer metallization is employed to route the center or peripheral die bond pads to the desired package I/O pattern. The first-level interconnects are sputtered TiN on the Al bond pads. A Ni/Au UBM is deposited on top of the TiN layer. On the UBM are package terminals which consist of inner high-lead solder, transferred copper land, and external eutectic solder bump. There are two generations in this family of CSPs. Although there are certain differences in the fabrication process, both versions of CSP use the same wiring technology for I/O redistribution at the wafer level. The earlier version of CSP is aimed at low-pin-count devices such as memory modules, whereas the later version is developed for high-pin-count applications such as ASICs. Depending on the pin count and the terminal pattern, the package bump pitch ranges from 0.5 to 1.0 mm.

The electrical and thermal performance of Mitsubishi's CSP have been evaluated by computational modeling. Good results were reported. Qualification tests were conducted to investigate the package reliability and the board-level solder joint reliability. In addition, a finite-element analysis was performed to predict the thermal fatigue life of solder joints. It was found that this CSP has the same level of package reliability as other conventional plastic packages. Also, the solder joint reliability of most CSP assemblies is comparable to that of TSOP. However, for very low-pin-count CSPs assembled on an organic substrate, the solder joint reliability is poor and underfill adhesive is needed to prolong the thermal fatigue life.

In summary, Mitsubishi's CSP has both "package" features and "bare-die" features. With the I/O redistribution capability, the standardization of footprints is very easy to implement. Therefore, this chip scale packaging technology has attracted considerable attention in the electronics industry.

33.8 References

1. M. Yasunaga, S. Baba, M. Matsuo, H. Matsushima, S. Nakao, and T. Tachikawa, "Chip Scale Package: A Lightly Dressed LSI Chip," *Proceedings of the IEEE/CPMT International Electronics Manufacturing Technology Symposium,* La Jolla, Calif., 1994, pp. 169–176.
2. S. Baba, Y. Tomita, M. Matsuo, H. Matsushima, N. Ueda, and O. Nakagawa, "Molded Chip Scale Package for High Pin Count," *Proceedings of the 46th ECTC,* Orlando, Fla., 1996, pp. 1251–1257.

National Semiconductor's μSMD

34.1 Introduction and Overview

National Semiconductor developed a new wafer-level chip scale package called μSMD [1]. Although this CSP was originally designed for low-pin-count applications, its form factor can accommodate higher pin counts up to 144. The package terminals of μSMD are solder bumps. A thin polyimide film may be coated on the bumped surface to enhance the solder joint reliability. The bump material may be high-lead or eutectic Sn/Pb solders. The bump pitch ranges from 0.8 to 0.5 mm. The current lineup for μSMD is 8 and 14 I/Os. The assembly of μSMD can make use of the existing surface-mount facilities. Underfill encapsulation of μSMD on PCB is not required. The board-level solder joint reliability has been investigated by experimental testing and computational simulation. Both thermal and mechanical loading were studied. Good solder joint reliability was reported.

34.2 Design Concepts and Package Structure

National Semiconductor's μSMD is a chip scale package with wafer-level fabrication process. This CSP resembles a flip-chip die, but does not require underfill for board-level assembly. The backside of the silicon chip of the μSMD is covered by a proprietary encapsulant for protection. On the other hand, the active chip face of the μSMD may be coated with a polyimide thin film to enhance the interconnect reliability. The thickness of this polyimide film is about 4 to 4.5 μm.

The original design for the μSMD was for packaging low-pin-count devices. Nevertheless, because of its superior form factor, this CSP can accommodate higher pin counts up to 144. The Sn/Pb solder bumps are the board-level interconnects of the μSMD. The solder may be high-

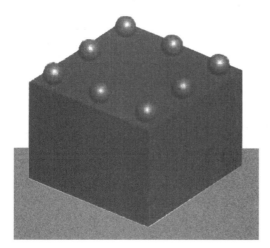

Figure 34.1   8-I/O μSMD.

Figure 34.2   14-I/O μSMD.

lead or eutectic. The bump pitch may range from 0.8 to 0.5 mm. The assembly of the μSMD on PCB can make use of the existing surface-mount facilities. After assembly, the bump height is 3 mils. The current lineup for the μSMD is 8 and 14 I/Os as shown in Figs. 34.1 and 34.2, respectively. Their footprints comply with JEDEC MO-211. The typical outline of the μSMD is 57×73 mils.

34.3   Material Issues and Manufacturing Process

The μSMD developed by National Semiconductor is a wafer-level chip scale package with backside encapsulation. The major packaging materials include a proprietary encapsulant, an optional polyimide thin film, and the Sn/Pb solder bumps. Unlike most other CSPs, the μSMD is encapsulated at the backside of the silicon chip. The purpose of this backside encapsulation is to protect the IC device. On the other hand, a polyimide thin film may be coated on the active chip face of the μSMD. This polyimide coating is optional and its function is to enhance the interconnect reliability. The thickness of this thin film is about 4 to 4.5 μm.

The package terminals of the μSMD are Sn/Pb solder bumps. Both high-lead (5Sn/95Pb) and eutectic (63Sn/37Pb) solders may be used. The typical bump height in a μSMD-PCB assembly is 3 mils. In principle, the high-lead solder has lower elastic modulus and, hence, may yield less inelastic strain during thermal cycling. However, the trade-off is that the reflow temperature of high-lead solder is much higher than that of eutectic solder. Therefore, in view of components and processing costs, the eutectic solder is still favored in general.

The fabrication of the μSMD involves several procedures. After the standard IC manufacturing process, wafer repassivation is performed. The next step is solder bump deposition followed by laser inspection. Once the bump characteristics are verified, backside encapsulation will be coated. After laser marking and singulation, the μSMD packages may be shipped in tape or reel form.

34.4   Solder Joint Reliability

The solder joint reliability of the μSMD-PCB assembly has been investigated comprehensively by National Semiconductor [1]. Both thermal and mechanical loading were studied. The first experimental evaluation with thermal loading was a −40 to 125°C thermal shock test (2 cycles per hour). The μSMD specimens were mounted on a PCB with 6-mil solder mask-defined pads. Four kinds of configurations, as shown in Table 34.1, were tested. The results of cumulative failure are presented in Table 34.2. In addition, the typical failure modes of different configurations are shown in Fig. 34.3. From these experimental results, it was concluded that the eutectic solder bumps together with polyimide film give the best solder joint reliability.

In addition to experimental testing, an elastoplastic analysis was

TABLE 34.1   Various μSMD Configurations for Solder Joint Reliability Evaluation

	Polyimide coating (PI)	No polyimide coating (NP)
Eutectic bump (EU)	PIEU	NPEU
High lead bump (HL)	PIHL	NPHL

TABLE 34.2   Cumulative Failure of NS' μSMD under Thermal Shock Loading (TMSL)

TMSL—40 ~ 125°C 2 cy./hr	SMD NPHL	SMD PIHL	SMD NPEU	SMD PIEU
250 X	0/37	16/34	39/39	1/40
500 X	37/37	34/34		9/40
750 X				34/40
1000 X				40/40

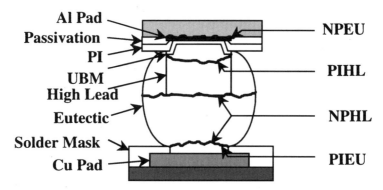

Figure 34.3   Summary of μSMD failure modes under thermal shock loading.

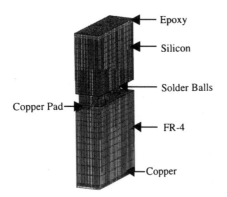

Figure 34.4   Finite element model for solder joint reliability analysis.

performed with the same thermal shock loading. The finite element model is illustrated in Fig. 34.4. The used material properties are given in Table 34.3. In particular, the constitutive relations of eutectic and high-lead solder are shown in Figs. 34.5 and 34.6, respectively [2]. It should be noted that the thin polyimide film was not modeled in the finite element analysis. The maximum equivalent plastic strain

TABLE 34.3   Material Properties Used for Finite-Element
Modeling of the μSMD

Material	Modulus (GPa)	CTE (ppm/°C)
Silicon	129	2.9
Copper	128	16.9
Eutectic solder	33.8	24.5
High-lead solder	23.5	25.7
FR4	15.4 (E_X, E_Y)	17.6($\alpha_{x,y}$)
	6.7 (E_Z)	64.1 (α_z)

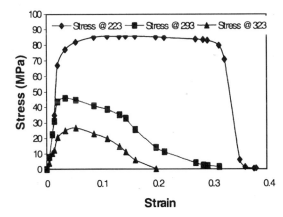

Figure 34.5   Constitutive relation of eutectic (63Sn/37Pb)
solder [2].

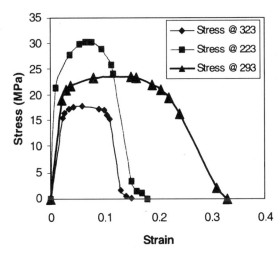

Figure 34.6   Constitutive relation of high-lead (5Sn/95Pb)
solder [2].

in the solder bumps was used as an index for comparison. The results from computational analysis are given in Table 34.4. The trend in this table agrees with that shown in Table 34.2.

Once the best configuration (eutectic solder bump with polyimide film) was identified, temperature cycling tests were performed for further reliability investigation. The test vehicle is shown in Fig. 34.7. Both 8-I/O and 14-I/O μSMDs were tested on a standard four-layer FR4 PCB (62 mils thick). The testing temperature profiles were 0 to 100°C and −40 to 125°C. Both profiles had a frequency of 1 cycle per hour. The various test configurations are shown in Table 34.5. The experimental results are given in Table 34.6 and Fig. 34.8. It can be seen

TABLE 34.4 Comparison of Computational Analysis Results for the μSMD

Model	Equivalent plastic strain induced in the solder bump
Eutectic bump model	0.066
High-lead bump model	0.038

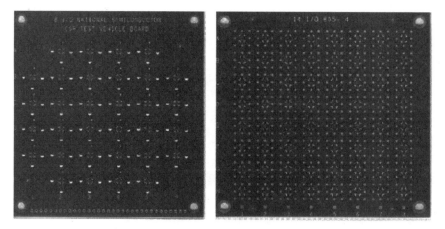

Figure 34.7 μSMD test vehicle boards for temperature cycling tests.

TABLE 34.5 Various Configurations for Temperature Cycling Tests of the μSMD

Test vehicle	I/O	I/O pitch	Max. DNP	Standoff	Test condition
0805DC1	14	0.4	1	0.100	−40°C to 125°C, 1 CPH
0605DC1	8	0.5	0.7	0.125	−40°C to 125°C, 1 CPH
0605DC1	8	0.5	0.7	0.125	0°C to 100°C, 1 CPH

CPH = cycle per hour

TABLE 34.6   Temperature Cycling Test Results of the μSMD

μSMD assembly	Test condition	0 cycles	500 cycles	800 cycles	1000 cycles	2300 cycles
8 I/O	0–100°C	0/62	0/62	0/62	0/62	0/62
8 I/O	−40–125°C	0/32	0/32	0/32	6/61	N/A
14 I/O	−40–125°C	0/83	8/83	58/83	70/83	N/A

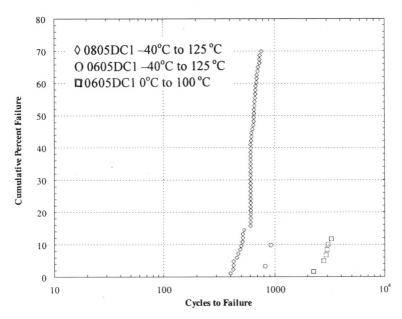

Figure 34.8   μSMD solder joint failure distribution under temperature cycling.

that the solder bumps of the 0605DC1 test vehicle subjected to the lower (0 to 100°C, 1 cycle per hour) temperature condition perform the best since they also have the taller standoff height (0.125 mm). Most of the solder bumps of the 0805DC1 test vehicle perform the worst, since they are subjected to the higher temperature condition (−40 to 125°C, 1 cycle per hour) and have the shorter standoff height (0.1 mm).

In addition to thermal loading, mechanical loading was applied to evaluate the solder joint reliability. The mechanical evaluations include the drop (shock) test, bending test, and vibration test. For the drop test, 8-I/O μSMD assemblies were dropped two times in the three mutually exclusive axes from a height of 750 mm onto a non-cushioning, vinyl tile surface. The test results showed that all 20 specimens passed this test.

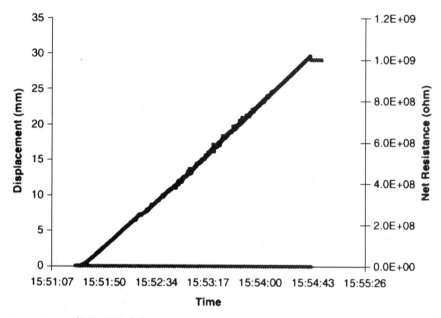

Figure 34.9 μSMD-PCB deflection and net resistance.

The second test was a three-point bending. The test board had a span of 100 mm. Static loading was applied at the center with a speed of 9.45 mm/min. The testing results for an 8-I/O μSMD specimen are given in Fig. 34.9. It can be seen that no solder joint failure occurred even with a deflection of 25 mm. For such board size, this deflection magnitude is beyond most manufacturing, shipping, handling, and re-work conditions [3].

For the vibration test, both random and sinusoidal excitations were used. It was found that on 8-I/O μSMD could survive 1 h of sinusoidal vibration at 20G followed by 3 h of sinusoidal vibration at 40G without any failure. In addition, this specimen also passed 3 h of $2G_{rms}$ random vibration with frequencies ranging from 20 to 2000 Hz.

34.5 Summary and Concluding Remarks

The μSMD is a wafer-level chip scale package developed by National Semiconductor in 1998. The backside of the silicon chip of this CSP is covered by a proprietary encapsulant for protection. On the other hand, the active chip face of the μSMD may be coated with a poly-imide thin film to enhance the interconnect reliability. The thickness of this polyimide film is about 4 to 4.5 μm.

The package terminals of the μSMD are solder bumps. Although this CSP was originally designed for low-pin-count applications, its form factor can accommodate higher pin counts up to 144. The bump material may be high-lead or eutectic Sn/Pb solders. The bump pitch ranges from 0.8 to 0.5 mm. The current μSMD family has 8 and 14 I/Os. The footprints comply with JEDEC MO-211.

The assembly of the μSMD on PCB can make use of the existing surface-mount facilities. Underfill encapsulation is not required. The board-level solder joint reliability has been investigated comprehensively by experimental testing and computational simulation. Both thermal and mechanical loading were studied. Experimental results indicated that the μSMD has good solder joint reliability.

34.6 References

1. L. Nguyen, N. Kelkar, T. Kao, A. Prabhu, and H. Takiar, "Wafer Level Chip Scale Packaging-Solder Joint Reliability," *Proceedings of IMAPS International Symposium on Microelectronics,* San Diego, November 1998.
2. SRC Cindas Database, *Semiconductor Research Corp.,* Raleigh, N.C.
3. J. H. Lau, "Solder Joint Reliability of Flip Chip and Plastic Ball Grid Array Assemblies Under Thermal, Mechanical, and Vibrational Conditions," *IEEE Transactions on CPMT-Part B,* vol. 19, no. 4, November 1996, pp. 728–735.

35

Sandia National Laboratories' Mini Ball Grid Array Package (mBGA)

35.1 Introduction and Overview

The mini ball grid array (mBGA) is a chip scale packaging technology developed by Sandia National Laboratories in 1995 [1]. The CSP package is a flip-chip die with I/O redistribution at the wafer level. The redistribution is implemented with two dielectric layers and two conductive layers. The fine-pitch peripheral die bond pads are reconfigured to area-array pads with much larger pitch. The minimum diameter and pitch of area-array pads in mBGA packages are 10 and 20 mils, respectively. The package terminals may be Au bumps or solder bumps. For the former, conductive adhesive should be used for the second-level interconnection. Otherwise, the conventional SMT is applied for board-level assembly. In addition to the CSP, two types of multichip testing modules (MCM-D and MCM-L) were developed to characterize mBGA packages. Experimental results indicate that good performance and reliability can be achieved. However, for the MCM-L configuration, underfill encapsulant is required. A cost analysis was also conducted to compare various packages. It was shown that the packaging cost of the mBGA is comparable to that of the TSOP for low-pin-count ICs and is much lower than that of the PQFP and PBGA for high-I/O devices.

35.2 Design Concepts and Package Structure

Sandia's mBGA is a wafer-level CSP with a flip-chip configuration. The cross section of the package structure is illustrated in Fig. 35.1. This CSP uses two polymer dielectric layers together with two metal conduc-

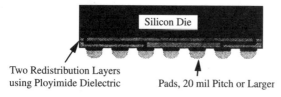

Two Redistribution Layers
using Ployimide Dielectric   Pads, 20 mil Pitch or Larger

Figure 35.1 Cross-sectional view of the mBGA package structure.

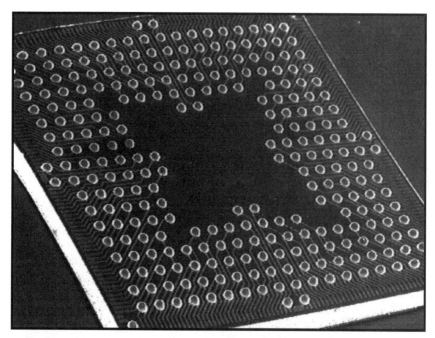

Figure 35.2 Bottom view of the mBGA package (ASIC with 275 I/Os).

tive layers to redistribute the peripheral wire bond pads on the die to an area-array format, as shown in Fig. 35.2. So far nine different die types, as listed in Table 35.1, have been processed into mBGA packages.

The first-level interconnection of the mBGA is established by sputtering metal on the Al bond pads. The routing traces are formed by photomasking with a fan-in pattern. The typical design rule is 2 mils for lines/spaces/vias and 10 mils for the diameter of terminal pads. For most devices, the pitch of area-array pads is 30 mils. However, for ASICs with dense wire-bond pads (5 mils), a 20-mil area-array pitch is required. The package I/Os of the mBGA are either Au bumps or solder bumps. If Au bumps are adopted, conductive adhesive should be used for board-level interconnection. The solder bumps may be

TABLE 35.1   mBGA Processed Die Types

Die type	Description	Number of pads	Pad pitch (in)
MT5C2568	Micron memory	34	0.030
PDSP16515	Plessey DSP	128	0.030
LCA100106	LSI logic ASIC	275	0.020
L8095	ASIC	298	0.020
LCA	Daisy chain die[†]	266	0.020
SA3470	Microcontroller[*]	52	0.030
X4010	Daisy chain die[†]	190	0.030
ATC03Quad	Assembly test chip[*]	224	0.030
ATC04	Assembly test chip[*]	35	0.030

[*]Silicon dies designed and fabricated at Sandia.
[†]Silicon dies fabricated at Sandia.

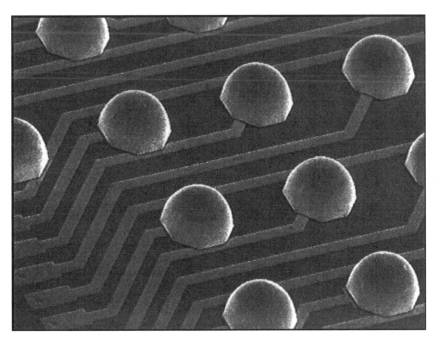

Figure 35.3   Metal traces and solder bumps of the mBGA (20-mil bump pitch).

made of 40Pb/60Sn or 95Pb/5Sn solder. Conventional SMT procedures are employed for the assembly of solder bumps. The micrograph of typical solder bumps of the mBGA package is shown in Fig. 35.3.

Because of the flip-chip configuration and the compact size, mBGA packages can be closely tiled together on a substrate to yield a very high-density multichip module (MCM). During the development of the mBGA packages, two types of testing MCMs were designed as well for the evaluation of performance. The first type was an MCM-D, which had five layers of metal/polyimide dielectric on a ceramic substrate [2]. Four mBGA packages were mounted on the substrate, as illustrated in Fig. 35.4. Three of them were ATC03 quad modules, which are Sandia's assembly test chips containing temperature sensors, strain sensors, and heaters. In fact, each ATC03 quad module is a chip set consisting of four square ATC03 dies. The die size of each ATC03 is 0.25 in. The fourth chip was an X4010 daisy chain testing die with 190 leads. The whole MCM-D had 862 I/Os and a module size of 1.26×1.26 in. The corresponding silicon density and interconnection density were 66 percent and 694 I/Os per in^2, respectively.

The second type of testing MCM for mBGA packages was an MCM-L, which used a 1.8×1.8 in FR-4 laminate as the substrate [3]. There were two test vehicles in this category, as shown in Fig. 35.5. Test vehicle #1 consisted of four daisy chain dies. Three of them were LCA chips, and the fourth one was an X4010 (see Table 35.1). The former had a die size of 0.41×0.41 in, whereas the latter was 0.554×0.563 in. This MCM-L had a total number of 988 I/Os. This test vehicle was designed mainly for solder joint reliability tests under thermal shock

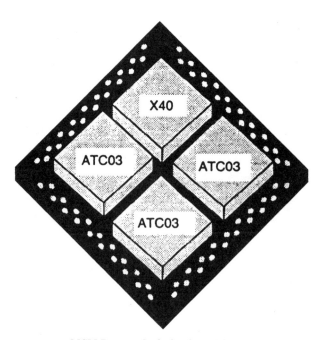

Figure 35.4 MCM-D test vehicle for the mBGA.

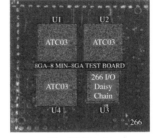

Test Vehicle #1   Test Vehicle #2

Figure 35.5 MCM-L test vehicles for the mBGA.

environment. Test vehicle #2 was composed of three ATC03 quad modules and one LCA chip. In total, there were 938 I/Os. This MCM-L was developed to test the thermal performance of mBGA packages.

35.3 Material Issues

The mBGA package is essentially a flip-chip die with I/O redistribution at the wafer level. The redistribution is implemented with two layers of polymer dielectric and another two layers of conductive metallization. The thickness of each dielectric layer is 3 µm. The material may be either polyimide (PI) or benzocyclobutene (BCB). For the former, DuPont 2525 or 2611 is used. If PI is adopted as the dielectric, an additional step, surface plasma etching, may be required in order to improve the adhesion strength between the dielectric and metallization layers. On the other hand, BCB adheres to Cu without requiring aid from other means. Also, it provides better planarization and absorbs less moisture. Therefore, BCB is recommended for the fabrication of mBGA packages.

The basic metallurgy for I/O redistribution in the mBGA is titanium (Ti), tungsten (W), nickel (Ni), and copper (Cu). The first three metals act as diffusion barriers, and the last one is the main conductor for circuit routing. It should be noted that Ti, W, and Cu (seed) are deposited by sputtering, but Cu (trace) and Ni are by plating. The package terminals of mBGA may be Au bumps or solder bumps. Both 40Pb/60Sn and 95Pb/5Sn solder can be used for package bumping. If Au bumps are employed, a conductive adhesive such as 3M's S40B should be used for the second-level interconnection. Otherwise, the conventional SMT is applied for board-level assembly.

For the MCM-D application, the substrate materials are silicon and alumina. There are five metal/dielectric pairs laminated on the top of this substrate. These layers are for one power, one ground, two signal planes, and the mounting pads. The metallurgy of the conductive layers is Ti, W, Ni, and Cu. However, the surface of all the mounting pads is finished with Au. For the MCM-L configuration, the substrate is made of polyclad FR-4 Tetra II plus #370G material. The glass transition temperature T_g and the dielectric constant are 180°C and 3.9, respectively. The CTE in the planar directions is 14 ppm/°C, while that in the thickness direction is 45 ppm/°C. Since the bulk of the mBGA is silicon chip, the effective CTE of the package is rather small. Because of the substantial CTE mismatch between mBGA packages and the FR-4 substrate, the flip-chip assembly in MCM-L requires underfill encapsulation. The underfill adhesives adopted for mBGA MCM-L are HYSOL 4511 or HYSOL 4520.

35.4 Manufacturing Process

The mBGA is a wafer-level CSP. The packaging process is performed on the whole wafer instead of on a singulated die. The fabrication of the mBGA involves conventional semiconductor technologies such as dielectric coating, metal deposition, photomasking, and chemical etching. The manufacturing flow for the solder-bumped mBGA is illustrated in Fig. 35.6. The actual wafer redistribution and bumping processes for the mBGA were performed by APTOS (Milpitas, California) and GE (Schenectady, New York). A typical wafer processed with mBGA technology is shown in Fig. 35.7. After the wafer-level process, the mBGA dies are singulated by mechanical sawing for the next level assembly.

The assembly processing conditions of the mBGA for MCMs depend on the bump materials and substrate types. For MCM-D, if Au bumps are adopted, conductive adhesive should be used for interconnection. The adhesive sheet is cured at 180 to 190°C for 3 min at a pressure of 115 lb per die. If the bumps are 40Pb/60Sn solder, the prefluxed assembly needs to be heated to 235°C to reflow the solder. For 95Pb/5Sn solder bumps, the reflow temperature is 355°C, and a nitrogen environment is required.

For MCM-L application, only 40Pb/60Sn solder bumps are used. The FR-4 substrate is prefluxed with Alpha 5003 before the pick-and-place of mBGA dies. The solder bumps are reflowed at 220°C for 1 min. The assembly is rinsed in IONOX solvent and dried in ACCEL cleaning equipment. After 1.5 h of soaking in isopropyl alcohol, the cleaned assembly is underfilled with HYSOL 4511 or 4520. It should be noted that the height of the solder joints may be controlled by using different bond-pad finishings on the laminate substrate. If the bond pads are finished with Au flash, then the solder joint height will be 3 mils, as shown in Fig. 35.8a. If 6-mil-thick solder is deposited on the pads instead of Au flash, then 7-mil-tall joints, as presented in Fig. 35.8b, can be achieved. The latter configuration is implemented

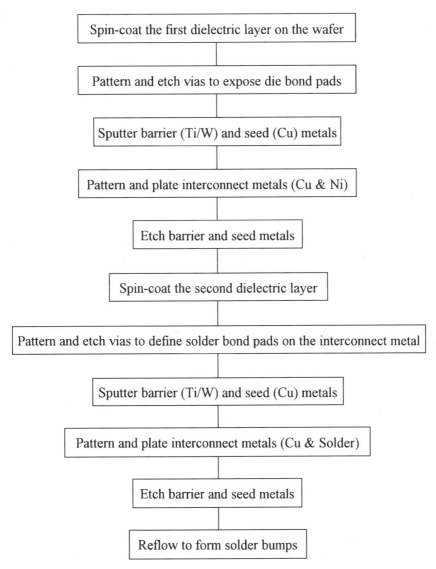

Figure 35.6 Manufacturing flow for Sandia's mBGA.

with precision pad technology (PPT), which was developed by MAS-TEK, Inc. In the PPT process, the substrate is coated with a 3-mil-thick solder mask. The openings on the bond pads are 1.5 times the pad area. Solder paste is screen-printed in the opening areas and re-flowed with the screen still in place. With this process, a flat solder layer (6 mils thick) is deposited on top of the substrate bond pads.

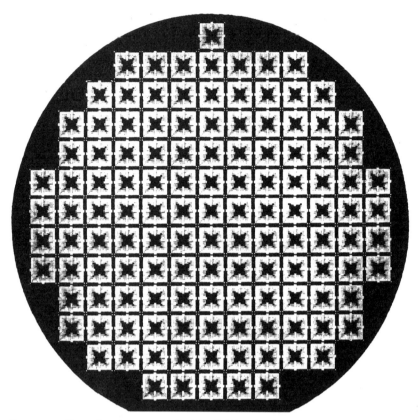

Figure 35.7   Typical wafer processed with mBGA technology.

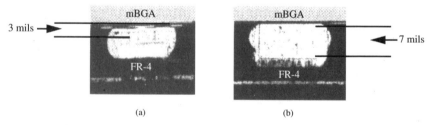

Figure 35.8   Cross section of solder joints in the mBGA MCM-L assembly: (*a*) bond pad with Au flash; (*b*) bond pad with 6-mil-high solder by PPT.

35.5   Electrical and Thermal Performance

The electrical performance of mBGA packages was investigated with an HP82000 IC tester and an AMT M300 burn-in system. The specimens were clamped by a universal testing fixture on a pin grid array (PGA) alumina chip carrier. The electrical interconnection was estab-

TABLE 35.2   Comparison of Maximum Operation Frequency (F_{max}) between mBGA and Conventional Wire-Bonded Packages

Die type	Package	F_{max} at 0°C (MHz)		F_{max} at 70°C (MHz)	
		4.5 V	5.5 V	4.5 V	5.5 V
MT5C2568	mBGA	72.5	80.6	67.6	75.7
	Conventional	71.8	78.7	66.1	74.2
PDSP16515	mBGA	86.3	97.2	75.3	83.2
	Conventional	88.0	98.9	77.3	85.0
LCA100106	mBGA	27.3	32.1	21.2	25.4
	Conventional	25.6	30.8	20.0	24.1

lished by a z-axis conductive interposer inserted between the mBGA package and the chip carrier substrate. The tested items included dc parameters, ac timing, and maximum operation frequency using full functional vector sets. A total of 300 mBGA dies were tested over the voltage range of 4.5 to 5.5 V and over the temperature range of 0 to 70°C. From the test results given in Table 35.2, the performance of the mBGA at maximum frequency F_{max} is close to that of certain wire-bonded conventional packages. In addition, after the electrical tests, all specimens were subjected to a burn-in test at 125°C for 168 h. No observable degradation was found afterwards.

A comprehensive study was conducted to evaluate the thermal performance of mBGA packages in the MCM applications. An MCM-D, as shown in Fig. 35.4, was investigated under four cooling conditions, namely, natural air–cooled without heat sink, natural air–cooled with heat sink (fan off), forced air–cooled with heat sink (air flow at 1.2 ft^3/min), and liquid-cooled with heat sink (water flow at 0.018 ft^3/min). A minimum of 1.0 W per ATC03 die was initially provided, and then the power was gradually increased. The die temperature was monitored by the sensing diodes in the ATC03. Incremental measurements were made during the power-up period, and before each reading was taken, 20 min was allowed for the module to reach a steady state. From the results shown in Fig. 35.9, it is observed that the MCM-D of the mBGA was able to dissipate about 6 W/in^2 when a temperature rise (die temperature minus room temperature) of 65°C was reached. With the aid of forced air cooling and liquid cooling, more power could be dissipated. In summary, if 65°C is considered to be the allowable die temperature rise, a natural air–cooled mBGA MCM-D without heat sink can survive under a module power density of 6 W/in^2 or less. However, for higher power density, a heat sink and forced air or liquid cooling should be used.

The thermal performance of the MCM-L test vehicle #2 (see Fig. 35.5) with 7-mil-high solder joints was also evaluated by experimental

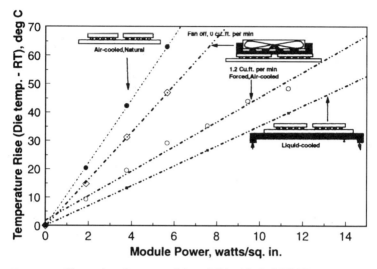

Figure 35.9 Thermal performance of the mBGA with the MCM-D test vehicle.

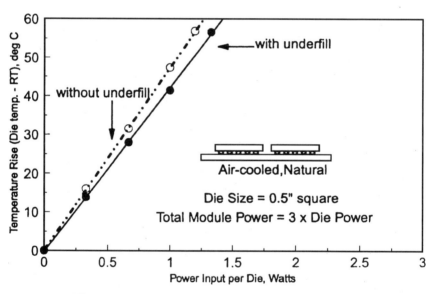

Figure 35.10 Effect of underfill on the thermal performance of the mBGA MCM-L.

study. The testing procedures were similar to those for the MCM-D. The results, shown in Fig. 35.10, reveal the effect of underfill (HYSOL 4511). It can be seen that the existence of underfill encapsulant did improve the heat-dissipation capability of the MCM-L as a result of the direct conduction effect. Figure 35.11 presents the comparison of thermal performance among various cooling schemes for the mBGA

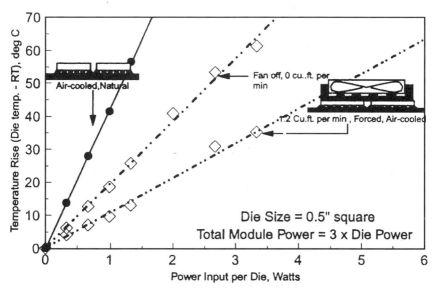

Figure 35.11   Thermal performance of the mBGA with MCM-L test vehicle #2.

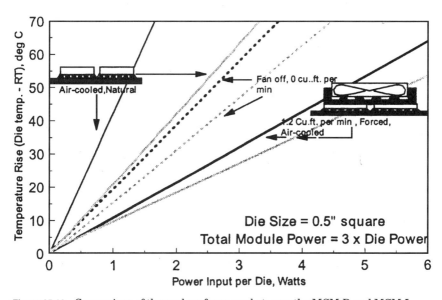

Figure 35.12   Comparison of thermal performance between the MCM-D and MCM-L.

MCM-L. All specimens were underfilled. The abscissa of this plot indicates the power generated by each ATC03 quad module. It can be seen that the heat sink and forced air cooling can substantially enhance the thermal performance. Figure 35.12 shows the comparison between the MCM-L (with underfill) and MCM-D (without underfill).

Obviously the latter has much better thermal performance, which is due to the higher thermal conductivity of the ceramic substrate.

35.6 Bump Shear Strength and Assembly Joint Reliability

Since the package interconnects of the mBGA are rather small, the reliability of bumps before and after assembly is a major concern. A bump shear test was conducted on unassembled mBGA packages using the Dage Precima ball shear tester. The solder bumps were sheared at a height of 12 μm from the base surface, as shown in Fig. 35.13a. The possible failure modes are illustrated in Fig. 35.13b. Of course, solder shear is the desired failure mode, as this correlates with higher joint reliability. To investigate the effect of the dielectric layer on the bump shear strength, specimens using PI and BCB as the dielectric were tested. Furthermore, for some mBGA packages with a PI dielectric, a modified process with additional surface plasma etching was applied to improve the adhesion strength between the dielectric and metallization layers. The results of bump shear are given in Table 35.3. Because of the high shear strength (with relatively low standard deviation) and the elimination of the additional surface treatment, the BCB dielectric is recommended.

An experimental study was performed to investigate the joint reliability of the mBGA MCM-D and MCM-L assemblies. For the former, MCM-D specimens with various joint and substrate materials were sub-

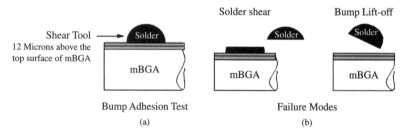

Figure 35.13 Bump shear test for the mBGA: (a) test configuration; (b) failure modes.

TABLE 35.3   Comparison of Bump Shear Strength among Various Dielectric Materials and Processes

Dielectric material/process	Number of wafers	Average shear force (g force)	Average standard deviation	Failure mechanism
Polyimide—old process	4	149	39	Mostly liftoff
Polyimide—modified process	4	206	12	Solder shear
BCB—recommended process	1	197	9	Solder shear

jected to temperature cycling between 0 and 100°C. The period of each cycle was 30 min, and the specimens were tested for 1000 cycles. During the thermal cycling test, the electrical connection of all joints was monitored. The testing results are given in Table 35.4. All specimens passed the thermal fatigue test without failure except for one particular module, which was broken as a result of mishandling. It should be noted that there is no underfill encapsulation in the mBGA MCM-D assembly.

Test vehicle #1 shown in Fig. 35.5 was used to investigate the solder joint reliability of the mBGA MCM-L assembly. Two testing programs were conducted to study the effects of various parameters. In the first testing program, the specimens were subjected to air-to-air thermal shock between 0 and 100°C. For comparison, various specimen configurations with different joint heights and underfills (including without underfill) were used for testing. However, all the mBGA packages used PI for the dielectric layers, and no plasma etching was performed. The experimental results are given in Table 35.5. It was observed that without underfill, the solder joints failed almost immediately after the test began, although a taller joint height did improve the solder joint reliability to a certain extent. However, once underfilled, the solder joint fatigue life of the MCM-L was enormously prolonged and became insensitive to the joint height and the underfill material type.

TABLE 35.4   Results of Solder Joint Thermal Cycling Tests for mBGA MCM-D Assembly

Module no.	Substrate	Die pad	Substrate pad	Joint material	Passed/ failed
29	Silicon	Gold	Gold	S40B adhesive	Passed
43	Silicon	60/40 solder	60/40 solder	60/40 solder	Passed
39	Silicon	60/40 solder	60/40 solder	60/40 solder	Passed
13	Silicon	60/40 solder	95/5 solder	60/40–95/5 solder	Passed
47	Silicon	95/5 solder	95/5 solder	95/5 solder	*
c14	Alumina	60/40 solder	95/5 solder	60/40–95/5 solder	Passed

Temperature cycled, 0 to 100°C, 1000 cycles.
*Module 47 ran for 700 cycles only and broke during handling.

TABLE 35.5   Results of Air-to-Air Thermal Shock (0 and 100°C) Tests for mBGA MCM-L Assembly (Program I)

Substrate	Joint height	Underfill	Number of modules	Total number of dies	Total number of joints	Number of cycles survived
FR-4	3 mils	No underfill	1	4	988	<10
FR-4	7 mils	No underfill	3	12	2964	<40
FR-4	3 mils	HYSOL 4511	3	12	2964	2000+
FR-4	7 mils	HYSOL 4511	2	8	1976	2000+
FR-4	7 mils	HYSOL 4520	1	4	988	2000+

TABLE 35.6   Results of Air-to-Air Thermal Shock (−55 and 125°C) Tests for mBGA MCM-L Assembly (Program II)

Substrate	mBGA process	Joint height	Underfill	Number of modules	Total number of dies	Total number of joints	Number of cycles survived
FR-4	Polyimide— old process	7 mils	HYSOL 4511	2	8	1976	5 dies failed within 500 cycles 1st die failed after 300 cycles
FR-4	Polyimide— modified process	7 mils	HYSOL 4511	2	8	1976	All dies survived 500 + cycles

Therefore, it may be concluded that underfill encapsulation is essential for the solder joint reliability of an mBGA MCM-L assembly.

In the second testing program, all specimens had 7-mil-high solder joints and were underfilled with HYSOL 4511. Although PI was used as the dielectric for all mBGA packages, plasma etching was applied to 50 percent of the specimens to enhance the adhesion strength of the solder bumps (modified process). All specimens were subjected to thermal shock between −55 and 125°C according to MIL-STD 883. The test results are given in Table 35.6. It can be seen that the modified process led to better solder joint reliability. Therefore, the PI dielectric with plasma treatment should be the preferred process for mBGA packages. On the other hand, from the results of bump shear tests, it was found that a BCB dielectric can also provide good adhesion strength for the solder bumps. However, no solder joint reliability data were reported for mBGA packages using BCB as the dielectric layers.

35.7   Applications and Advantages

The mBGA package is a flip-chip CSP which redistributes the peripheral wire bond pads on the die to area-array bumps. The package I/Os range from 34 to 298 pins. As shown in Table 35.1, so far there are nine IC devices packaged by mBGA technology. Among them are memory modules, DSPs, microcontrollers, and ASICs. Also, because of the low package profile and compact size, the mBGA is very suitable for MCM applications. To date, the MCM-D and MCM-L mBGA packages have been developed. Experimental studies have shown that good thermal performance and assembly reliability can be achieved.

In addition to superior form factors, there are other advantages associated with mBGA packages. The redistribution capability can accommodate almost all kinds of peripheral die bond pads from various wafer suppliers and turn them into an area array with a common footprint.

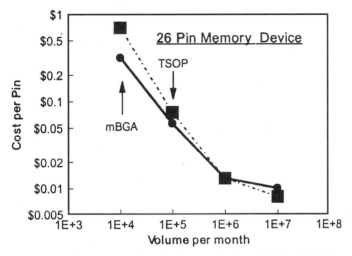

Figure 35.14 Comparison of packaging cost for the mBGA and TSOP.

Also, the mBGA is able to manage most custom-designed patterns and footprints. Compared to C4 flip chips, mBGA packages can provide bigger solder bumps with much larger bump pitch. Therefore, the requirements for surface-mount assembly, such as pick-and-place accuracy and placement tolerance, are much looser. Furthermore, since mBGA packages can be fully tested at the wafer level, this packaging technology can provide a known good die (KGD) solution to DCA and MCM.

A cost analysis was conducted to compare the packaging cost of the mBGA with that of other conventional packages. As shown in Fig. 35.14, the cost of the mBGA and TSOP is almost the same if the production volume is higher than 10,000 per month. For high-pin-count devices with large production volume, the packaging cost of the mBGA is much lower than that of the PQFP and PBGA packages, as indicated in Fig. 35.15. Besides, just like other wafer-level CSPs, the mBGA can easily cope with die shrink and wafer expand. As a result, an even greater cost efficiency may be achieved.

35.8 Summary and Concluding Remarks

Sandia's mBGA is a CSP with wafer-level I/O redistribution. The basic configuration is a flip-chip die. The fine-pitch peripheral die bond pads are reconfigured by two dielectric layers and two conductive layers to area-array bumps with much larger pad size and bump pitch. The package terminals may be Au bumps or solder bumps. For the former, conductive adhesive should be used for the second-level interconnection. Otherwise, the conventional SMT is applied for board-level assembly. In addition to the CSP, two types of multichip modules (MCM-D and MCM-L) were developed as testing vehicles to

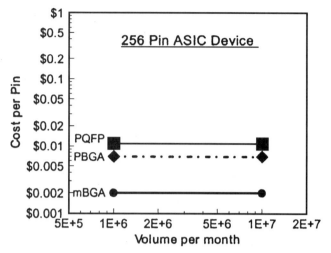

Figure 35.15 Comparison of packaging cost among the mBGA, PBGA, and PQFP.

characterize mBGA packages. Experimental results indicate that good thermal performance and assembly reliability can be achieved. It should be noted that for the mBGA MCM-L assembly, underfill encapsulation is essential for solder joint reliability.

The mBGA package was developed for IC devices with low to medium pin counts. The package I/Os vary from 34 to 298 pins. So far there are nine ICs processed with the mBGA package. Typical devices include memory modules, DSPs, microcontrollers, and ASICs. In addition to superior form factors, the mBGA packaging technology has several other advantages. The redistribution capability can manage both common footprints and custom-designed patterns. Also, the relatively big solder bumps with larger bump pitch can relax the requirements for surface-mount assembly. Furthermore, since mBGA packages can be fully tested at the wafer level, this packaging technology can provide KGDs for DCA and MCM. From a cost analysis for various packages, it was determined that the packaging cost of the mBGA is comparable to that of the TSOP for low-pin-count ICs and is much lower than that of the PQFP and PBGA for high-I/O devices in large-volume production.

35.9 References

1. R. Chanchani, K. Treece, and P. Dressendorfer, "Mini Ball Grid Array (mBGA) Technology," *Proceedings of NEPCON-West '95,* Anaheim, Calif., 1995, pp. 938–945.
2. R. Chanchani, K. Treece, and P. Dressendorfer, "A New Mini Ball Grid Array (mBGA) Multichip Module Technology," *International Journal of Microcircuits and Electronic Packaging,* vol. 18, no. 3, 1995, pp. 185–192.
3. R. Chanchani, K. Treece, and P. Dressendorfer, "Mini Ball Grid Array (mBGA) Assembly on MCM-L Boards" *Proceedings of the 47th ECTC,* San Jose, Calif., 1997, pp. 656–663.

36

ShellCase's
Shell-PACK/Shell-BGA

36.1 Introduction and Overview

In 1995, ShellCase Ltd. released a new wafer-level chip scale package that had the same name as the company [1]. Later on this CSP was called SlimCase in order to highlight the ultra-thin feature of this package [2]. More recently a new version was added to this CSP family, and the names have been updated to Shell-PACK [3] and Shell-BGA [4], respectively. Although ShellCase's products are classified as wafer-level CSPs, their package structures are quite different from those of all others in this category. In Shell-PACK and Shell-BGA, the I/Os are not really redistributed "on the wafer." Instead, the wafer is sandwiched between two glass plates and the I/O redistribution is performed on one of these glass plates. The major features of ShellCase's CSPs are small package thickness and fine terminal pitch. The Shell-PACK has peripheral flat lands with a pitch of 0.3 mm, and the Shell-BGA has area-array solder bumps with a pitch of 0.5 mm. The thickness of both CSPs is 0.3 to 0.5 mm. Because they are so slim, ShellCase's CSPs are a very appealing option for smart card applications.

36.2 Design Concepts and Package Structure

Shell-PACK and Shell-BGA are wafer-level CSPs developed by ShellCase. These two packages basically share the same structure. The only difference is the format of the package terminals. Shell-PACK has peripheral flat lands with a pitch of 0.3 mm, whereas Shell-BGA has area-array solder bumps with a pitch of 0.5 mm. The bottom views of both CSPs are shown in Fig. 36.1 for comparison.

(a)

(b)

Figure 36.1   Bottom views of ShellCase's CSPs: (*a*) Shell-PACK, (*b*) Shell-BGA.

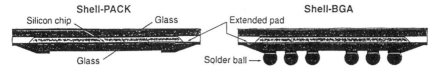

Figure 36.2   Cross-sectional views of the package structure for ShellCase's CSPs.

As illustrated in the package structure cross section shown in Fig. 36.2, the silicon chip is sandwiched between two glass plates. Adhesive epoxy is used to bond the IC and the glass plates together. The perimeter of the IC is also encapsulated by epoxy for environmental protection. The package thickness of ShellCase's CSPs is 0.3 to 0.5 mm. The package size is about 100 μm (maximum) larger than the die size. The first-level interconnects of Shell-PACK and Shell-BGA are extended die pads which are attached to the metallization deposited on one of the glass plates. These patterned metal leads wrap around the glass plate from the sides to the bottom to redistribute the package I/Os. Since this metallization is gold or solder, depending on the package design, the flat lands at the bottom of the package can be either directly assembled to the next level of substrate (Shell-PACK) or used for solder-ball attachment (Shell-BGA) by the standard SMT process.

36.3   Material Issues

ShellCase's CSPs are wafer-level chip scale packages. However, unlike those of the other CSPs in the same category, the captioned packages have rigid substrates for protection and I/O redistribution. These substrates are thin glass plates. Other materials such as aluminum nitride (ALN) may be considered for better thermal management. The silicon chip is sandwiched between the glass plates. The bonding is done by epoxy adhesive. The same epoxy is also used to encapsulate the perimeter of the IC for environmental protection.

The I/O redistribution of ShellCase's CSPs is performed by the thin-film metallization deposited on one of the glass plates. This metallization is patterned to wrap around the glass substrate from the sides to the bottom to make the desired routing traces. The ends of these metal leads are in contact with the cross section of the extended die pads at the periphery of the package to form the first-level interconnects. The other ends are flat lands at the bottom of the package for direct assembly on the PCB (Shell-PACK) or for the placement of solder balls (Shell-BGA). It should be noted that the aforementioned metallization is plated with Ni/Au or Pb/Sn solder. Therefore, the standard SMT process can be applied for board-level assembly or solder-ball attachment.

36.4   Manufacturing Process

Shell-PACK and Shell-BGA are classified as wafer-level CSPs because some of their manufacturing processes are performed on the wafer. The detailed fabrication procedures are given in Figs. 36.3 and 36.4. From these two flowcharts, it can be seen that Shell-PACK and Shell-BGA basically share the same packaging technology. The only differences are the patterning of the bottom metallization and the attachment of the solder balls.

The fabrication of ShellCase's CSPs starts with bonding of the active face of the silicon wafer to a glass substrate by epoxy adhesive. Then a mechanical thinning process is performed on the backside of the wafer to reduce the wafer thickness to 100 μm. The next step is to singulate the ICs by etching away the silicon inside the scribe streets between dies. Also, the whole wafer is further etched to make it 50 μm thick. At this stage, the die bond pads are exposed to the outside as a result of the etching of the silicon. It should be noted that this process requires oversized bond pads which extend from the regular die areas into the scribe streets between dies. If a conventional wafer is used, additional metallization should be deposited and patterned before the packaging process in order to fulfill this requirement.

The subsequent procedure is to attach a second glass substrate to the bottom of the wafer for further protection. Additional epoxy needs to be applied for the bonding of the glass plate and the encapsulation of etched scribe streets. Then the sandwiched wafer is flipped over and deep notches over the scribe streets are made in the first glass plate to expose the extended die bond pads. A thin-film metallization is deposited to cover the whole wafer. At the bottom of the deep notches, the metal layer is in contact with the cross section of the extended die bond pads to form the first-level interconnection. The next step is to pattern the deposited metal layer using photolithography to form the desired routing traces and bottom lands. For Shell-BGA, an extra procedure is needed to attach the solder balls. Finally, the completed CSPs are singulated from the wafer by mechanical dicing.

36.5   Performance and Reliability

Since ShellCase's CSPs have a very short electrical passage, their electrical performance should be relatively good. An experimental measurement was conducted to characterize the parasitics of Shell-PACK. The mutual inductance was found to be less than 1 nH (at 1

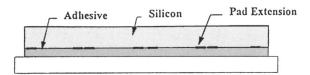

Adhesive   Silicon   Pad Extension

a. Attach the silicon to glass with adhesive

b. Thin the silicon to 100 μm thick by grinding

c. Etch the silicon wafer to isolate ICs and then
   etch ICs to 50 μm thick

d. Apply adhesive to the back side of ICs and
   attach the second glass substrate

e. Draw grooves from the front side so as to
   reveal the pad cross-section

f. Coat the wafer with a metal layer

g. Define electrical leads by photolithography

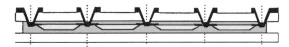

h. Separate the packaged dies by dicing

Figure 36.3 Manufacturing process for Shell-PACK.

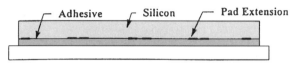

a. **Attach the silicon to glass with adhesive**

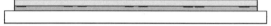

b. **Thin the silicon to 100 μm thick by grinding**

c. **Etch the silicon wafer to isolate ICs and then etch ICs to 50 μm thick**

d. **Apply adhesive to the back side of ICs and attach the second glass substrate**

e. **Draw grooves from the front side so as to reveal the pad cross-section**

f. **Coat the wafer with a metal layer**

g. **Define electrical leads by photolithography**

h. **Deposit solder bumps**

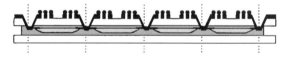

i. **Separate the packaged dies by dicing**

Figure 36.4 Manufacturing process for Shell-BGA.

GHz), and the lead capacitance was less than 0.07 pF [1]. Such low electrical parasitics reduce the high-frequency delay and improve the ground bounce. A comparison of parasitics among various packages is given in Table 36.1. From this comparison, it can be seen that the Shell-PACK has superior electrical performance.

The thermal performance of Shell-PACK was also evaluated. An experimental measurement showed that the thermal resistance θ_{ja} under natural convection was 40°C/W and 15°C/W for the package with and without heat sink, respectively. Since the package structure of the Shell-PACK and Shell-BGA is almost the same, it is expected that their electrical and thermal performance should be similar.

The package reliability of ShellCase's CSPs is currently under evaluation. The comprehensive qualification program shown in Table 36.2 has been planned and is under way. However, the test results are not yet available. Since the glass cover plates provide robustness to the package, ShellCase's CSPs are easy to burn in for known good die (KGD). A schematic diagram of this is presented in Fig. 36.5. With this testing fixture, the package can be tested at speed before the board-level assembly.

The solderability of Shell-PACK has been investigated. A standard

TABLE 36.1   Comparison of Electrical Performance among Various Types of Packages

	PQFP	PGA	TBG	BGA	Shell-PACK
Capacitance	1 pF	10 pF	2.4 pF	10 pF	<0.07 pF
Inductance	6 nH	7 nH	5.5 nH	5 nH	<1 nH
Ground bounce	90%	100%	90%	85%	37%

TABLE 36.2   Qualification Testing Program for ShellCase's CSPs

Test	Conditions
THB	85°C/85% RH, 5-V bias
Endurance	125°C, 50 mA/lead, $J = 3.3 \cdot 10^4$ A/cm^2
Thermal cycling	$-40-+85$°C, air to air
Popcorn	JEDEC A112
Shear	Mil-STD.883D, Method 2019.5
Pull	Mil-STD.883D, Method 2011.7, F
Bend	Internal procedure
Heat dissipation	5 W

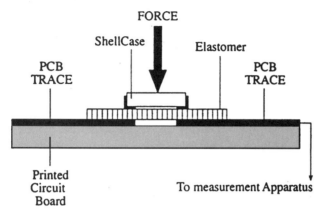

Figure 36.5 Schematic diagram for electrical testing of Shell-PACK.

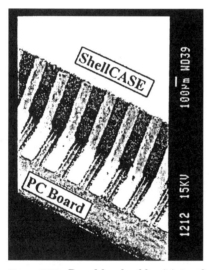

Figure 36.6 Board-level solder joints of Shell-PACK.

SMT process was applied to mount the CSP to a conventional FR-4 board. It was confirmed that even with pitch as fine as 0.3 mm, the CSP can be assembled to the PCB without any problem. A micrograph of some typical solder joints of Shell-PACK is shown in Fig. 36.6. The board-level solder joint reliability of ShellCase's CSPs is currently under evaluation together with the package qualification testing program.

36.6   Applications and Advantages

Shell-PACK and Shell-BGA are wafer-level chip scale packages with a compact footprint. Their potential applications include PCMCIA cards, camcorders, cellular phones, and other portable equipment and devices [5]. In particular, because of the ultra-thin package and the robustness provided by the covering glass plates, the Shell-PACK is very suitable for smart card applications [6]. A comparison of the thickness of a Shell-PACK package and a regular credit card is given in Fig. 36.7. From this figure, it can be seen that the Shell-PACK can fit nicely in a credit card without any glob-top encapsulation. In addition to conventional smart cards, the Shell-PACK may be used for contactless smart cards, as illustrated in Fig. 36.8. ShellCase's CSPs are SMT-compatible, and this feature makes the assembly of the coil and die relatively simple. Furthermore, another type of noncontact smart cards communicates by optics instead of electromagnetic transmission. Because the Shell-PACK is covered by transparent glass plates, this CSP is also suitable for applications in smart cards with optical link. On the other hand, since Shell-BGA can accommodate more I/Os, this package may be used for ASIC applications. Certain industrial customers are currently evaluating ShellCase's CSPs. Their configurations are summarized in Table 36.3.

Figure 36.7   Shell-PACK for smart card application.

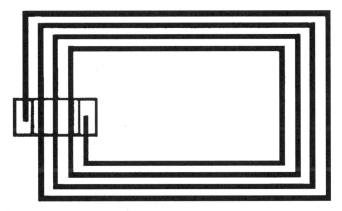

Figure 36.8   Schematic diagram for the contactless smart card module assembly.

TABLE 36.3   ShellCase's CSPs under Customers' Evaluation

Size	Size (mm × mm)	Type	Number of I/Os	Device
Large	10 × 10	NA	120	Test die
Medium	7 × 7	Shell-BGA	48	ASIC
Small	2.8 × 2.3	Shell-PACK	6	EEPROM

Figure 36.9   Compact features of ShellCase's CSPs.

Compared to conventional packages, the most significant advantage of ShellCase's CSPs is the compact form factors. This dimensional merit can be easily identified from the comparisons illustrated in Fig. 36.9. In addition, the IC is firmly bonded between two glass covering plates. This sandwich structure not only gives the package substantial robustness for environmental protection, but also provides great security to prevent reverse engineering. Based on the aforementioned features and the compatibility with SMT, ShellCase's CSPs have become a very appealing option to the manufacturers of both conventional and contactless smart cards.

36.7 Summary and Concluding Remarks

Shell-PACK and Shell-BGA are wafer-level CSPs developed by ShellCase. Unlike other chip scale packages in the same category, ShellCase's CSPs employ two glass plates to sandwich the IC, and the I/O redistribution is performed on one of the glass plates. Small package thickness and fine terminal pitch are the major features of ShellCase's CSPs. The Shell-PACK has peripheral flat lands with a pitch of 0.3 mm, whereas the Shell-BGA has area-array solder bumps with a pitch of 0.5 mm. The thickness of both CSPs is 0.3 to 0.5 mm. Because of the compact dimensions and rather short routing traces, ShellCase's CSPs have superior electrical and thermal performance.

The potential applications of Shell-PACK and Shell-BGA include PCMCIA cards, camcorders, cellular phones, and other portable devices. Compared to conventional packages, the most significant advantage of ShellCase's CSPs is the compact form factors. Furthermore, the IC is firmly bonded between two glass covering plates. This sandwich structure not only gives the package substantial robustness for environmental protection, but also provides great security to prevent reverse engineering. Based on these advantages and the compatibility with SMT, ShellCase's CSPs have become a very appealing option to the manufacturers of both conventional and contactless smart cards.

36.8 References

1. A. Badihi and E. Por, "ShellCase—A True Miniature Integrated Circuit Package," *Proceedings of the International Flip Chip, Ball Grid Array, Advanced Packaging Symposium,* San Jose, Calif., February 1995, pp. 244–252.
2. A. Badihi, N. Schlomovich, M. de-la-Vega, N. Karligano, and Z. Baron, "SlimCase— An Ultra Thin Chip Size Integrated Circuit Package," *Proceedings of the ICEMCM '96,* Denver, Colo., April 1996, pp. 234–238.
3. G. Zilber, M. de-la-Vega, N. Schlomovich, N. Karligano, U. Tropp, and A. Badihi, "Shell-PACK—A Thin Chip Size Integrated Circuit Package," *Proceedings of Semicon West,* San Jose, Calif., June 1996.
4. G. Zilber, H. Gershetman, N. Schlomovich, N. Natan, N. Karligano, U. Tropp, and A. Badihi, "Shell-BGA—A Thin Chip Size Integrated Circuit Package," *Proceedings of the ISHM Workshop on CSP,* Whistler, B.C., August 1996.
5. A. Badihi, "Thin-Film, Related Processes Used for Unique, Ultra-Thin CSP," *Chip Scale Review,* 1997, pp. 32–33.
6. G. Zilber, N. Schlomovich, M. Yarden, I. Schweki, and A. Badihi, "Ultra-Thin Chip Size Package Enables a Simpler Smart Card Assembly," 1997.

Index

Accelerated soft error rate (ASER) test, 136
Accelerated thermal cycling (ATC), 118–119, 121–123
Advanced Solder Mask (ASM), 351
Allied Signal, 163
Amkor/Anam (*see* ChipArray package)
Application-specific ICs (ASICs), 2, 161
ASER (accelerated soft error rate) test, 136
ASICs (*see* Application-specific ICs)
ASM (Advanced Solder Mask), 351
ASMAT, 219–221, 225, 231, 232
Assembly process, chip, 4–5
ATC (*see* Accelerated thermal cycling)

Ball grid arrays (BGAs), 1, 141, 158, 160–163, 167–171, 185, 186, 188
  (*See also* Fine-pitch ball grid array package; Micro BGA and quad flat nonleaded package)
BCC package (*see* Bump chip carrier package)
µBGA (*see* Micro-ball grid array)
BGAs (*see* Ball grid arrays)
BLP package (*see* Bottom-leaded plastic package)
Bottom-leaded plastic (BLP) package, 43–44, 107–127
  applications/advantages of, 124–126
  design of, 107–110
  electrical/thermal performance of, 115–118
  manufacturing process for, 112–115
  materials in, 110–112
  qualifications/reliability of, 118–124
Bump chip carrier (BCC) package, 61–70
  applications/advantages of, 69–70
  design/structure of, 61–63
  electrical/thermal performance of, 66–67
  manufacturing process for, 64–66
  materials in, 63
  qualifications/reliability of, 67–69

C-BLP, 108, 109, 114–115, 120, 123, 124
Ceramic/plastic fine-pitch BGA package (C/P-FBGA), 439–453
  applications/advantages of, 452
  design/structure of, 439–441
  manufacturing process for, 443–448
  materials in, 441–443
  qualifications/reliability of, 448–452
Chip scale packages (CSPs), 2
ChipArray package (Amkor/Anam), 305–312
  applications/advantages of, 311
  design/structure of, 305–309
  electrical/thermal performance of, 310–311
  manufacturing process for, 309–310
  materials in, 307
  reliability of, 311
Chip-on-board (COB), 61
Chip-on-flex chip scale package (COF-CSP), 157–172
  applications/advantages of, 169–172
  design of, 157–163
  electrical/thermal performance of, 168–169
  manufacturing process for, 164–168
  materials in, 163–164
  qualifications/reliability of, 169
ChipScale, Inc. (*see* Micro SMT package)
COB (chip-on-board), 61
Coefficient of thermal expansion (CTE), 90–91, 123, 207
COF-CSP (*see* Chip-on-flex chip scale package)
Cost factors (*see* Solder-bumped flip-chip technology, cost factors in; Wire-bonding technology, cost factors in)
C/P-FBGA (*see* Ceramic/plastic fine-pitch BGA package)
Crystallization, 21
CSP (*see* Chip-on-flex chip scale package; Enhanced Flex CSP; EPIC chip scale package; Lead-on-chip chip scale package; Memory chip scale package; Mitsubishi chip scale package; Sharp chip scale package)

CTE (*see* Coefficient of thermal expansion)
Curing, 21–23
Customized-lead-frame-based CSPs (*see* Bottom-leaded plastic package; Bump chip carrier package; Lead-on-chip chip scale package; Memory chip scale package; Micro BGA and quad flat nonleaded package; Micro stud array package; Small outline no-lead/C-lead package)

DCA (*see* Direct chip attach)
Die:
  wafer-bumping cost per, 7–9
  wire-bonding cost per, 11, 13–14
Differential scanning calorimeter (DSC), 19, 21–23
Digital signal processors (DSPs), 143
Direct chip attach (DCA), 1, 2
DMA (dynamic mechanical analysis), 19
DRAM (dynamic random-access memories), 2
DSC (*see* Differential scanning calorimeter)
DSPs (digital signal processors), 143
DuPont, 163
Dynamic mechanical analysis (DMA), 19
Dynamic random-access memories (DRAM), 2

Electrical/thermal performance:
  bottom-leaded plastic (BLP) package, 115–118
  bump chip carrier (BCC) package, 66–67
  ChipArray package (Amkor/Anam), 310–311
  chip-on-flex chip scale package (COF-CSP), 168–169
  enhanced Flex CSP, 150–151
  EPIC chip scale package, 473–474
  fine-pitch ball grid array (FPBGA) package, 213
  lead-on-chip chip scale package (LOC-CSP), 93–94
  memory chip scale package (MCSP), 136–137
  micro SMT (MSMT) package, 461–462
  micro stud array (MSA) package, 103–104
  micro-ball grid array (μBGA), 266, 269–273

Electrical/thermal performance (*Cont.*):
  mini ball grid array (mBGA) package (Sandia National Laboratories), 536–539
  mini-ball grid array (mini-BGA) package (IBM), 346
  Mitsubishi chip scale package (CSP), 512–514
  MN-PAC, 365–370
  NuCSP, 328–329
  plastic chip carrier (PCC) package, 403–404
  sharp chip scale package, 240–241
  shell-PACK/Shell-BGA, 548, 551
  slightly larger than IC carrier (SLICC), 384–389
  small outline no-lead/C-lead package (SON/SOC), 49–52
  three-dimensional memory module (3DM), 422–424
  *Ultra*CSP, 482–483
  underfill materials, 34, 35
EMC (*see* Encapsulation molding compound)
Encapsulants, underfill, 34–36
Encapsulation molding compound (EMC), 48, 90–91, 112, 237
Enhanced Flex CSP, 143–156
  applications/advantages of, 155
  design of, 143–146
  electrical/thermal performance of, 150–151
  manufacturing process for, 148–150
  materials in, 145–149
  qualifications/reliability of, 151–155
EPIC chip scale package, 467–475
  applications/advantages of, 474
  design/structure of, 467–470
  electrical/thermal performance of, 473–474
  manufacturing process for, 471–473
  materials in, 471
  reliability of, 474
EPIC Technologies, Inc. (*see* EPIC chip scale package)
EPS (*see* NuCSP)

FC/PBGA (*see* Flip chip-plastic ball grid array)
Fiber-push connection (FPC), 183, 192–193
Film redistribution layer (FRL), 2

Fine-pitch ball grid array (FPBGA) package, 201–218
applications/advantages of, 216–218
design of, 201–206
electrical/thermal performance of, 213
manufacturing process for, 208–213
materials in, 206–208
qualifications/reliability of, 213–216
Flash memories, 2
Flexible interposer, CSPs with (*see* Chip-on-flex chip scale package; Enhanced Flex CSP; Fine-pitch ball grid array package; *flex*PAC; Memory chip scale package; Memory devices, chip scale package for; Micro-ball grid array; Micro-star BGA; Molded chip size package; Sharp chip scale package)
*flex*PAC, 183–198
applications/advantages of, 196–198
design of, 183–187
manufacturing process for, 188–195
materials in, 186–188
qualifications/reliability of, 195–197
Flip Chip Technologies (FCT) (*see* *Ultra*CSP)
Flip chip-plastic ball grid array (FC/PBGA), 349–357
design/structure of, 349–351
manufacturing process for, 351–355
materials in, 351
qualifications/reliability of, 353–356
FLOTHERM software, 423
Flow rate, underfill, 30–31
Fourier transform infrared spectroscopy (FTIR), 132
FPBGA (*see* Fine-pitch ball grid array package)
FPC (*see* Fiber-push connection)
FRL (film redistribution layer), 2
FTIR (Fourier transform infrared spectroscopy), 132
Fujitsu (*see* Bump chip carrier package; Micro BGA and quad flat nonleaded package; Small outline no-lead/C-lead package; *Super*CSP)

General Electric (*see* Chip-on-flex chip scale package)
Glass transition temperature (T_g), 19, 34, 36
GREENFIELD software, 49, 423

High-density interconnect (HDI), 2, 157–158
Hitachi (*see* Memory devices, chip scale package for)
Hitachi Cable (*see* Lead-on-chip chip scale package; Micro stud array package)

IBM (*see* Flip chip-plastic ball grid array; Mini-ball grid array package)
IBSS (*see* Interpenetrating polymer buildup structure system)
Inner lead bonding (ILB), 173, 174
Interpenetrating polymer buildup structure system (IBSS), 2
IZM (*see* *flex*PAC)

JACS-Pak (*see* Slightly larger than IC carrier)
JEDEC standard, 107, 108, 349–350, 459
JEIDA Type II memory cards, 57, 58

Known good die (KGD), 1–3, 70, 158
Kyocera, 439

Land grid array (LGA), 359, 411, 430–431
Lead-on-chip chip scale package (LOC-CSP), 87–95
applications/advantages of, 94–95
design of, 87–90
electrical performance of, 93–94
manufacturing process for, 91–93
materials in, 89–91
qualifications/reliability of, 94
Lead-on-chip (LOC) configuration, 43–44, 148, 293–294
LG Semicon, 43–44
(*See also* Bottom-leaded plastic package)
LGA (*see* Land grid array)
Linear variable differential transducer (LVDT), 24
LOC configuration (*see* Lead-on-chip configuration)
LOC-CSP (*see* Lead-on-chip chip scale package)
Lockheed-Martin, 157
Low-cost solder-bumped NuCSP (*see* NuCSP)
LVDT (*see* Linear variable differential transducer)

Manufacturing process:
  bottom-leaded plastic (BLP) package, 112–115
  bump chip carrier (BCC) package, 64–66
  ceramic/plastic fine-pitch BGA package (C/P-FBGA), 443–448
  ChipArray package (Amkor/Anam), 309–310
  chip-on-flex chip scale package (COF-CSP), 164–168
  enhanced Flex CSP, 148–150
  EPIC chip scale package, 471–473
  fine-pitch ball grid array (FPBGA) package, 208–213
  *flex*PAC, 188–195
  flip chip-plastic ball grid array (FC/PBGA), 351–355
  lead-on-chip chip scale package (LOC-CSP), 91–93
  memory chip scale package (MCSP), 134–136, 297–299
  memory devices, chip scale package for, 177–179
  micro BGA and quad flat nonleaded package (QFN), 75–79
  micro SMT (MSMT) package, 460–461
  micro stud array (MSA) package, 100–104
  micro-ball grid array (μBGA), 265–268
  micro-star BGA (μStar BGA), 285
  mini ball grid array (mBGA) package (Sandia National Laboratories), 534–535
  mini-ball grid array (mini-BGA) package (IBM), 341–346
  Mitsubishi chip scale package (CSP), 505–512
  MN-PAC, 363–365
  molded chip size package (MCSP), 225–226
  NuCSP, 323–328
  plastic chip carrier (PCC) package, 402–403
  sharp chip scale package, 238–240
  shell-PACK/Shell-BGA, 548–550
  slightly larger than IC carrier (SLICC), 383–384
  small outline no-lead/C-lead package (SON/SOC), 48–50
  μSMD, 521
  *Super*CSP (SCSP), 489–491
  three-dimensional memory module (3DM), 418–422

Manufacturing process (*Cont.*):
  transformed grid array (TGA) package, 432–433
  *Ultra*CSP, 482, 483
Matsushita Electronics Corporation (MEC), 313
  (*See also* MN-PAC)
Maxwell-3D, 50
mBGA (*see* Mini ball grid array package)
MCM-F technology (*see* Multichip-on-flex technology)
MCMs (*see* Multichip modules)
MCSP (*see* Memory chip scale package; Molded chip size package)
MEC (*see* Matsushita Electronics Corporation)
Mechanical performance (of underfill), 32–34
Memory chip scale package (MCSP), 129–141, 293–303
  applications/advantages of, 140–141, 302
  design of, 129–131
  design/structure of, 293–295
  electrical/thermal performance of, 136–137
  manufacturing process for, 134–136, 297–299
  materials in, 131–133, 294–297
  qualifications/reliability of, 137–140, 299–302
Memory devices, chip scale package for, 173–181
  applications/advantages of, 180–181
  design of, 173–176
  manufacturing process for, 177–179
  materials in, 176–177
  qualifications/reliability of, 179–180
MF-LOC (*see* Multiframe lead-over-chip)
Micro BGA and quad flat nonleaded package (QFN), 71–85
  applications/advantages of, 83–85
  design/structure of, 71–74
  manufacturing process for, 75–79
  materials in, 74–75
  qualifications/reliability of, 80–82
Micro SMT (MSMT) package, 455–465
  applications/advantages of, 462–464
  design/structure of, 455–459
  electrical/thermal performance of, 461–462
  manufacturing process for, 460–461
  materials in, 459–460
  qualifications/reliability of, 462

Micro stud array (MSA) package, 97–105
    applications/advantages of, 104–105
    design of, 97–100
    electrical/thermal performance of,
        103–104
    manufacturing process for, 100–104
    qualifications/reliability of, 104
Micro-ball grid array (μBGA), 143, 144,
        155, 173, 174, 176, 259–281
    applications/advantages of, 276–281
    design/structure of, 259–263
    electrical/thermal performance of, 266,
        269–273
    manufacturing process for, 265–268
    materials in, 263–265
    qualifications/reliability of, 271–276
Micro-star BGA (μStar BGA), 143, 144,
        151, 283–292
    applications/advantages, 291
    design/structure of, 283–284
    manufacturing process for, 285
    materials in, 284–285
    qualifications/reliability of, 285–291
Mini ball grid array (mBGA) package
        (Sandia National Laboratories),
        529–544
    applications/advantages of, 542–543
    design/structure of, 529–553
    electrical/thermal performance of,
        536–539
    manufacturing process for, 534–535
    materials in, 533–534
    reliability of, 540–542
Mini-ball grid array (mini-BGA) package
        (IBM), 337–348
    applications/advantages of, 347
    design/structure of, 337–340
    manufacturing process for, 341–346
    materials in, 339–341
    reliability of, 346
    thermal performance of, 346
Mitsubishi chip scale package (CSP),
        495–518
    applications/advantages of, 516–517
    design/structure of, 496–499
    electrical/thermal performance of,
        512–514
    manufacturing process for, 505–512
    materials in, 400–505
    reliability of, 514–516
MN-PAC, 359–375
    applications/advantages of, 371–372
    design/structure of, 359–361

MN-PAC (Cont.):
    electrical/thermal performance of,
        365–370
    manufacturing process for, 363–365
    materials in, 361–363
    qualifications/reliability of, 368–374
Modulus, underfill, 26–28
Moisture content, 28–30
Molded chip size package (MCSP), 219–232
    applications/advantages of, 231–232
    design/structure of, 219–220
    manufacturing process for, 225–226
    materials in, 220–225
    qualifications/reliability of, 226–231
Motorola (see Slightly larger than IC car-
        rier)
MSA package (see Micro stud array pack-
        age)
MSMT (see Micro SMT package)
Multichip modules (MCMs), 157, 158,
        407, 532–534, 537–544
Multichip-on-flex (MCM-F) technology, 157
Multiframe lead-over-chip (MF-LOC),
        43–44, 48–49, 88, 129

National Semiconductor (see Plastic chip
        carrier package)
Near-die-size packages (NDSP), 305–306
NEC (see Fine-pitch ball grid array pack-
        age; Three-dimensional memory mod-
        ule)
Nitto Denko (see Molded chip size package)
NuCSP, 313–335
    applications/advantages of, 334
    design of, 320–323
    design/structure of, 313–315
    manufacturing process for, 323–328
    materials in, 315–320
    mechanical/electrical performance of,
        328–329
    reliability of, 329–334
    solder-bump characterizations, 317–318
    solder-bump height measurements,
        318–319
    solder-bump strength measurements,
        319–320

PacTech GmbH, 183
Pad:
    wafer-bumping cost per, 9–12
    wire-bonding cost per, 14–15

Parasitic Parameter-3D, 50
PCBs (printed circuit boards), 1
PCC package (*see* Plastic chip carrier package)
PCT (pressure cooker testing), 151
Plasma-etched redistribution layers (PERL), 1–2
Plastic chip carrier (PCC) package, 399–406
    applications/advantages of, 404–405
    design/structure of, 399–401
    manufacturing process for, 402–403
    materials in, 401–402
    qualifications/reliability of, 404
    thermal performance of, 403–404
Pressure cooker testing (PCT), 151
Printed circuit boards (PCBs), 1

QFN (*see* Micro BGA and quad flat non-leaded package)

Redistribution layer (RDL), 477
Rigid substrate, CSPs with (*see* Ceramic/plastic fine-pitch BGA package; ChipArray package; Flip chip-plastic ball grid array; Mini-ball grid array package; MN-PAC; NuCSP; Plastic chip carrier package; Slightly larger than IC carrier; Three-dimensional memory module; Transformed grid array package)

SAM (*see* Scanning acoustic microscopy)
Sandia National Laboratories (*see* Mini ball grid array)
SBB (*see* Solder-ball bumper; Stud bump bonding)
S-BLP, 108–111, 115, 124
Scanning acoustic microscopy (SAM), 118, 195
SCC (*see* Substrate chip carrier)
SCSP (*see* *Super*CSP)
SER simulation (SSER), 136
Sharp chip scale package, 233–257
    applications/advantages of, 255–257
    design/structure of, 233–237
    electrical/thermal performance of, 240–241
    manufacturing process for, 238–240
    materials in, 237–238

Sharp chip scale package (*Cont.*):
    qualifications/reliability of, 241–256
    solder joint reliability of, 244–255
ShellCase Ltd., 545
Shell-PACK/Shell-BGA, 545–546
    applications/advantages of, 553–555
    design/structure of, 535–537
    electrical/thermal performance of, 548, 551
    manufacturing process for, 548–550
    materials in, 547–548
    reliability of, 551–552
Shindo, 284
Shrink small outline packages (SSOP), 61
SLC (surface laminar circuits), 2
Slightly larger than IC carrier (SLICC), 313, 377–397
    design/structure of, 378–381
    electrical/thermal performance of, 384–389
    manufacturing process for, 383–384
    materials in, 381–383
    qualifications/reliability of, 389–397
Small outline no-lead/C-lead package (SON/SOC), 43–58
    applications/advantages of, 54, 57
    design/structure of, 43–47
    electrical/thermal performance of, 49–52
    manufacturing process for, 48–50
    materials in, 48
    qualifications/reliability of, 52–57
SMC (*see* Surface-mount component)
μSMD, 519–527
    design/structure of, 519–520
    manufacturing process for, 521
    materials in, 520–521
    reliability of, 521–526
SMM (*see* Sumitomo Metal Mining)
SOC (*see* Small outline no-lead/C-lead package)
SOJ, 129
Solder joint reliability (Sharp chip scale package), 244–255
Solder-ball bumper (SBB), 183
Solder-bumped flip-chip technology, cost factors in, 2–18
    assembly process, 4–5
    die, wafer-bumping cost per, 7–9
    major equipment needed, 5
    materials, 5
    pad, wafer-bumping cost per, 9–12
    wafer, 5–7

Solder-bumped flip-chip technology, cost factors in (*Cont.*):
wire-bonding technology, comparison with, 15–18
SON (*see* Small outline no-lead/C-lead package)
Sony Corporation (*see* Transformed grid array package)
SRAM (*see* Static random-access memories)
SSER (*see* SER simulation)
SSOP (*see* Shrink small outline packages)
μStar BGA (*see* Micro-star BGA)
Static random-access memories (SRAM), 2
Stud bump bonding (SBB), 313, 359
Substrate chip carrier (SCC), 399
Sumitomo Metal Mining (SMM), 284
*Super*CSP (SCSP), 487–494
design/structure of, 487–488
manufacturing process for, 489–491
materials in, 487–489
qualifications/reliability of, 490, 492–493
Surface laminar circuits (SLC), 2
Surface-mount component (SMC), 118–119

T_g (*see* Glass transition temperature)
TAB tape, 284
Tangent delta (tan δ), 28
Tape ball grid array (TBGA), 143–145
Tape carrier package (TCP), 173
Tape-LOC, 43, 44, 88, 130
TAPI, 219, 220, 223–228, 231–232
TBGA (*see* Tape ball grid array)
TCE (*see* Thermal coefficient of expansion)
TCMT (Tessera-compliant mounting tape), 265
TCP (*see* Tape carrier package)
TDR (time-domain reflectometory) method, 66
Technical University of Berlin (TU-Berlin), 183, 188
Tessera (*see* Micro-ball grid array)
Tessera-compliant mounting tape (TCMT), 265
Texas Instruments (TI) Japan (*see* Memory chip scale package; Micro-star BGA)
TGA (*see* Thermal gravimetric analysis)
TGA package (*see* Transformed grid array package)

Thermal coefficient of expansion (TCE), 19, 22, 24–26, 34, 36
Thermal gravimetric analysis (TGA), 19, 28
Thermal mechanical analysis (TMA), 19, 22
Thermal performance (*see* Electrical/thermal performance)
Thin small outline package (TSOP), 43, 50–52, 54–58, 107, 129, 240–241, 409
Thin-zero outline package (TZOP), 158
Three-dimensional memory module (3DM), 407–427
applications/advantages of, 424–426
design/structure of, 407–415
electrical/thermal performance of, 422–424
manufacturing process for, 418–422
materials in, 414–418
qualifications/reliability of, 423–424
Three-dimensional packaging module (3DPM), 43, 47, 58
3DM (*see* Three-dimensional memory module)
3DPM (*see* Three-dimensional packaging module)
3M (*see* Enhanced Flex CSP)
TI Japan (*see* Texas Instruments Japan)
Time-domain reflectometory (TDR) method, 66
TMA (*see* Thermal mechanical analysis)
Toshiba Corporation (*see* Ceramic/plastic fine-pitch BGA package)
Transformed grid array (TGA) package, 429–437
applications/advantages of, 435–437
design/structure of, 429–431
manufacturing process for, 432–433
materials in, 431–432
qualifications/reliability of, 433–436
TSOP (*see* Thin small outline package)
TU-Berlin (*see* Technical University of Berlin)
TZOP (thin-zero outline package), 158

UBM (*see* Under-bump metallurgy)
*Ultra*CSP, 477–486
applications/advantages of, 485–486
design/structure of, 477–480
electrical/thermal performance of, 482–483

*Ultra*CSP (*Cont.*):
  manufacturing process for, 482,
    483
  materials in, 479, 480
  reliability of, 482, 484–485
UMC (*see* United Microelectronics
  Corporation)
Under-bump metallurgy (UBM), 187,
  189–190, 317, 353, 362, 442, 443,
  479–481, 488–490, 495
Underfill materials, 19–37
  curing conditions, 21–23
  electrical performance of, 34, 35
  encapsulants, 34–36
  flow rate, 30–31
  mechanical performance of, 32–34
  moisture content, 28–30
  properties, 22, 24–30
  storage modulus, 26–28
  tangent delta, 28
  TCE, 22, 24–26
United Microelectronics Corporation
  (UMC), 315
USON, 88, 107

Wafer-level redistribution CSPs (*see* EPIC
  chip scale package; Micro SMT package; Mini ball grid array; Mitsubishi
  chip scale package; Shell-PACK/Shell-
  BGA; μSMD; *Super*CSP; *Ultra*CSP)
Wafers, 5–7
Wire-bonding technology, cost factors in,
  2–18
  assembly process, 4–5
  die, wire-bonding cost per, 11, 13–14
  materials, 5
  pad, wire-bonding cost per, 14–15
  solder-bumped flip-chip technology,
    comparison with, 15–18
  wafer, 5–7

XPS (*see* X-ray photoemission spectroscopy)
X-ray photoemission spectroscopy (XPS),
  131–132

YAG laser, 240

ABOUT THE AUTHORS

JOHN H. LAU is the president of Express Packaging Systems, Inc., in Palo Alto, California. His current interests cover a broad range of electronics packaging and manufacturing technology.

Prior to founding EPS in November, 1995, Lau worked for Hewlett-Packard Company, Sandia National Laboratory, Bechtel Power Corporation, and Exxon Production and Research Company. With more than twenty-eight years of R&D and manufacturing experience in the electronics, petroleum, nuclear, and defense industries, he has authored and co-authored over one hundred peer-reviewed technical publications, and is the author and editor of twelve books: *Solder Joint Reliability; Handbook of Tape Automated Bonding; Thermal Stress and Strain in Microelectronics Packaging; The Mechanics of Solder Alloy Interconnects; Handbook of Fine Pitch Surfface Mount Technology; Chip On Board Technologies for Multichip Modules; Ball Grid Array Technology; Flip Chip Technologies; Solder Joint Reliability of BGA, CSP, Flip Chip and Fine Pitch SMT Assemblies; Electronics Packaging: Design, Materials, Process, and Reliability; Chip Scale Package (CSP): Design, Materials, Process, Reliability, and Applications;* and the forthcoming *Direct Chip Attach (DCA): Design, Materials, Process, Reliability, and Applications.*

John served as one of the associate editors of the *IEEE Transactions on Components, Packaging, and Manufacturing Technology* and *ASME Transactions, Journal of Electronic Packaging.* He also served as general chairman, program chairman, and session chairman, and invited speaker of several *IEEE, ASME, ASM, MRS, ISHM, SEMI, NEPCON,* and *SMI International* conferences. He received a few awards from ASME, and IEEE for best papers and technical achievements, and is an IEEE Fellow. He is listed in *American Men and Women of Science* and *Who's Who in America.*

John received his PH.D in Theorectical and Applied Mechanics from the University of Illinois, a M.A.S. degree in Structural Engineering from the University of British Columbia, a second M.S. degree in Engineering Mechanics from the University of Wisconsin, and a third M.S. degree in Management Science from Fairleigh Dickinson University. He also has a B.E. degree in Civil Engineering from National Taiwan University.

SHI-WEI RICKY LEE received his B.S. degree in Mechanical Engineering from National Taiwan University in 1981. After two years of military service, he joined the Yue Loong Motor Engineering Center as Structural Testing Engineer. In 1986, he went to the United States for post-graduate studies. He received his M.S. degree in Engineering Mechanics from Virginia Polytechnic Institute & State Univeristy (VPI & SU) and Ph.D. in Aeronautics and Astronautics from Purdue University in 1987 and 1992, respectively. Through years of intensive research, he has developed an expertise in computational modeling and experimental methods. Before taking the teaching position at the Hong Kong Univeristy of Science & Technology (HKUST) in 1993, he spent one year at Purdue University as Post-Doctoral Research Associate and Visiting Assistant Professor.

Currently, Ricky is Assistant Professor in the Department of Mechanical Engineering, HKUST, and serves as Vice-Chair of the Hong Kong Section, ASME International. He also sits on the Editorial Advisory Board of three international journals: *Smart Materials & Structures, Circuit World,* and *Soldering & Surface Mount Technology.* In 1997–1998, he served as Guest Editor for *Smart Materials & Structures,* and published a special issue on *Piezoelectric Motors/Actuators & Their Applications.* Ricky is very active in professional societies. He is a member of Tau Beta Pi (TBΠ), ASME, IEEE-CPMT, and SMTA. He has served as Track Organizer and Session Chair for many international conferences and symposiums. Furthermore, he is quite keen on continuing education for professional development. He has organized several workshops and short courses, and has been invited to deliver lectures and seminars around the world.

Ricky's recent research interests cover electronic packaging and surface mount technology, smart materials and mechanics for sensors and actuators, composite grid structures and structural repairs. He has contributed to more than fifty publications in refereed journals, conference proceedings, and technical monographs in these areas. He also filed one U.S. patent application for his new invention of a piezoelectric motor. Recently, Ricky co-authored with John Lau a book on electronic packaging: *Chip Scale Package (CSP): Design, Materials, Process, Reliability, and Applications.* In addition to CSP, his research activities in electronic packaging include reliability modeling of SMT solder joints and development of a low-cost process for high density substrates.